CONVEXITY AND OPTIMIZATION IN $\mathbb{R}^n$

PURE AND APPLIED MATHEMATICS

A Wiley-Interscience Series of Texts, Monographs, and Tracts

A complete list of the titles in this series appears at the end of this volume.

CONVEXITY AND OPTIMIZATION IN $\mathbb{R}^n$

LEONARD D. BERKOVITZ
Purdue University

A Wiley-Interscience Publication
JOHN WILEY & SONS, INC.

This book is printed on acid-free paper. ∞

Published simultaneously in Canada.

For ordering and customer service, call 1-800-CALL-WILEY.

Library of Congress Cataloging-in-Publication Data:

Berkovitz, Leonard David, 1924–
Convexity and optimization in R [superscript n] / Leonard D. Berkovitz.
p. cm.--(Pure and applied mathematics)
"A Wiley-Interscience publication."
Includes bibliographical references and index.
ISBN 0-471-35281-0 (cloth : alk. paper)
1. Convex sets. 2. Mathematical optimization. I. Title. II. Pure and applied mathematics (John Wiley & Sons : Unnumbered)

QA640.B46 2001
516'.08--dc21

2001045391

Printed in the United States of America.

10 9 8 7 6 5 4 3 2 1

To my wife, Anna

CONTENTS

PREFACE

This book presents the mathematics of finite-dimensional optimization, featuring those aspects of convexity that are useful in this context. It provides a basis for the further study of convexity, of more general optimization problems, and of numerical algorithms for the solution of finite-dimensional optimization problems. The intended audience consists of beginning graduate students in engineering, economics, operations research, and mathematics and strong undergraduates in mathematics. This was the audience in a one-semester course at Purdue, MA 521, from which this book evolved.

Ideally, the prerequisites for reading this book are good introductory courses in real analysis and linear algebra. In teaching MA 521, I found that while the mathematics students had the real analysis prerequisites, many of the other students who took the course because of their interest in optimization did not have this prerequisite. Chapter I is for those students and readers who do not have the real analysis prerequisite; in it I present those concepts and results from real analysis that are needed. Except for the Weierstrass theorem on the existence of a minimum, the "heavy" or "deep" theorems are stated without proof. Students without the real variables prerequisite found the material difficult at first, but most managed to assimilate it at a satisfactory level. The advice to readers for whom this is the first encounter with the material in Chapter I is to make a serious effort to master it and to return to it as it is used in the sequel.

To address as wide an audience as possible, I have not always presented the most general result or argument. Thus, in Chapter II I chose the "closest point" approach to separation theorems, rather than more generally valid arguments, because I believe it to be more intuitive and straightforward for the intended audience. Readers who wish to get the best possible separation theorem in finite dimensions should read Sections 6 and 7 of Chapter II. In proving the Fritz John Theorem, I used a penalty function argument due to McShane rather than more technical arguments involving linearizations. I limited the discussion of duality to Lagrangian duality and did not consider Fenchel duality, since the latter would require the development of more mathematical machinery.

The numbering system and reference system for theorems, lemmas, remarks, and corollaries is the following. Within a given chapter, theorems, lemmas, and remarks are numbered consecutively in each section, preceded by the section number. Thus, the first theorem of Section 1 is Theorem 1.1, the second, Theorem 1.2, and so on. The same applies to lemmas and remarks. Corollaries are numbered consecutively within each section without a reference to the section number. Reference to a theorem in the same chapter is given by the theorem number. Reference to a theorem in a chapter different from the current one is given by the theorem number preceded by the chapter number in Roman numerals. Thus, a reference in Chapter IV to Theorem 4.1 in Chapter II would be Theorem II.4.1. References to lemmas and remarks are similar. References to corollaries within the same section are given by the number of the corollary. References to corollaries in a different section of the same chapter are given by prefixing the section number to the corollary number; references in a different chapter are given by prefixing the chapter number in Roman numerals to the preceding.

I thank Rita Saerens and John Gregory for reading the course notes version of this book and for their corrections and suggestions for improvement. I thank Terry Combs for preparing the figures. I also thank Betty Gick for typing seemingly innumerable versions and revisions of the notes for MA 521. Her skill and cooperation contributed greatly to the success of this project.

LEONARD D. BERKOVITZ

West Lafayette, Indiana

I

TOPICS IN REAL ANALYSIS

1. INTRODUCTION

The serious study of convexity and optimization problems in $\mathbb{R}^n$ requires some background in real analysis and in linear algebra. In teaching a course based on notes from which this text evolved, the author and his colleagues assumed that the students had an undergraduate course in linear algebra but did not necessarily have a background in real analysis. The purpose of this chapter is to provide the reader with most of the necessary background in analysis. Not all statements will be proved. The reader, however, is expected to understand the definitions and the theorems and is expected to follow the proofs of statements whenever the proofs are given. The bad news is that many readers will find this chapter to be the most difficult one in the text. The good news is that careful study of this material will provide background for many other courses and that subsequent chapters should be easier. If necessary, the reader should return to this chapter when encountering this material later on.

2. VECTORS IN $\mathbb{R}^n$

By euclidean n-space, or $\mathbb{R}^n$, we mean the set of all n-tuples $\mathbf{x} = (x_1, \ldots, x_n)$, where the x_i, $i = 1, \ldots, n$ are real numbers. Thus, $\mathbb{R}^n$ is a generalization of the familiar two- and three-dimensional spaces $\mathbb{R}^2$ and $\mathbb{R}^3$. The elements $\mathbf{x}$ of $\mathbb{R}^n$ are called *vectors* or *points*. We will often identify the vector $\mathbf{x} = (x_1, \ldots, x_n)$ with the $n \times 1$ matrix

$$\begin{pmatrix} x_1 \\ \vdots \\ x_n \end{pmatrix}$$

and abuse the use of the equal sign to write

$$\mathbf{x} = \begin{pmatrix} x_1 \\ \vdots \\ x_n \end{pmatrix}.$$

In this case we shall call $\mathbf{x}$ a column vector. We shall also identify $\mathbf{x}$ with the $1 \times n$ matrix $(x_1, \ldots, x_n)$ and write $\mathbf{x} = (x_1, \ldots, x_n)$. In this case we shall call $\mathbf{x}$ a row vector. It will usually be clear from the context whether we consider $\mathbf{x}$ to be a row vector or a column vector. When there is danger of confusion, we will identify $\mathbf{x}$ with the column vector and use the transpose symbol, which is a superscript t, to denote the row vector. Thus $\mathbf{x}^t = (x_1, \ldots, x_n)$.

We define two operations, vector addition and multiplication by a scalar. If $\mathbf{x} = (x_1, \ldots, x_n)$ and $\mathbf{y} = (y_1, \ldots, y_n)$ are two vectors, we define their sum $\mathbf{x} + \mathbf{y}$ to be

$$\mathbf{x} + \mathbf{y} = (x_1 + y_1, \ldots, x_n + y_n).$$

For any scalar, or real number, α we define $\alpha\mathbf{x}$ to be

$$\alpha\mathbf{x} = (\alpha x_1, \ldots, \alpha x_n).$$

We assume that the reader is familiar with the properties of these operations and knows that under these operations $\mathbb{R}^n$ is a vector space over the reals.

Another important operation is the *inner product*, or *dot product*, of two vectors, denoted by $\langle \mathbf{x}, \mathbf{y} \rangle$ or $\mathbf{x}^t\mathbf{y}$ and defined by

$$\langle \mathbf{x}, \mathbf{y} \rangle = \sum_{i=1}^{n} x_i y_i.$$

Again, we assume that the reader is familiar with the properties of the inner product. We use the inner product to define the norm $\|\cdot\|$ or length of a vector $\mathbf{x}$ as follows:

$$\|\mathbf{x}\| = \langle \mathbf{x}, \mathbf{x} \rangle^{1/2} = \left[\sum_{i=1}^{n} x_i^2 \right]^{1/2}.$$

This norm is called the *euclidean norm.* In $\mathbb{R}^2$ and $\mathbb{R}^3$ the euclidean norm reduces to the familiar length. It is straightforward to show that the norm has the following properties:

$$\|\mathbf{x}\| \geqslant 0 \quad \text{for all } \mathbf{x} \text{ in } \mathbb{R}^n, \tag{1}$$

$$\|\mathbf{x}\| = 0 \quad \text{if and only if } \mathbf{x} \text{ is the zero vector } \mathbf{0} \text{ in } \mathbb{R}^n, \tag{2}$$

$$\|\alpha\mathbf{x}\| = |\alpha| \, \|\mathbf{x}\| \quad \text{for all real numbers } \alpha \text{ and vectors } \mathbf{x} \text{ in } \mathbb{R}^n. \tag{3}$$

The norm has two additional properties, which we will prove:

For all vectors $\mathbf{x}$ and $\mathbf{y}$ in $\mathbb{R}^n$

$$|\langle \mathbf{x}, \mathbf{y} \rangle| \leqslant \|\mathbf{x}\| \, \|\mathbf{y}\| \tag{4}$$

with equality holding if and only if one of the vectors is a scalar multiple of the other.

For all vectors **x** and **y** in $\mathbb{R}^n$

$$\|\mathbf{x} + \mathbf{y}\| \leqslant \|\mathbf{x}\| + \|\mathbf{y}\|, \tag{5}$$

with equality holding if and only if one of the vectors is a *nonnegative* scalar multiple of the other.

The inequality (4) is known as the Cauchy–Schwarz inequality. The reader should not be discouraged by the slickness of the proof that we will give or by the slickness of other proofs. Usually, the first time that a mathematical result is discovered, the proof is rather complicated. Often, many very smart people then look at it and constantly improve the proof until a relatively simple and clever argument is found. Therefore, do not be intimidated by clever arguments. Now to the proof.

Let **x** and **y** be any pair of fixed vectors in $\mathbb{R}^n$. Let λ be any real scalar. Then

$$\begin{aligned} 0 \leqslant \|\mathbf{x} + \lambda\mathbf{y}\|^2 &= \langle \mathbf{x} + \lambda\mathbf{y}, \mathbf{x} + \lambda\mathbf{y} \rangle \\ &= \|\mathbf{x}\|^2 + 2\lambda\langle \mathbf{x}, \mathbf{y} \rangle + \|\mathbf{y}\|^2\lambda^2 \quad \text{for all } \lambda. \end{aligned} \tag{6}$$

This says that the quadratic in λ on the right either has a double real root or has no real roots. Therefore, by the quadratic formula, its discriminant must satisfy

$$4\langle \mathbf{x}, \mathbf{y} \rangle^2 - 4\|\mathbf{x}\|^2\|\mathbf{y}\|^2 \leqslant 0.$$

Transposing and taking square roots give

$$|\langle \mathbf{x}, \mathbf{y} \rangle| \leqslant \|\mathbf{x}\|\,\|\mathbf{y}\|.$$

Equality holds if and only if the quadratic has a double real root, say $\lambda = \kappa$. But then, from (6), $\|\mathbf{x} + \kappa\mathbf{y}\| = 0$. Hence by (2) $\mathbf{x} + \kappa\mathbf{y} = 0$, and $\mathbf{x} = -\kappa\mathbf{y}$.

To prove (5), we write

$$\begin{aligned} \|\mathbf{x} + \mathbf{y}\|^2 = \langle \mathbf{x} + \mathbf{y}, \mathbf{x} + \mathbf{y} \rangle &= \|\mathbf{x}\|^2 + 2\langle \mathbf{x}, \mathbf{y} \rangle + \|\mathbf{y}\|^2 \\ &\leqslant \|\mathbf{x}\|^2 + 2\|\mathbf{x}\|\,\|\mathbf{y}\| + \|\mathbf{y}\|^2 = (\|\mathbf{x}\| + \|\mathbf{y}\|)^2, \end{aligned} \tag{7}$$

where the inequality follows from the Cauchy–Schwarz inequality. If we now take square roots, we get the triangle inequality. We obtain equality in (7), and hence in the triangle inequality, if and only if $\langle \mathbf{x}, \mathbf{y} \rangle$ is nonnegative and either **y** is a scalar multiple of **x** or **x** is a scalar multiple of **y**. But under these circumstances, for $\langle \mathbf{x}, \mathbf{y} \rangle$ to be nonnegative, the multiple must be nonnegative. The inequality (5) is called the triangle inequality because when it is applied to vectors in $\mathbb{R}^2$ or $\mathbb{R}^3$ it says that the length of a third side of a triangle is less than or equal to the sum of the lengths of the other two sides.

Exercise 1.1. For any vectors $\mathbf{x}$ and $\mathbf{y}$ in $\mathbb{R}^n$, show that $\|\mathbf{x}+\mathbf{y}\|^2 + \|\mathbf{x}-\mathbf{y}\|^2 = 2\|\mathbf{x}\|^2 + 2\|\mathbf{y}\|^2$. Interpret this relation as a statement about parallelograms in $\mathbb{R}^2$ and $\mathbb{R}^3$.

In elementary courses it is shown that in $\mathbb{R}^2$ and $\mathbb{R}^3$ the cosine of the angle θ, $0 \leqslant \theta \leqslant \pi$, between two nonzero vectors $\mathbf{x}$ and $\mathbf{y}$ is given by the formula

$$\cos\theta = \frac{\langle \mathbf{x}, \mathbf{y}\rangle}{\|\mathbf{x}\|\,\|\mathbf{y}\|}.$$

Thus, $\mathbf{x}$ and $\mathbf{y}$ are orthogonal if and only if $\langle \mathbf{x}, \mathbf{y}\rangle = 0$. In $\mathbb{R}^n$, the Cauchy–Schwarz inequality allows us to give meaning to the notion of angle between nonzero vectors as follows. Since $|\langle \mathbf{x}, \mathbf{y}\rangle|| \leqslant \|\mathbf{x}\|\,\|\mathbf{y}\|$, the absolute value of the quotient

$$\frac{\langle \mathbf{x}, \mathbf{y}\rangle}{\|\mathbf{x}\|\,\|\mathbf{y}\|} \tag{8}$$

is less than or equal to 1, and thus (8) is the cosine of an angle between zero and π. We define this angle to be the angle between $\mathbf{x}$ and $\mathbf{y}$. We say that $\mathbf{x}$ and $\mathbf{y}$ are *orthogonal* if $\langle \mathbf{x}, \mathbf{y}\rangle = 0$.

3. ALGEBRA OF SETS

Given two sets A and B, the set A is said to be a *subset* of B, or to be *contained* in B, and written $A \subseteq B$, if every element of A is also an element of B. The set A is said to be a *proper subset* of B if A is a subset of B and there exists at least one element of B that is not in A. This is written as $A \subset B$. Two sets A and B are said to be *equal*, written $A = B$, if $A \subseteq B$ and $B \subseteq A$. The *empty set*, or *null* set, is the set that has no members. The empty set will be denoted by $\varnothing$. The notation $x \in S$ will mean "x belongs to the set S" or "x is an element of the set S."

Let $\{S_\alpha\}_{\alpha\in A}$ be a collection of sets indexed by an index set A. By the *union* of the sets $\{S_\alpha\}_{\alpha\in A}$, denoted by $\bigcup_{\alpha\in A} S_\alpha$, we mean the set consisting of all elements that belong to at least one of the S_α. Note that if not all of the S_α are empty, then $\bigcup_{\alpha\in A} S_\alpha$ is not empty. By the *intersection* of the sets $\{S_\alpha\}_{\alpha\in A}$, denoted by $\bigcap_{\alpha\in A} S_\alpha$, we mean the set of points s with the property that s belongs to *every* set S_α. Note that for a collection $\{S_\alpha\}_{\alpha\in A}$ of nonempty sets, the intersection can be empty.

Let $\mathfrak{X}$ be a set, say $\mathbb{R}^n$ for example, and let S be a subset of $\mathfrak{X}$. By the *complement* of S (relative to $\mathfrak{X}$), written cS, we mean the set consisting of those points in $\mathfrak{X}$ that do not belong to S. Note that $c\mathfrak{X}$ is the empty set. By convention, we take the complement of the empty set to be $\mathfrak{X}$. We thus always have $c(cS) = S$. Also, if $A \subset B$, then $cA \supset cB$. The reader should convince himself or herself of the truth of the following lemma, known as *de Morgan's*

cS = the complement of the set S

law, either by writing out a proof or drawing Venn diagrams for some simple cases.

LEMMA 3.1. *Let $\{S_\alpha\}_{\alpha \in A}$ be a collection of subsets of a set $\mathfrak{X}$. Then*

$$\bigcup_{\alpha \in A} S_\alpha = c\left[\bigcap_{\alpha \in A} (cS_\alpha)\right],$$

$$\bigcap_{\alpha \in A} S_\alpha = c\left[\bigcup_{\alpha \in A} (cS_\alpha)\right].$$

4. METRIC TOPOLOGY OF $\mathbb{R}^n$

Let $\mathfrak{X}$ be a set. A function d that assigns a real number to each pair of points (x, y) with $x \in \mathfrak{X}$ any $y \in \mathfrak{X}$ is said to be a *metric*, or *distance function*, on $\mathfrak{X}$ if it satisfies the following:

$$d(x, y) \geqslant 0, \quad \text{with equality holding if and only if } x = y, \tag{1}$$

$$d(x, y) = d(y, x), \tag{2}$$

$$d(x, y) \leqslant d(x, z) + d(z, y) \quad \text{for all } x, y, z \text{ in } \mathfrak{X}. \tag{3}$$

The last inequality is called the triangle inequality.

In $\mathbb{R}^n$ we define a function d on pairs of points $\mathbf{x}, \mathbf{y}$ by the formula

$$d(\mathbf{x}, \mathbf{y}) = \|\mathbf{x} - \mathbf{y}\| = \left(\sum_{i=1}^{n} (x_i - y_i)^2\right)^{1/2}. \tag{4}$$

Exercise 4.1. Use the properties of the norm to show that the function d defined by (4) is a metric, or distance function, on $\mathbb{R}^n$.

A set $\mathfrak{X}$ with metric d is called a *metric space*. Although metric spaces occur in many areas of mathematics and applications, we shall confine ourselves to $\mathbb{R}^n$. In $\mathbb{R}^2$ and $\mathbb{R}^3$ the function defined in (4) is the ordinary euclidean distance.

We now present some important definitions to the reader, who should master them and try to improve his or her understanding by drawing pictures in $\mathbb{R}^2$.

The (open) *ball* centered at $\mathbf{x}$ with radius $r > 0$, denoted by $B(\mathbf{x}, r)$, is defined to be the set of points $\mathbf{y}$ whose distance from $\mathbf{x}$ is *less than* r. We write this in symbols as

$$B(\mathbf{x}, r) = \{\mathbf{y} : \|\mathbf{y} - \mathbf{x}\| < r\}.$$

The *closed ball* $\overline{B(\mathbf{x}, r)}$ with center at $\mathbf{x}$ and radius $r > 0$ is defined by

$$\overline{B(\mathbf{x}, r)} = \{\mathbf{y} : \|\mathbf{y} - \mathbf{x}\| \leqslant r\}.$$

$B(x,r)$ = the open ball

$\overline{B(x,r)}$ = the closed ball

S' = the limit points of a set S

Let $S \subseteq \mathbb{R}^n$. A point $\mathbf{x}$ is an *interior point* of S if there exists an $r > 0$ such that $B(\mathbf{x}, r) \subset S$. A set S need not have any interior points. For example, consider the set of points in the plane of the form $(x_1, 0)$ with $0 < x_1 < 1$. This is the interval $0 < x_1 < 1$ on the x_1-axis. It has no interior points, considered as a set in $\mathbb{R}^2$. On the other hand, if we consider this as a set in $\mathbb{R}^1$, every point is an interior point. This leads to the important observation that for a given set, whether or not it has interior points may depend on in which euclidean space $\mathbb{R}^n$ we take the set to be lying, or *embedded*.

If the set of interior points of a set S is not empty, then we call the set of interior points the *interior* of S and denote it by $\operatorname{int}(S)$.

A set S is said to be *open* if all points of S are interior points. Thus an equivalent definition of an open set is that it is equal to its interior. If we go back to the definition of interior point, we can restate the definition of an open set as follows. A set S is open if for every point $\mathbf{x}$ in S there exists a positive number $r > 0$, which may depend on $\mathbf{x}$, such that the ball $B(\mathbf{x}, r)$ is contained in S. The last definition, while being wordy, is the one that the reader should picture mentally.

A point $\mathbf{x}$ is said to be a *limit point* of a set S if for every $\varepsilon > 0$ there exists a point $\mathbf{x}_\varepsilon \neq \mathbf{x}$ such that $\mathbf{x}_\varepsilon$ belongs to S and $\mathbf{x}_\varepsilon \in B(\mathbf{x}, \varepsilon)$. The point $\mathbf{x}_\varepsilon$ will in general depend on ε. A set need not have any limit points, and a limit point of a set need not belong to the set. An example of a set without a limit point is the set $\mathbb{N}$ of positive integers, considered as a set in $\mathbb{R}^1$. For an example of a limit point that does not belong to a set, consider the set $S \equiv \{x : x = 1/n,\ n = 1, 2, 3, \ldots\}$ in $\mathbb{R}^1$. Zero is a limit point of the set, yet zero does not belong to S. We shall denote the set of limit points of a set S by S'.

Exercise 4.2. (a) Sketch the graph of $y = \sin(1/x)$, $x > 0$.

(b) Consider the graph as a set in $\mathbb{R}^2$ and find the limit points of this set.

$\bar{S}$ = the closure of a set S

The *closure* of a set S, denoted by $\bar{S}$, is defined to be the set $S \cup S'$; that is, $\bar{S} = S \cup S'$. A set S is said to be *closed* if S contains all of its limit points; that is, $S' \subset S$. A set S is closed if and only if $S = \bar{S}$. To see this, note that $S = \bar{S}$ and the definitions $\bar{S} = S \cup S'$ imply that $S = S \cup S'$. Hence $S' \subset S$. On the other hand, if $S' \subset S$, then $S \cup S' = S$, and so $\bar{S} = S \cup S' = S$.

A set can be neither open nor closed. In $\mathbb{R}^2$ consider $B(\mathbf{0}, 1)$, the ball with center at the origin and radius 1. It should be intuitively clear that all points $\mathbf{x}$ in $\mathbb{R}^2$ with $\|\mathbf{x}\| = 1$ are limit points of $B(\mathbf{0}, 1)$. Now consider the set

Do this

$$S = B(\mathbf{0}, 1) \cup \{\mathbf{x} = (x_1, x_2) : \|\mathbf{x}\| = 1,\ x_1 \geqslant 0\}.$$

(The reader should sketch this set.) Points $\mathbf{x} = (x_1, x_2)$ with $\|\mathbf{x}\| = 1$ and $x_1 \geqslant 0$ are not interior points of S since for such an $\mathbf{x}$, no matter how small we choose $\varepsilon > 0$, the ball $B(\mathbf{x}, \varepsilon)$ will not belong to S. Hence S is not open. Also, S is not closed since points $\mathbf{x} = (x_1, x_2)$ with $\|\mathbf{x}\| = 1$ and $x_1 < 0$ are limit points of S, yet they do not belong to S.

It follows from our definitions that $\mathbb{R}^n$ itself is both open and closed. By convention, we will also take the empty set to be open and closed.

THEOREM 4.1. *Complements of closed sets are open. Complements of open sets are closed.*

Proof. If $S = \varnothing$, where $\varnothing$ denotes the empty set, or if $S = \mathbb{R}^n$, then there is nothing to prove. Let $S \neq \varnothing$, $S \neq \mathbb{R}^n$ and let S be closed. Let $\mathbf{x}$ be any point in cS. Then $\mathbf{x} \notin S$ and $\mathbf{x}$ is not a limit point of S. If we recall the definition of "$\mathbf{x}$ is a limit point of S," we see that the statement "$\mathbf{x}$ is not a limit point of S" means that there is *an* $\varepsilon_0 > 0$ such that all points $\mathbf{y} \neq \mathbf{x}$ with $\mathbf{y} \in B(\mathbf{x}, \varepsilon_0)$ are in cS. Since $\mathbf{x} \in cS$, this means that $B(\mathbf{x}, \varepsilon_0) \subset cS$, and so cS is open. Now let $S = \varnothing$, $S \neq \mathbb{R}^n$ and let S be open. Let $\mathbf{x}$ be a limit point of cS. Then for *every* $\varepsilon > 0$ there exists a point $\mathbf{x}_\varepsilon \neq \mathbf{x}$ in cS such that $\mathbf{x}_\varepsilon \in B(\mathbf{x}, \varepsilon)$. Since $\mathbf{x}_\varepsilon \in cS$, the ball $B(\mathbf{x}, \varepsilon)$ does not belong to S. Thus $\mathbf{x}$ cannot belong to the open set S, for then we would be able to find an ε_0 such that $B(\mathbf{x}, \varepsilon_0)$ did belong to S. Hence $\mathbf{x} \in cS$, so cS is closed.

Exercise 4.3. Show that for $\mathbf{x} \in \mathbb{R}^n$ and $r > 0$ the set $B(\mathbf{x}, r)$ is open; that is, show that an open ball is open.

Exercise 4.4. Show that for $\mathbf{x} \in \mathbb{R}^n$ and $r > 0$ the closed ball $\overline{B(\mathbf{x}, r)}$ is closed.

Exercise 4.5. Show that any finite set of points $\mathbf{x}_1, \ldots, \mathbf{x}_k$ in $\mathbb{R}^n$ is closed.

Exercise 4.6. Show that in $\mathbb{R}^n$ no point $\mathbf{x}$ with $\|\mathbf{x}\| = 1$ is an interior point of $B(\mathbf{0}, 1)$.

Exercise 4.7. Show that for any set S in $\mathbb{R}^n$ the set $\bar{S}$ is closed.

Exercise 4.8. Show that for any set S the closure $\bar{S}$ is equal to the intersection of all closed sets containing S.

THEOREM 4.2. (*i*) *Let $\{O_\alpha\}_{\alpha \in A}$ be a collection of open sets. Then $\bigcup_{\alpha \in A} O_\alpha$ is open.*
(*ii*) *Let $O_1, \ldots, O_n$ be a finite collection of open sets. Then $\bigcap_{i=1}^n O_i$ is open.*
(*iii*) *Let $\{F_\alpha\}_{\alpha \in A}$ be a collection of closed sets. Then $\bigcap_{\alpha \in A} F_\alpha$ is closed.*
(*iv*) *Let $F_1, \ldots, F_n$ be a finite collection of closed sets. Then $\bigcup_{i=1}^n F_i$ is closed.*

Note that an infinite collection of open sets need not have an intersection that is open. For example, in $\mathbb{R}^1$, for each $n = 1, 2, 3, \ldots$, let

$$O_n = (1 - 1/n, 1 + 1/n) = \{x \in \mathbb{R}^1 : 1 - 1/n < x < 1 + 1/n\}.$$

In $\mathbb{R}^1$ each O_n is open, and $\bigcap_{n=1}^\infty O_n = \{1\}$, a point. By Exercise 4.5, the set consisting of the point $x = 1$ is closed. Similarly, the union of an infinite number of closed sets need not be closed. To see this, for each $n = 1, 2, 3, \ldots$ let $F_n = [0, 1 - 1/(n+1)] = \{x \in \mathbb{R}^1 : 0 \leqslant x \leqslant 1 - 1/(n+1)\}$. Each F_n is closed, and

$\bigcup_{n=1}^{\infty} F_n = [0, 1) = \{x \in \mathbb{R}^1 : 0 \leqslant x < 1\}$, which is not closed since it does not contain the limit point $x = 1$.

We now prove the theorem. To establish (i), we take an arbitrary element $\mathbf{x}$ in $\bigcup_{\alpha \in A} O_\alpha$. Since $\mathbf{x}$ is in the union, it must belong to at least one of the sets $\{O_\alpha\}_{\alpha \in A}$, say O_{α_0}. Since O_{α_0} is open, there exists an $r_0 > 0$ such that $B(\mathbf{x}, r_0) \subseteq O_{\alpha_0}$. Hence $B(\mathbf{x}, r_0) \subseteq \bigcup_{\alpha \in A} O_\alpha$, and thus the union is open. To establish (ii), let $\mathbf{x}$ be an arbitrary point in $\bigcap_{i=1}^{n} O_i$. Then $\mathbf{x} \in O_i$ for each $i = 1, \ldots, n$. Since each O_i is open, for each $i = 1, \ldots, n$ there exists an $r_i > 0$ such that $B(\mathbf{x}, r_i) \subseteq O_i$. Let $r = \min\{r_i : i = 1, \ldots, n\}$. Then $B(\mathbf{x}, r) \subseteq B(\mathbf{x}, r_i) \subseteq O_i$ for each $i = 1, \ldots, n$. Hence $B(\mathbf{x}, r) \subset \bigcap_{i=1}^{n} O_i$, and thus the intersection is open.

The most efficient way to establish (iii) is to write

$$\bigcap_{\alpha \in A} F_\alpha = \bigcap_{\alpha \in A} c(cF_\alpha) = c\left[\bigcup_{\alpha \in A} (cF_\alpha)\right],$$

where the second equality follows from Lemma 3.1. Since F_α is closed, cF_α is open and so is $\bigcup_{\alpha \in A} (cF_\alpha)$. Hence the complement of this set is closed, and therefore so is $\bigcap_{\alpha \in A} F_a$. To establish (iv), write

$$\bigcup_{i=1}^{n} F_i = \bigcup_{i=1}^{n} c(cF_i) = c\left[\bigcap_{i=1}^{n} (cF_i)\right].$$

Again, cF_i is open for each $i = 1, \ldots, n$. Hence the intersection $\bigcap_{i=1}^{n} (cF_i)$ is open and the complement of the intersection is closed.

The reader who wishes to test his or her mastery of the relevant definitions should prove statements (iii) and (iv) of the theorem directly.

5. LIMITS AND CONTINUITY

A *sequence* in $\mathbb{R}^n$ is a function that assigns to each positive integer k a vector or point $\mathbf{x}_k$ in $\mathbb{R}^n$. We usually write the sequence as $\{\mathbf{x}_k\}_{k=1}^{\infty}$ or $\{\mathbf{x}_k\}$, rather than in the usual function notation. Examples of sequences in $\mathbb{R}^2$ are

$$\begin{aligned}
&\text{(i)} \quad \{\mathbf{x}_k\} = \{(k, k)\}, \\
&\text{(ii)} \quad \{\mathbf{x}_k\} = \left\{\left(\cos\frac{k\pi}{2}, \sin\frac{k\pi}{2}\right)\right\}, \\
&\text{(iii)} \quad \{\mathbf{x}_k\} = \left\{\left(\frac{(-1)^k}{2^k}, \frac{1}{2^k}\right)\right\}, \\
&\text{(iv)} \quad \{\mathbf{x}_k\} = \left\{\left((-1)^k - \frac{1}{k}, (-1)^k - \frac{1}{k}\right)\right\}.
\end{aligned} \tag{1}$$

The reader should plot the points in these sequences.

A sequence $\{\mathbf{x}_k\}$ is said to have a *limit* $\mathbf{y}$ or to *converge* to $\mathbf{y}$ if for *every* $\varepsilon > 0$ there is a positive integer $K(\varepsilon)$ such that whenever $k > K(\varepsilon)$, $\mathbf{x}_k \in B(\mathbf{y}, \varepsilon)$. We write

$$\lim_{k\to\infty} \mathbf{x}_k = \mathbf{y} \quad \text{or} \quad \mathbf{x}_k \to \mathbf{y}.$$

The sequences (i), (ii), and (iv) in (1) do not converge, while the sequence (iii) converges to $\mathbf{0} = (0, 0)$. Sequences that converge are said to be *convergent*; sequences that do not converge are said to be *divergent*. A sequence $\{\mathbf{x}_k\}$ is said to be *bounded* if there exists an $M > 0$ such that $\|\mathbf{x}_k\| < M$ for all k. A sequence that is not bounded is said to be unbounded. The sequence (i) is unbounded; the sequences (ii), (iii), and (iv) are bounded.

The following theorem lists some basic properties of convergent sequences.

THEOREM 5.1. (*i*) *The limit of a convergent sequence is unique.*

(*ii*) *Let* $\lim_{k\to\infty} \mathbf{x}_k = \mathbf{x}_0$, *let* $\lim_{k\to\infty} \mathbf{y}_k = \mathbf{y}_0$, *and let* $\lim_{k\to\infty} \alpha_k = \alpha$, *where* $\{\alpha_k\}$ *is a sequence of scalars. Then* $\lim_{k\to\infty} (\mathbf{x}_k + \mathbf{y}_k)$ *exists and equals* $\mathbf{x}_0 + \mathbf{y}_0$, *and* $\lim_{k\to\infty} \alpha_k \mathbf{x}_k$ *exists and equals* $\alpha\mathbf{x}_0$.

(*iii*) *A convergent sequence is bounded.*

Exercise 5.1. Prove Theorem 5.1.

Let $\{\mathbf{x}_k\}$ be a sequence in $\mathbb{R}^n$ and let $r_1 < r_2 < r_3 < \cdots < r_m < \cdots$ be a strictly increasing sequence of positive integers. Then the sequence $\{\mathbf{x}_{r_m}\}_{m=1}^{\infty}$ is called a *subsequence* of $\{\mathbf{x}_k\}$. Thus, $\{\mathbf{x}_{2k}\}_{k=1}^{\infty} = \{(2k, 2k)\}$ is a subsequence of (i). The sequence $\{\mathbf{x}_{2k}\}_{k=1}^{\infty} = \{(\cos k\pi, \sin k\pi)\}_{k=1}^{\infty}$ is a subsequence of (ii). Each of the sequences $\{\mathbf{x}_{2k}\} = \{((-1)^{2k} - 1/2k, (-1)^{2k} - 1/2k)\} = \{(1 - 1/2k, 1 - 1/2k)\}$ and $\{\mathbf{x}_{2k+1}\} = \{(-1 - 1/(2k+1), -1 - 1/(2k+1))\}$ is a subsequence of (iv). The reader should plot the various subsequences.

We can now formulate a very useful criterion for a point to be a limit point of a set. (The reader should not confuse the notions of limit and limit point. They are different.)

LEMMA 5.1. *A point* $\mathbf{x}$ *is a limit point of a set* S *if and only if there exists a sequence* $\{\mathbf{x}_k\}$ *of points in* S *such that, for each* k, $\mathbf{x}_k \neq \mathbf{x}$ *and* $\mathbf{x}_k \to \mathbf{x}$.

Proof. Let $\mathbf{x}$ be a limit point of S. Then for every positive integer k there is a point $\mathbf{x}_k$ in S such that $\mathbf{x}_k \neq \mathbf{x}$ and $\mathbf{x}_k \in B(\mathbf{x}, 1/k)$. For every $\varepsilon > 0$, there is a positive integer $K(\varepsilon)$ satisfying $1/K(\varepsilon) < \varepsilon$. Hence for $k > K(\varepsilon)$, we have $1/k < \varepsilon$ and $B(\mathbf{x}, 1/k) \subset B(\mathbf{x}, \varepsilon)$. Hence the sequence $\{\mathbf{x}_k\}$ converges to $\mathbf{x}$. Conversely, let there exist a sequence $\{\mathbf{x}_k\}$ of points in S with $\mathbf{x}_k \neq \mathbf{x}$ and $\mathbf{x}_k \to \mathbf{x}$. Let $\varepsilon > 0$ be arbitrary. Since $\mathbf{x}_k \to \mathbf{x}$, there exists a positive integer $K(\varepsilon)$ such that, for $k > K(\varepsilon)$, $\mathbf{x}_k \in B(\mathbf{x}, \varepsilon)$. Since by hypothesis $\mathbf{x}_k \neq \mathbf{x}$ for all k, it follows that we have satisfied the requirements of the definition that $\mathbf{x}$ is a limit point of S.

Remark 5.1. Let $\{\mathbf{x}_k\}$ be a sequence of points belonging to a set S and converging to a point $\mathbf{x}$. Then $\mathbf{x}$ must belong to $\bar{S}$. If $\mathbf{x}$ is in S, there is nothing to prove. If $\mathbf{x}$ were not in S, then since $\mathbf{x}_k \in S$ for each k, it follows that $\mathbf{x}_k \neq \mathbf{x}$. Lemma 5.1 then implies that $\mathbf{x}$ is a limit point of S. Thus $\mathbf{x} \in \bar{S}$. If S is a closed set, then since $S = \bar{S}$, the point $\mathbf{x}$ belongs to S.

Remark 5.2. Let S be a set in $\mathbb{R}^n$. If $\mathbf{x}$ belongs to $\bar{S}$, then there is a sequence of points $\{\mathbf{x}_k\}$ in S such that $\mathbf{x}_k \to \mathbf{x}$. To see this, note that if $\mathbf{x}$ is in $\bar{S}$, then $\mathbf{x}$ belongs either to S or to S'. If $\mathbf{x} \in S'$, the assertion follows from Lemma 5.1. If $\mathbf{x} \in S$ and $\mathbf{x} \notin S'$, we may take $\mathbf{x}_k = \mathbf{x}$ for all positive integers k.

Let S be a subset (proper or improper) of $\mathbb{R}^n$. Let $\mathbf{f}$ be a function with domain S and range contained in $\mathbb{R}^m$. We denote this by $\mathbf{f}: S \to \mathbb{R}^m$. Thus, $\mathbf{f} = (f_1, \ldots, f_m)$, where each f_i is a real-valued function.

Let $\mathbf{s}$ be a limit point of S. We say that "*the limit of* $\mathbf{f}(\mathbf{x})$ *as* $\mathbf{x}$ *approaches* $\mathbf{s}$ *exists and equals* $\mathbf{L}$" if for every $\varepsilon > 0$ there exists a $\delta > 0$, which may depend on ε, such that for all $\mathbf{x} \in B(\mathbf{s}, \delta) \cap S$, $\mathbf{x} \neq \mathbf{s}$, we have $\mathbf{f}(\mathbf{x}) \in B(\mathbf{L}, \varepsilon)$. We write

$$\lim_{\mathbf{x} \to \mathbf{s}} \mathbf{f}(\mathbf{x}) = \mathbf{L}.$$

It is not hard to show that, if $\mathbf{L} = (L_1, \ldots, L_m)$, $\lim_{\mathbf{x} \to \mathbf{s}} \mathbf{f}(\mathbf{x}) = \mathbf{L}$ if and only if, for each $i = 1, \ldots, m$, $\lim_{\mathbf{x} \to \mathbf{s}} f_i(\mathbf{x}) = L_i$.

It is also not hard to show that *if a limit exists it is unique* and that the algebra of limits, as stated in the next theorem, holds.

THEOREM 5.2. *Let* $\lim_{\mathbf{x} \to \mathbf{s}} \mathbf{f}(\mathbf{x}) = \mathbf{L}$ *and let* $\lim_{\mathbf{x} \to \mathbf{s}} \mathbf{g}(\mathbf{x}) = \mathbf{M}$. *Then*

$$\lim_{\mathbf{x} \to \mathbf{s}} [\mathbf{f}(\mathbf{x}) + \mathbf{g}(\mathbf{x})]$$

exists and equals $\mathbf{L} + \mathbf{M}$. *Also* $\lim_{\mathbf{x} \to \mathbf{s}} [\alpha \mathbf{f}(\mathbf{x})]$ *exists and equals* $\alpha \mathbf{L}$.

A useful sequential criterion for the existence of a limit is now given.

THEOREM 5.3. $\operatorname{Lim}_{\mathbf{x} \to \mathbf{s}} \mathbf{f}(\mathbf{x}) = \mathbf{L}$ *if and only if for every sequence* $\{\mathbf{x}_k\}$ *of points in* S *such that* $\mathbf{x}_k \neq \mathbf{s}$ *for all* k *and* $\mathbf{x}_k \to \mathbf{s}$ *it is true that* $\lim_{k \to \infty} \mathbf{f}(\mathbf{x}_k) = \mathbf{L}$.

Proof. Let $\lim_{\mathbf{x} \to \mathbf{s}} \mathbf{f}(\mathbf{x}) = \mathbf{L}$ and let $\{\mathbf{x}_k\}$ be an arbitrary sequence of points in S with $\mathbf{x}_k \neq \mathbf{s}$ for all k and with $\mathbf{x}_k \to \mathbf{s}$. Let $\varepsilon > 0$ be given. Since $\lim_{\mathbf{x} \to \mathbf{s}} \mathbf{f}(\mathbf{x}) = \mathbf{L}$, there exists a $\delta(\varepsilon) > 0$ such that, for all $\mathbf{x} \neq \mathbf{s}$ and in $B(\mathbf{s}, \delta) \cap S$, $\mathbf{f}(\mathbf{x}) \in B(\mathbf{L}, \varepsilon)$. Since $\mathbf{x}_k \to \mathbf{s}$, there exists a positive integer $K(\delta(\varepsilon))$ such that whenever $k > K(\delta(\varepsilon))$, $\mathbf{x}_k \in B(\mathbf{s}, \delta)$. Thus, for $k > K(\delta(\varepsilon))$, $\mathbf{f}(\mathbf{x}_k) \in B(\mathbf{L}, \varepsilon)$, and so $\mathbf{f}(\mathbf{x}_k) \to \mathbf{L}$.

We now prove the implication in the other direction. Suppose that $\lim_{\mathbf{x} \to \mathbf{s}} \mathbf{f}(\mathbf{x})$ does not equal $\mathbf{L}$. Then there exists an $\varepsilon_0 > 0$ such that for every $\delta > 0$ there exists a point $\mathbf{x}_\delta$ in S satisying the conditions $\mathbf{x}_\delta \neq \mathbf{s}$, $\mathbf{x}_\delta \in B(\mathbf{s}, \delta)$, and $\mathbf{f}(\mathbf{x}_\delta) \notin B(\mathbf{L}, \varepsilon_0)$. Take δ through the sequence of values $\delta = 1/k$, where k runs through the positive integers. We then get a sequence $\{\mathbf{x}_k\}$, with $\mathbf{x}_k \in S$, $\mathbf{x}_k \neq \mathbf{s}$, and $\mathbf{x}_k \to \mathbf{s}$, with the further property that $\mathbf{f}(\mathbf{x}_k) \notin B(\mathbf{L}, \varepsilon_0)$. Thus $\mathbf{f}(\mathbf{x}_k)$ does not converge to $\mathbf{L}$, and the theorem is proved.

Let $\mathbf{x}_0$ be a point of S that is also a limit point of S; that is, $\mathbf{x}_0 \in S \cap S'$. We say that $\mathbf{f}$ is *continuous* at $\mathbf{x}_0$ if

$$\lim_{\mathbf{x} \to \mathbf{x}_0} \mathbf{f}(\mathbf{x}) = \mathbf{f}(\mathbf{x}_0).$$

If $\mathbf{x}_0 \in S$ and $\mathbf{x}_0$ is not a limit point of S, we will take $\mathbf{f}$ to be continuous at $\mathbf{x}_0$ by convention.

From the corresponding property of limits it follows that $\mathbf{f}$ is continuous at $\mathbf{x}_0$ if and only if each of the real-valued functions $f_1, \ldots, f_m$ is continuous at $\mathbf{x}_0$. It follows from Theorem 5.2 that if $\mathbf{f}$ and $\mathbf{g}$ are continuous at $\mathbf{s}$, then so are $\mathbf{f} + \mathbf{g}$ and $\alpha\mathbf{f}$.

The function $\mathbf{f}$ is said to be *continuous on* S if it is continuous at every point of S.

Exercise 5.2. Show that if $\lim_{\mathbf{x} \to \mathbf{s}} \mathbf{f}(\mathbf{x}) = \mathbf{L}$, then $\mathbf{L}$ is unique.

Exercise 5.3. Prove Theorem 5.2.

Exercise 5.4. Let f be a real-valued function defined on a set S in $\mathbb{R}^n$. Show that if f is continuous at a point $\mathbf{x}_0$ in S and if $f(\mathbf{x}_0) < 0$, then there exists a $\delta > 0$ such that $f(\mathbf{x}) < 0$ for all $\mathbf{x}$ in $B(\mathbf{x}_0, \delta) \cap S$.

Exercise 5.5. Show that the real-valued function $f : \mathbb{R}^n \to \mathbb{R}^1$ defined by $f(\mathbf{x}) = \|\mathbf{x}\|$ is continuous on $\mathbb{R}^n$ (i.e., show that the norm is a continuous function). *Hint:* Use the triangle inequality to show that, for any pair of vectors $\mathbf{x}$ and $\mathbf{y}$, $|\,\|\mathbf{x}\| - \|\mathbf{y}\|\,| \leqslant \|\mathbf{x} - \mathbf{y}\|$.

6. BASIC PROPERTY OF REAL NUMBERS

A set S in $\mathbb{R}^n$ is said to be *bounded* if there exists a positive number M such that $\|\mathbf{x}\| \leqslant M$ for all $\mathbf{x} \in S$. Another way of stating this is $S \subset B(\mathbf{0}, M)$. The number M is said to be a *bound* for the set. Note that if a set is bounded, then there are infinitely many bounds.

We now shall deal with sets in $\mathbb{R}^1$ only. In $\mathbb{R}^1$ a set S being bounded means that, for every x in S, $|x| \leqslant M$, or $-M \leqslant x \leqslant M$. A set S in $\mathbb{R}^1$ is said to be *bounded above* if there exists a real number A such that $x \leqslant A$ for all x in S. A set S is *bounded below* if there exists a real number B such that $x \geqslant B$ for all x in S. The number A is said to be an *upper bound* of S, the number B a *lower bound*.

A number U is said to be a *least upper bound* (l.u.b.), or *supremum* (sup), of a set S

(i) if U is an upper bound of S and

(ii) if U' is another upper bound of S and then $U' \geqslant U$.

If a set has a least upper bound, then the least upper bound is unique. To see this, let U_1 and U_2 be two least upper bounds of a set S. Then since U_2 is an upper bound and U_1 is a least upper bound, $U_2 \geqslant U_1$. Similarly, $U_1 \geqslant U_2$, and so $U_1 = U_2$.

Condition (ii) states that no number $A < U$ can be an upper bound of S. Therefore condition (ii) can be replaced by the equivalent statement:

(ii′) For every $\varepsilon > 0$ there is a number x_ε in S with $U \geqslant x_\varepsilon > U - \varepsilon$.

Let S be the set of numbers $1 - 1/n$, $n = 1, 2, 3, \ldots$. Then one is the l.u.b. of this set. Note that one does not belong to S. Now consider the set $S_2 = S \cup \{2\}$. The number 2 is the l.u.b. of this set and belongs to S_2. The reader should be clear about the distinction between an upper bound and a supremum, or l.u.b.

A number L is said to be a *greatest lower bound* (g.l.b.), or *infimum* (inf), of a set S

(i) if L is a lower bound of S and

(ii) if L' is another lower bound of S and then $L' \leqslant L$.

Observations analogous to those made following the definition of l.u.b. hold for the definition of g.l.b. We leave their formulation to the reader.

Exercise 6.1. Let $L =$ g.l.b. of a set S. Show that there exists a sequence of points $\{x_k\}$ in S such that $x_k \to L$. Show that the sequence $\{x_k\}$ can be taken to be nonincreasing, that is, $x_{k+1} \leqslant x_k$ for every k. Does L have to be a limit point of S?

Exercise 6.2. Let S be a set in $\mathbb{R}^1$. We define $-S$ to be $\{x: -x \in S\}$. Thus $-S$ is the set that we obtain by replacing each element x in S by the element $-x$. Show that S is bounded below if and only if $-S$ is bounded above. Show that α is the g.l.b. of S if and only if $-\alpha$ is the l.u.b. of $-S$.

We can now state the basic property of the real numbers $\mathbb{R}^1$, which is sometimes called the *completeness property*.

Basic Property of the Reals

Every set $S \subset \mathbb{R}^1$ that is bounded $\begin{pmatrix}\text{above}\\ \text{below}\end{pmatrix}$ has a $\begin{pmatrix}\text{l.u.b.}\\ \text{g.l.b.}\end{pmatrix}$.

It follows from Exercise 6.2 that every set in $\mathbb{R}^1$ that is bounded above has a l.u.b. if and only if every set that is bounded below has a g.l.b. Thus, our statement of the completeness property can be interpreted as two equivalent statements.

Not every number system has this property. Consider the rational numbers $\mathbb{Q}$. They possess all of the algebraic and order properties of the reals but do not possess the completeness property. We will outline the argument showing that the rationals are not complete. For details see Rudin [1976]. Recall that a rational number is a number that can be expressed as a quotient of integers and recall that $\sqrt{2}$ is not rational. (The ancient Greeks knew that $\sqrt{2}$ is not rational.) Consider the set S of rational numbers x defined by $S = \{x : x \text{ rational}, x^2 < 2\}$. Clearly, 2 is an upper bound. It can be shown that for any x in S there is an x' in S such that $x' > x$. Thus no element of S can be an upper bound for S, let alone a l.u.b. It can also be shown that for any rational number y not in S that is an upper bound of S there is another rational number y' that is not in S, that is, also an upper bound of S and satisfies $y' < y$. Thus no rational number not in S can be a l.u.b. of S. Hence, there is no l.u.b. of S in the *rationals*. The number $\sqrt{2}$ in the *reals* is the candidate for the title of l.u.b., but $\sqrt{2}$ is not in the system of rational numbers.

Exercise 6.3. Let A and B be two bounded sets of real numbers with $A \subseteq B$. Show that

$$\sup\{a : a \in A\} \leqslant \sup\{b : b \in B\},$$
$$\inf\{a : a \in A\} \geqslant \inf\{b : b \in B\}.$$

Exercise 6.4. Let A and B be two sets of real numbers.

(i) Show that if A and B are bounded above, then

$$\sup\{(a + b) : a \in A, b \in B\} = \sup\{a : a \in A\} + \sup\{b : b \in B\}.$$

(ii) Show that if A and B are bounded below, then

$$\inf\{(a + b) : a \in A, b \in B\} = \inf\{a : a \in A\} + \inf\{b : b \in B\}.$$

Exercise 6.5. Let a and b be real numbers and let $I = \{x : a \leqslant x \leqslant b\}$. A real-valued function f is said to have a *right-hand limit* ρ at a point x_0, where $a \leqslant x_0 < b$, if for every $\varepsilon > 0$ there exists a $\delta(\varepsilon)$ such that whenever $0 < x - x_0$

$< \delta(\varepsilon)$ the inequality $|f(x) - \rho| < \varepsilon$ holds. We shall write

$$f(x_0+) = \lim_{x \to x_0+} f(x) = \rho.$$

(i) Formulate a definition for a left-hand limit at a point x_0, where $a < x_0 \leqslant b$. If λ is the left-hand limit, we shall write

$$f(x_0-) = \lim_{x \to x_0-} f(x) = \lambda.$$

(ii) Show that f has a limit at a point x_0, where $a < x_0 < b$, if and only if f has right- and left-hand limits at x_0 and these limits are equal.

Exercise 6.6. Let I be as in Exercise 6.5. A real-valued function f defined on I is said to be nondecreasing if for any pair of points x_1 and x_2 in I with $x_2 > x_1$ we have $f(x_2) \geqslant f(x_1)$.

(i) Formulate the definition of nonincreasing function.
(ii) Show that if f is a nondecreasing function on I, then f has a right-hand limit at every point x, where $a \leqslant x < b$, and left hand limit at every point $a < x \leqslant b$.
(iii) Show that if I is an open interval $(a, b) = \{x : a < x < b\}$ and if f is nondecreasing and bounded below on I, then $\lim_{x \to a+} f(x)$ exists. If f is bounded above, then $\lim_{x \to b-} f(x)$ exists.

7. COMPACTNESS

A property that is of great importance in many contexts is that of compactness. The definition that we shall give is one that is applicable in very general contexts as well as in $\mathbb{R}^n$. We do this so that the reader who takes more advanced courses in analyis will not have to unlearn anything. After presenting the definition of compactness, we will state, without proof, two theorems that give necessary and sufficient conditions for a set S in $\mathbb{R}^n$ to be compact. We frequently shall use the properties stated in these theorems when working with compact sets.

Let S be a set in $\mathbb{R}^n$. A collection of open sets $\{O_\alpha\}_{\alpha \in A}$ is said to be an *open cover* of S if every $\mathbf{x} \in S$ is contained in some O_α. A set S in $\mathbb{R}^n$ is said to be *compact* if for every open cover $\{O_\alpha\}_{\alpha \in A}$ of S there is a *finite* collection of sets $O_1, \ldots, O_m$ from the original collection $\{O_\alpha\}_{\alpha \in A}$ such that the finite collection $O_1, \ldots, O_m$ is also an open cover of S. In mathematical jargon this property is stated as "Every open cover has a finite subcover."

To illustrate, consider the set S in $\mathbb{R}^1$ defined by $S = (0, 1) = \{x : 0 < x < 1\}$. For each positive integer k, let $O_k = (0, 1 - 1/k) = \{x : 0 < x < 1 - 1/k\}$. Then $\{O_k\}_{k=1}^{\infty}$ is an open cover of S, but no finite collection of sets in the cover will cover S. Thus, $(0, 1)$ is *not* compact. It is difficult to give a meaningful example of a compact set based on the definition, since we must show that *every* open cover has a finite subcover.

THEOREM 7.1. *In* $\mathbb{R}^n$ *a set* S *is compact if and only if every sequence* $\{\mathbf{x}_k\}$ *of points in* S *has a subsequence* $\{\mathbf{x}_{k_j}\}$ *that converges to a point in* S *(Bolzano–Weierstrass property).*

THEOREM 7.2. *In* $\mathbb{R}^n$ *a set* S *is compact if and only if* S *is closed and bounded.*

We refer the reader to Bartle and Sherbert [1999] or Rudin [1976] for proofs of these theorems.

We emphasize that the criteria for compactness given in Theorems 7.1 and 7.2 are not necessarily valid in contexts more general than $\mathbb{R}^n$.

Theorem 7.2 provides an easily applied criterion for compactness. Using Theorem 7.2, we see immediately that $S = (0, 1) = \{x : 0 < x < 1\}$ is not compact in $\mathbb{R}^1$. (S is not closed.) We also see that $S_1 = [0, 1] = \{x : 0 \leqslant x \leqslant 1\}$ is compact, since it is closed and bounded.

We conclude with another characterization of compactness that is valid in more general contexts than $\mathbb{R}^n$. A set S is said to have the *finite-intersection property* if for *every* collection of closed sets $\{F_\alpha\}$ such that $\bigcap F_\alpha \subseteq cS$ there is a finite subcollection $F_1, \dots, F_k$ such that $\bigcap_{i=1}^k F_i \subseteq cS$. We shall illustrate the definition by exhibiting a set that fails to have the finite-intersection property. The set $(0,1) = \{x : 0 < x < 1\}$ in $\mathbb{R}^1$ does not have the finite-intersection property. Take $F_n = [1 - 1/n, \infty)$, $n = 2, 3, 4, \dots$. Then $\bigcap_{n=2}^\infty F_n = [1, \infty) \subset c(0, 1)$, yet the intersection of any finite subcollection of the F_n's has nonempty intersection with $(0, 1)$ and so is not contained in $c(0, 1)$.

THEOREM 7.3. *A set* S *is compact if and only if it has the finite-intersection property.*

The proof is an exercise in the use of Lemma 3.1. Let S have the finite-intersection property. Let $\{O_\alpha\}$ be an open cover of S. Then $S \subseteq \bigcup O_\alpha = c\bigcap(cO_\alpha)$, and so $cS \supseteq c(c\bigcap(cO_\alpha)) = \bigcap(cO_\alpha)$. Each set cO_α is closed, so there exist a finite number of sets $O_1, \dots, O_k$ such that $cS \supseteq \bigcap_{i=1}^k (cO_i)$. Hence $S \subseteq c(\bigcap_{i=1}^k (cO_i)) = \bigcup_{i=1}^n O_i$, and so $\{O_\alpha\}$ has a finite subcover $O_1, \dots, O_k$. Thus we have shown that S is compact. Now let S be compact. Let $\{F_\alpha\}$ be a collection of closed sets such that $\bigcap F_\alpha \subseteq cS$. Then $c(\bigcup(cF_\alpha)) \subseteq cS$, and so $\bigcup(cF_\alpha) \supseteq S$. Thus $\{(cF_\alpha)\}$ is an open cover of S. Since S is compact, there exist a finite subcollection of sets $F_1, \dots, F_k$ such that $\bigcup_{i=1}^k (cF_i) \supseteq S$. Therefore $\bigcap_{i=1}^k F_i = c\bigcup_{i=1}^k (cF_i) \subseteq cS$, and thus S has the finite-intersection property.

Remark 7.1. Note that the finite-intersection property can be stated in the following equivalent form. Every collection of closed sets $\{F_\alpha\}$ such that $(\bigcap_\alpha F_\alpha) \cap S = \phi$ has a finite subcollection $F_1, \dots, F_k$ such that $(\bigcap_{i=1}^k F_i) \cap S = \phi$.

8. EQUIVALENT NORMS AND CARTESIAN PRODUCTS

A norm on a vector space $\mathfrak{D}$ is a function $\nu(\cdot)$ that assigns a real number $\nu(\mathbf{v})$ to every vector $\mathbf{v}$ in the space and has the properties (1)–(3) and (5) listed in Section 2, with $\|\cdot\|$ replaced by $\nu(\cdot)$. Once a norm has been defined, one can define a function ρ by the formula $\rho(\mathbf{v}, \mathbf{w}) = \nu(\mathbf{v} - \mathbf{w})$. Properties (1)–(3) and (5) of the norm imply that the function ρ satisfies the properties (1)–(3) of a metric and therefore is a metric, or distance function. Once a distance has been defined, open sets are defined in terms of the distance, as before. Convergence is defined in terms of open sets.

To illustrate these ideas, we give an example of another norm in $\mathbb{R}^n$, defined by

$$\nu(\mathbf{x}) = \|\mathbf{x}\|_\infty = \max\{|x_1|, \ldots, |x_n|\}.$$

The corresponding distance is defined by

$$\rho(\mathbf{x}, y) = \|\mathbf{x} - \mathbf{y}\|_\infty = \max\{|x_1 - y_1|, \ldots, |x_n - y_n|\}.$$

Since a norm determines the family of open sets and since the family of open sets determines convergence, the question arises, "Given two norms, ν_1 and ν_2, do they determine the same family of open sets?" The answer, which is not difficult to show, is, "If there exist positive constants m and M such that for all $\mathbf{x}$ in $\mathbb{R}^n$,

$$m\nu_2(\mathbf{x}) \leqslant \nu_1(\mathbf{x}) \leqslant M\nu_2(\mathbf{x}), \tag{1}$$

then a set is open under the ν_1 norm if and only if it is open under the norm ν_2." Two norms ν_1 and ν_2 that satisfy (1) are said to be *equivalent norms*.

An extremely important fact is that in $\mathbb{R}^n$ *all norms are equivalent*. For a proof see Fleming [1977]. In optimization theory, and in other areas, it may happen that the analysis of a problem or the development of a computational algorithm is more convenient in one norm than in another. The equivalence of the norms guarantees that the convergence of an algorithm is intrinsic, that is, is independent of the norm.

Let $A_1, \ldots, A_m$ be m sets with $A_i \subseteq R^{n_i}$, $i = 1, \ldots, m$. By the *cartesian product* of these sets, denoted by

$$\prod_{i=1}^{m} A_i \quad \text{or} \quad A_1 \times A_2 \times \cdots \times A_m,$$

we mean the set A in $\mathbb{R}^{n_1+n_2+\cdots n_m}$ consisting of all possible points in $\mathbb{R}^{n_1+\cdots+n_m}$ obtained by taking the first n_1 components from a point in A_1, the next n_2 components from a point in A_2, the next n_3 components from a point in A_3, and so forth, until the last n_m components, which are obtained by taking a point in A_m. For example, $\mathbb{R}^n$ can be considered to be the cartesian product of $\mathbb{R}^1$ taken n times with itself:

$$\mathbb{R}^n = \underbrace{\mathbb{R}^1 \times \mathbb{R}^1 \times \cdots \times \mathbb{R}^1}_{n \text{ times}}.$$

If A_1 is the set in $\mathbb{R}^2$ consisting of points on the circumference of the circle with center at the origin and radius 1 and if $A_2 = [0, 1]$ in $\mathbb{R}^1$, then $A_1 \times A_2$ is the set in $\mathbb{R}^3$ that is the lateral surface of a cylinder of height 1 whose base is the closed disk in the (x_1, x_2) plane with center at the origin and radius equal to 1.

Let $\mathfrak{X} = \mathbb{R}^{n_1} \times \cdots \times \mathbb{R}^{n_m}$. Let $\mathbf{x}$ be a vector in $\mathfrak{X}$. Then $\mathbf{x} = (\mathbf{x}_1, \ldots, \mathbf{x}_m)$, where $\mathbf{x}_i = (x_{i1_i}, \ldots, x_{in_i})$, $i = 1, \ldots, m$. If for vectors $\mathbf{x}$ and $\mathbf{y}$ in $\mathfrak{X}$ and scalars $\lambda \in \mathbb{R}^1$ we define

$$(\mathbf{x} + \mathbf{y}) = (\mathbf{x}_1 + \mathbf{y}_1, \ldots, \mathbf{x}_m + \mathbf{y}_m), \qquad \lambda\mathbf{x} = (\lambda\mathbf{x}_1, \ldots, \lambda\mathbf{x}_m)$$

and let $n = n_1 + n_1 + \cdots + n_m$, then $\mathfrak{X}$ can be identified with the usual vector space $\mathbb{R}^n$. The euclidean norm in $\mathfrak{X}$ would then be given by

$$\|\mathbf{x}\| = [\|\mathbf{x}_1\|^2 + \cdots + \|\mathbf{x}_m\|^2]^{1/2} = \left[\sum_{i=1}^{m} \sum_{j=1}^{n_i} (x_{ij})^2\right]^{1/2}. \tag{2}$$

On the other hand, in keeping with the definition used in more general contexts, the norm on $\mathfrak{X}$ is defined by

$$\|\mathbf{x}\|_\pi = \sum_{i=1}^{m} \|\mathbf{x}_i\|, \tag{3}$$

where $\|\mathbf{x}_i\|$ is the euclidean norm of $\mathbf{x}_i$ in $\mathbb{R}^{n_i}$, defined by $\|\mathbf{x}_i\| = \langle \mathbf{x}_i, \mathbf{x}_i \rangle^{1/2} = [\Sigma_{j=1}^{n_i} (x_{ij})^2]^{1/2}$. Since all norms on $\mathbb{R}^n$ are equivalent, the norm defined by (3) is equivalent to the euclidean norm defined by (2).

The following lemma is an immediate consequence of the definition of $\|\cdot\|_\pi$.

LEMMA 8.1. *A sequence $\mathbf{x}_k = (\mathbf{x}_{k1}, \ldots, \mathbf{x}_{km})$ of points in $\mathfrak{X}$ converges to a point $\mathbf{x}_0 = (\mathbf{x}_{01}, \ldots, \mathbf{x}_{0m})$ of $\mathfrak{X}$ in the metric determined by $\|\cdot\|_\pi$ if and only if for each $i = 1, \ldots, m$ the sequence $\{\mathbf{x}_{ki}\}$ converges to $\mathbf{x}_{0i}$ in the euclidean norm on $\mathbb{R}^{n_i}$.*

The next theorem can be proved by using Lemma 8.1 and Theorem 7.1 and by selecting appropriate subsequences.

THEOREM 8.1. *Let $A_1, \ldots, A_m$ be m sets with $A_i \subset \mathbb{R}^{n_i}$ and let each A_i be compact, $i = 1, \ldots, m$. Then $\Pi_{i=1}^{m} A_i$ is compact in $\mathbb{R}^n$, where $n = n_1 + n_2 + \cdots + n_m$.*

Exercise 8.1. (a) In $\mathbb{R}^2$ sketch the set $B_\infty(\mathbf{0}, 1) = \{\mathbf{x}: \|\mathbf{x}\|_\infty < 1\}$.

(b) Show that $\| \|_\infty$ is a norm [i.e., that it satisfies (1)–(3) and (5) of Section 2].

(c) Can you draw a sketch in $\mathbb{R}^2$ that would indicate that a set is open under the metric $d(\mathbf{x}, \mathbf{y}) = \|\mathbf{x} - \mathbf{y}\|$ if and only if it is open under the metric $\rho(\mathbf{x}, \mathbf{y}) = \|\mathbf{x} - \mathbf{y}\|_\infty$?

(d) Is it true that $\|\mathbf{x} + \mathbf{y}\|_\infty = \|\mathbf{x}\|_\infty + \|\mathbf{y}\|_\infty$ if and only if $\mathbf{y} = \kappa\mathbf{x}$, $\kappa > 0$?

Exercise 8.2. (a) Show that the function $\|\cdot\|_\pi$ defined by (3) is a norm on $\mathfrak{X}$; that is $\|\cdot\|_\pi$ satisfies (1)–(3) and (5) of Section 2.

(b) If $\mathfrak{X} = \mathbb{R}^2 = \mathbb{R}^1 \times \mathbb{R}^1$, sketch the set $B_\pi(0, 1) = \{\mathbf{x}: \|\mathbf{x}\|_\pi \leqslant 1\}$.

Exercise 8.3. Without using the fact that all norms in $\mathbb{R}^n$ are equivalent, show that $\|\cdot\|$ and $\|\cdot\|_\pi$ defined by (2) and (3), respectively, are equivalent norms on $\mathfrak{X}$.

Exercise 8.4. Show that the following functions are continuous:

(a) $f: \mathbb{R}^n \times \mathbb{R}^n \to \mathbb{R}^n$, where $f(\mathbf{x}, \mathbf{y}) = (\mathbf{x} + \mathbf{y})$ (addition is continuous);

(b) $f: \mathbb{R}^1 \times \mathbb{R}^n \to \mathbb{R}^n$, where $f(\lambda, \mathbf{x}) = \lambda\mathbf{x}$ (scalar multiplication is continuous); and

(c) $f: \mathbb{R}^n \times \mathbb{R}^n \to \mathbb{R}^1$, where $f(\mathbf{x}, \mathbf{y}) = \langle \mathbf{x}, \mathbf{y} \rangle$ (scalar product is continuous).

Exercise 8.5. Prove Theorem 8.1.

Exercise 8.6. Show that if Q is a positive-definite real symmetric matrix, then $v(\mathbf{x}) = \langle \mathbf{x}, Q\mathbf{x} \rangle^{1/2}$ defines a norm on $\mathbb{R}^n$. *Hint:* Look at Section 1.

9. FUNDAMENTAL EXISTENCE THEOREM

The basic problem in optimization theory is as follows: Given a set S (not necessarily in $\mathbb{R}^n$) and a real-valued function f defined on S, does there exist an element s_* in S such that $f(s_*) \leqslant f(s)$ for all s in S, and if so find it. The problem as stated is a minimization problem. The point s_* is said to be a *minimizer* for f, or to *minimize* f. We also say that "*f attains a minimum on S at s_**." The problem of "maximizing f over S," or finding an s^* in S such that $f(s^*) \geqslant f(s)$ for all s, can be reduced to the problem of minimizing the function $-f$ over S. To see this, observe that if $-f(s^*) \leqslant -f(s)$ for all s in S, then $f(s^*) \geqslant f(s)$ for all s in S, and conversely.

The first question that we address is, "Does a minimizer exist?" To point out some of the difficulties here, we look at some very simple examples in $\mathbb{R}^1$.

Let $S = \{x : x \geqslant 1\}$ and let $f(x) = 1/x$. It is clear that $\inf\{f(x) : x \geqslant 1\} = 0$, yet there is no $x_* \in S$ such that $f(x_*) = 0$. Thus a minimizer does not exist, or f does not attain its minimum on S. As a second example, let $S = [1, 2)$ and let $f(x) = 1/x$. Now, $\inf(f(x) : 1 \leqslant x < 2\} = \frac{1}{2}$, yet for no x_* in S do we have $f(x_*) = \frac{1}{2}$, Of course, if we modify the last example by taking $S = [1, 2] = \{x : 1 \leqslant x \leqslant 2\}$, then $x_* = 2$ is the minimizer. For future reference, note that if $S = (1, 2] = \{x : 1 < x \leqslant 2\}$, then $x_* = 2$ is still the minimizer.

We now state and prove the fundamental existence theorem for minimization of functions with domains in $\mathbb{R}^n$.

THEOREM 9.1. *Let S be a compact set in $\mathbb{R}^n$ and let f be a real-valued continuous function defined in S. The f attains a maximum and a minimum on S.*

Before taking up the proof, we point out that the theorem gives a sufficient condition for the existence of a maximizer and minimizer. It is not a necessary condition, as the example in $\mathbb{R}^1$, $S = (1, 2]$, $f(x) = 1/x$, shows. In more advanced courses the reader may encounter the notions of upper semicontinuity and lower semicontinuity and learn that continuity can be replaced by upper semicontinuity to ensure the existence of maximizers and that continuity can be replaced by lower semicontinuity to ensure the existence of minimizers. Finally, the reader should study the proof of this theorem very carefully and be sure to understand it. The proof of the theorem in $\mathbb{R}^n$ is the model for the proofs of the existence of minimizers in more general contexts than $\mathbb{R}^n$.

Proof. If f is continuous on S, then so is $-f$. If $-f$ attains a minimum on S, then f attains a maximum on S. Thus we need only prove that f attains a minimum.

We first prove that the set of values that f assumes on S is bounded below, that is, that the set $\mathscr{R} = \{y : y = f(x), x \in S\}$ is bounded below. To show this, we suppose that $\mathscr{R}$ were not bounded below. Then for each positive integer k there would exist a point $\mathbf{x}_k$ in S such that $f(\mathbf{x}_k) < -k$. Since S is compact, the sequence $\{\mathbf{x}_k\}$ of points in S has a subsequence $\{\mathbf{x}_{k_j}\}$ that converges to a point $\mathbf{x}_*$ in S. The continuity of f implies that $\lim_{j\to\infty} f(\mathbf{x}_{k_j}) = f(\mathbf{x}_*)$. Thus, by the definition of limit (taking $\varepsilon = 1$) we have that there exists a positive integer K such that, for $j > K$, $f(\mathbf{x}_{k_j}) > f(\mathbf{x}_*) - 1$. This contradicts the defining property of $\{\mathbf{x}_k\}$, namely that, for each k, $f(\mathbf{x}_k) < -k$.

Since the set $\mathscr{R}$ is bounded below, it has an infimum, say L. Therefore, by definition of infimum, for each positive integer k there exists a point $\mathbf{x}_k$ in S such that

$$L \leqslant f(\mathbf{x}_k) < L + \frac{1}{k}. \tag{1}$$

Since S is compact, the sequence $\{\mathbf{x}_k\}$ has a subsequence $\{\mathbf{x}_{k_j}\}$ that converges to a point $\mathbf{x}_*$ in S. The continuity of f implies that $\lim_{j\to\infty} f(\mathbf{x}_{k_j}) = f(\mathbf{x}_*)$. Therefore, if we replace k by k_j in (1) and let $j \to \infty$, we get that $L \leqslant f(x_*) \leqslant L$. Hence

$$f(\mathbf{x}_*) = L \leqslant f(\mathbf{x}) \quad \text{for all } \mathbf{x} \in S.$$

This proves the theorem.

If the reader understands the proof of Theorem 9.1 and understands the use of compactness, he or she should have no trouble in proving the following theorem. Actually, Theorem 9.1 is a corollary of Theorem 9.2.

THEOREM 9.2. *Let S be a compact subset of $\mathbb{R}^n$ and let $\mathbf{f} = (f_1, \ldots, f_m)$ be a continuous function defined on S with range in $\mathbb{R}^m$. Then the set*

$$\mathbf{f}(S) \equiv \mathscr{R} \equiv \{\mathbf{y} : \mathbf{y} = f(\mathbf{x}),\ \mathbf{x} \in S\}$$

is compact.

In mathematical jargon this result is stated as follows: The continuous image of a compact set is compact.

Exercise 9.1. Prove Theorem 9.2.

Exercise 9.2. Let A be a real $n \times n$ symmetric matrix with entries a_{ij} and let $Q(\mathbf{x})$ be the quadratic form

$$Q(\mathbf{x}) = \mathbf{x}^t A\mathbf{x} = \langle \mathbf{x}, A\mathbf{x} \rangle = \sum_{i,j=1}^{n} a_{ij} x_i x_j.$$

Show that there exists a constant $M > 0$ such that, for all $\mathbf{x}$, $Q(\mathbf{x}) \leqslant M\|\mathbf{x}\|^2$. Show that if $Q(\mathbf{x}) > 0$ for all $\mathbf{x} \neq \mathbf{0}$, then there exists a constant $m > 0$ such that, for all $\mathbf{x}$, $Q(\mathbf{x}) \geqslant m\|\mathbf{x}\|^2$. *Hint:* First show that the conclusion is true for $\mathbf{x} \in S(0, 1) \equiv \{\mathbf{x} : \|\mathbf{x}\| = 1\}$.

Exercise 9.3. Let f be a real-valued function defined on $\mathbb{R}^n$ such that

$$\lim_{\|\mathbf{x}\|\to\infty} f(\mathbf{x}) = +\infty.$$

By this is meant that for each $M > 0$ there exists an $R > 0$ such that if $\|\mathbf{x}\| \geqslant R$, then $f(\mathbf{x}) > M$. Such a function f is said to be *coercive*. Show that if f is continuous on $\mathbb{R}^n$ and is coercive, then f attains a minimum on $\mathbb{R}^n$.

10. LINEAR TRANSFORMATIONS

A *linear transformation* from $\mathbb{R}^n$ to $\mathbb{R}^m$ is a function T with domain $\mathbb{R}^n$ and range contained in $\mathbb{R}^m$ such that for every $\mathbf{x}, \mathbf{y}$ in $\mathbb{R}^n$ and every pair of scalars α, β

$$T(\alpha\mathbf{x} + \beta\mathbf{y}) = \alpha T(\mathbf{x}) + \beta T(\mathbf{y}).$$

Example 10.1. $T:\mathbb{R}^2 \to \mathbb{R}^2$: Rotate each vector in $\mathbb{R}^2$ through an angle θ.

Example 10.2. $T:\mathbb{R}^3 \to \mathbb{R}^2$ defined by

$$T(x_1, x_2, x_3) = (x_1, x_2, 0).$$

This is the projection of $\mathbb{R}^3$ onto the x_1x_2 plane.

Example 10.3. Let A be an $m \times n$ matrix. Define

$$T\mathbf{x} = A\mathbf{x}. \tag{1}$$

In the event that the range space is $\mathbb{R}^1$, that is, we have a linear transformation from $\mathbb{R}^n$ to $\mathbb{R}^1$, we call the transformation a *linear functional.* The following is an example of a linear functional L. Let $\mathbf{a}$ be a fixed vector in $\mathbb{R}^n$. Then

$$L(\mathbf{x}) = \langle \mathbf{a}, \mathbf{x} \rangle \tag{2}$$

defines a linear functional.

We now show that (1) and (2) are essentially the only examples of a linear transformation and linear functional, respectively. Recall that the *standard basis* in $\mathbb{R}^n$ is the basis consisting of the vectors $\mathbf{e}_1, \ldots, \mathbf{e}_n$, where $\mathbf{e}_i = (0, 0, \ldots, 0, 1, 0, \ldots, 0)$ and 1 is the ith entry.

THEOREM 10.1. *Let T be a linear transformation from $\mathbb{R}^n$ to $\mathbb{R}^m$. Then relative to the standard bases in $\mathbb{R}^n$ and $\mathbb{R}^m$ there exists an $m \times n$ matrix A such that $T\mathbf{x} = A\mathbf{x}$.*

Let $\mathbf{x} \in \mathbb{R}^n$; then $\mathbf{x} = \Sigma_{j=1}^n x_j\mathbf{e}_j$. Hence

$$T\mathbf{x} = T\left(\sum_{j=1}^n x_j\mathbf{e}_j\right) = \sum_{j=1}^n x_j(T\mathbf{e}_j). \tag{3}$$

For each $j = 1, \ldots, n$, $T\mathbf{e}_j$ is a vector in $\mathbb{R}^m$, so if $\mathbf{e}_1^*, \ldots, \mathbf{e}_m^*$ is the standard basis in $\mathbb{R}^m$,

$$T\mathbf{e}_j = \sum_{i=1}^m a_{ij}\mathbf{e}_i^*. \tag{4}$$

Substituting (4) into (3) and changing the order of summation give

$$T\mathbf{x} = \sum_{j=1}^{n} x_j \left(\sum_{i=1}^{m} a_{ij}\mathbf{e}_i^* \right) = \sum_{i=1}^{m} \left(\sum_{j=1}^{n} a_{ij}x_j \right) \mathbf{e}_i^*.$$

Therefore if A is the $m \times n$ matrix whose jth column is $(a_{1j} \dots a_{mj})^t$, then $T\mathbf{x}$ expressed as an m-vector $(y_1, \dots, y_m)^t$ is given by

$$\begin{pmatrix} y_1 \\ \vdots \\ y_m \end{pmatrix} = T\mathbf{x} = A\mathbf{x}.$$

From (4) we get that the jth column of A gives the coordinates of $T\mathbf{e}_j$ relative to $\mathbf{e}_i^*, \dots, \mathbf{e}_m^*$.

In the case of a linear functional L, the matrix A reduces to a row vector $\mathbf{a}$ in $\mathbb{R}^n$, so that

$$L(\mathbf{x}) = \langle \mathbf{a}, \mathbf{x} \rangle.$$

Exercise 10.1. Show that a linear functional L is a continuous mapping from $\mathbb{R}^n$ to $\mathbb{R}$.

11. DIFFERENTIATION IN $\mathbb{R}^n$

Let D be an open interval in $\mathbb{R}^1$ and let f be real-valued function defined on D. The function f is said to have *a derivative* or to be *differentiable at a point* x_0 in D if

$$\lim_{h \to 0} \frac{f(x_0 + h) - f(x_0)}{h} \tag{1}$$

exists. The limit is called the *derivative* of f at x_0 and is denoted by $f'(x_0)$.

The question now arises as to how one generalizes this concept to real-valued functions defined on open sets D in $\mathbb{R}^n$ and to functions defined on open sets in $\mathbb{R}^n$ with range in $\mathbb{R}^m$. In elementary calculus the notion of partial derivative for real-valued functions defined on open sets D in $\mathbb{R}^n$ is introduced. For each $i = 1, \dots, n$ the *partial derivative with respect to* x_i at a point $\mathbf{x}_0$ is defined by

$$\frac{\partial f}{\partial x_i}(\mathbf{x}_0) = \lim_{h \to 0} \frac{f(x_{01}, \dots, x_{0,i-1}, x_{0,i} + h, x_{0,i+1}, \dots, x_{0,n}) - f(x_{01}, \dots, x_{0,n})}{h} \tag{2}$$

provided the limit on the right exists. It turns out that the notion of partial derivative is not the correct generalization of the notion of derivative.

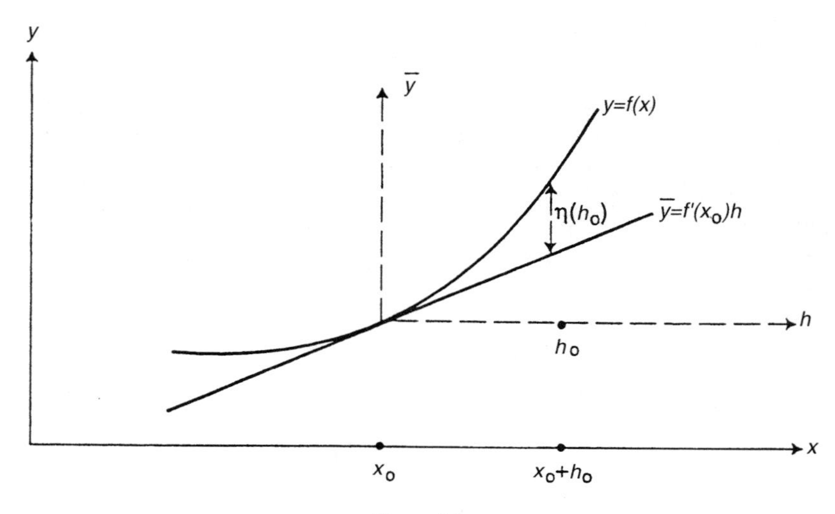

Figure 1.1.

To motivate the correct generalization of the notion of derivative to functions with domain and range in spaces of dimension greater than 1, we reexamine the notion of derivative for functions f defined on an open interval D in $\mathbb{R}^1$ with range in $\mathbb{R}^1$.

We rewrite (1) as

$$f'(x_0) = \frac{f(x_0 + h) - f(x_0)}{h} + \varepsilon(h),$$

where $\varepsilon(h) \to 0$ as $h \to 0$. This in turn can be rewritten as

$$f(x_0 + h) - f(x_0) = f'(x_0)h + \eta(h), \tag{3}$$

where $\eta(h)/h \to 0$ as $h \to 0$. The term $f'(x_0)h$ is a linear approximation to f near x_0 and it is a "good approximation for small h" in the sense that $\eta(h)/h \to 0$ as $h \to 0$. See Figure 1.1.

If we consider h to be an arbitrary real number, then $f'(x_0)h$ defines a linear functional L on $\mathbb{R}^1$ by the formula $L(h) = f'(x_0)h$. Thus if f is differentiable at x_0, there exists a linear functional (or linear transformation) on $\mathbb{R}^1$ such that

$$f(x_0 + h) - f(x_0) = L(h) + \eta(h), \tag{4}$$

where $\eta(h)/h \to 0$ as $h \to 0$. Conversely, let there exist a linear functional L on $\mathbb{R}^1$ such that (4) holds. Then $L(h) = ah$ for some real number a, and we may write

$$f(x_0 + h) - f(x_0) = ah + \eta(h).$$

If we divide by $h \neq 0$ and then let $h \to 0$, we get that $f'(x_0)$ exists and equals a. Thus we could have used (4) to define the notion of derivative and could have defined the derivative to be the linear functional L, which in this case is determined by the number a. The reader might feel that this is a very convoluted way of defining the notion of derivative. It has the merit, however, of being the definition that generalizes correctly to functions with domain in $\mathbb{R}^n$ and range in $\mathbb{R}^m$ for any $n \geqslant 1$ and any $m \geqslant 1$.

Let D be an open set in $\mathbb{R}^n$, $n \geqslant 1$, and let $\mathbf{f} = (f_1, \ldots, f_m)$ be a function defined on D with range in $\mathbb{R}^m$, $m \geqslant 1$. The function $\mathbf{f}$ is said to be *differentiable* at a point $\mathbf{x}_0$ in D, or to be *Fréchét differentiable* at $\mathbf{x}_0$, if there exists a linear transformation $T(\mathbf{x}_0)$ from $\mathbb{R}^n$ to $\mathbb{R}^m$ such that for all $\mathbf{h}$ in $\mathbb{R}^n$ with $\mathbf{h}$ sufficiently small

$$\mathbf{f}(\mathbf{x}_0 + \mathbf{h}) - \mathbf{f}(\mathbf{x}_0) = T(\mathbf{x}_0)\mathbf{h} + \boldsymbol{\eta}(\mathbf{h}), \tag{5}$$

where $\|\boldsymbol{\eta}(\mathbf{h})\|/\|\mathbf{h}\| \to 0$ as $\|\mathbf{h}\| \to 0$.

We now represent $T(\mathbf{x}_0)$ as a matrix relative to the standard bases $\mathbf{e}_1, \ldots, \mathbf{e}_n$ in $\mathbb{R}^n$ and $\mathbf{e}_1^*, \ldots, \mathbf{e}_m^*$ in $\mathbb{R}^m$. Let ε be a real number and let $\mathbf{h} = \varepsilon\mathbf{e}_j$ for a fixed j in the set $1, \ldots, n$. Then $\varepsilon\mathbf{e}_j = (0, \ldots, 0, \varepsilon, 0, \ldots, 0)$, where the ε occurs in the jth component. From (5) we have

$$\mathbf{f}(\mathbf{x}_0 + \varepsilon\mathbf{e}_j) - \mathbf{f}(\mathbf{x}_0) = T(\mathbf{x}_0)(\varepsilon\mathbf{e}_j) + \boldsymbol{\eta}(\varepsilon\mathbf{e}_j).$$

Hence for $\varepsilon \neq 0$

$$\frac{\mathbf{f}(\mathbf{x}_0 + \varepsilon\mathbf{e}_j) - \mathbf{f}(\mathbf{x}_0)}{\varepsilon} = T(\mathbf{x}_0)\mathbf{e}_j + \frac{\boldsymbol{\eta}(\varepsilon\mathbf{e}_j)}{\varepsilon}.$$

Since $\|\varepsilon\mathbf{e}_j\| = |\varepsilon|$ and for $i = 1, \ldots, m$,

$$\begin{aligned} f_i(\mathbf{x}_0 + \varepsilon\mathbf{e}_j) - f_i(\mathbf{x}_0) = f_i(&x_{01}, \ldots, x_{0,j-1}, x_{0j} + \varepsilon, x_{0,j+1}, \ldots, x_{0,n}) \\ &- f(x_{0,1}, \ldots, x_{0,n}), \end{aligned}$$

it follows on letting $\varepsilon \to 0$ that for $i = 1, \ldots, m$ the partial derivatives $\partial f_i(\mathbf{x}_0)/\partial x_j$ exist and

$$T(\mathbf{x}_0)\mathbf{e}_j = \begin{pmatrix} \dfrac{\partial f_1}{\partial x_j}(\mathbf{x}_0) \\ \vdots \\ \dfrac{\partial f_m}{\partial x_j}(\mathbf{x}_0) \end{pmatrix}.$$

Since the coordinates of $T(\mathbf{x}_0)\mathbf{e}_j$ relative to the standard basis in $\mathbb{R}^m$ are given by the jth column of the matrix representing $T(\mathbf{x}_0)$, the matrix representing $T(\mathbf{x}_0)$ is the matrix

$$\left(\frac{\partial f_i}{\partial x_j}(\mathbf{x}_0)\right).$$

If we define ∇f_i to be the row vector

$$\nabla f_i = \left(\frac{\partial f_i}{\partial x_1}, \ldots, \frac{\partial f_i}{\partial x_n}\right), \qquad i = 1, \ldots, n,$$

and $\nabla \mathbf{f}$ to be the matrix

$$\begin{pmatrix} \nabla f_1 \\ \vdots \\ \nabla f_m \end{pmatrix},$$

then we have shown that

$$T(\mathbf{x}_0) = \nabla \mathbf{f}(\mathbf{x}_0).$$

The matrix $\nabla \mathbf{f}(\mathbf{x}_0)$ is called the *Jacobian matrix* of $\mathbf{f}$ at $\mathbf{x}_0$. It is sometimes also written as $J(\mathbf{f}; \mathbf{x}_0)$. Note that we have shown that if $\mathbf{f}$ is differentiable at $\mathbf{x}_0$, then each component function f_i has all first partial derivatives existing at $\mathbf{x}_0$.

THEOREM 11.1. *Let D be an open set in $\mathbb{R}^n$ and let $\mathbf{f}$ be a function defined on D with range in $\mathbb{R}^m$. If $\mathbf{f}$ is differentiable at a point $\mathbf{x}_0$ in D, then $\mathbf{f}$ is continuous at $\mathbf{x}_0$.*

This follows immediately from (5) on letting $\mathbf{h} \to \mathbf{0}$.

We will now show that the existence of partial derivatives at a point does not imply differentiability. We shall consider real-valued functions f defined on $\mathbb{R}^2$. For typographical convenience we shall represent points in the plane as (x, y) instead of (x_1, x_2).

Example 11.1. Let

$$f(x, y) = \begin{cases} \dfrac{xy}{x^2 + y^2}, & (x, y) \neq (0, 0), \\ 0, & (x, y) = (0, 0). \end{cases}$$

Since $f(x, 0) = 0$ for all x, $\partial f/\partial x$ exists at $(0, 0)$ and equals 0. Similarly $\partial f/\partial y|_{(0,0)} = 0$. Along the line $y = kx$, the function f becomes

$$f(x, kx) = \frac{kx^2}{x^2 + k^2x^2} = \frac{k}{1 + k^2}, \qquad x \neq 0,$$

and so $\lim_{x\to 0} f(x, kx) = k/(1 + k^2)$. Thus $\lim_{(x,y)\to(0,0)} f(x, y)$ does not exist. Therefore f is not continuous at the origin and so by Theorem 11.1 cannot be differentiable at the origin.

One might now ask, what if the function is continuous at a point and has partial derivatives at the point? Must the function be differentiable at the point? The answer is no, as the next example shows.

Example 11.2. Let

$$f(x, y) = \begin{cases} \dfrac{xy}{\sqrt{x^2 + y^2}} & (x, y) \neq (0, 0), \\ 0, & (x, y) = 0. \end{cases}$$

As in Example 11.1, $\partial f/\partial x$ and $\partial f/\partial y$ exist at the origin and are both equal to zero. Note that

$$0 \leqslant (|x| - |y|)^2 = |x|^2 - 2|x|\,|y| + |y|^2$$

implies that $x^2 + y^2 \geqslant 2|x|\,|y|$. Hence for $(x, y) \neq 0$

$$|f(x, y)| = \frac{|xy|}{\sqrt{x^2 + y^2}} \leqslant \frac{1}{2}\frac{(x^2 + y^2)}{\sqrt{x^2 + y^2}} = \frac{1}{2}\sqrt{x^2 + y^2}.$$

Therefore $\lim_{(x,y)\to(0,0)} f(x, y) = 0$, and so f is continuous at $(0, 0)$.

If f were differentiable at $(0, 0)$, then for any $\mathbf{h} = (h_1, h_2)$

$$f(\mathbf{0} + \mathbf{h}) = f(\mathbf{0}) + \langle \nabla \mathbf{f}(\mathbf{0}), \mathbf{h} \rangle + \eta(\mathbf{h}),$$

where $\eta(\mathbf{h})/\|\mathbf{h}\| \to \mathbf{0}$ as $\mathbf{h} \to \mathbf{0}$. Since $\nabla f(\mathbf{0}) = ((\partial f/\partial x)(\mathbf{0}), (\partial f/\partial y)(\mathbf{0})) = (0, 0)$ and since $f(\mathbf{0}) = 0$, we would have

$$f(\mathbf{h}) = \eta(\mathbf{h}), \qquad \mathbf{h} \neq \mathbf{0}.$$

From the definition of f we get, writing (h_1, h_2) in place of (x, y),

$$\eta(\mathbf{h}) = \frac{h_1 h_2}{\sqrt{(h_1)^2 + (h_2)^2}}.$$

We require that $\eta(\mathbf{h})/\|\mathbf{h}\| \to 0$ as $\mathbf{h} \to \mathbf{0}$. But this is not true since

$$\frac{\eta(\mathbf{h})}{\|\mathbf{h}\|} = \frac{h_1 h_2}{(h_1)^2 + (h_2)^2},$$

and by Example 11.1, the right-hand side does not have a limit as $\mathbf{h} \to \mathbf{0}$.

A real-valued function f defined on an open set D in $\mathbb{R}^n$ is said to be of *class* $C^{(k)}$ on D if all of the partial derivatives up to and including those of order k exist and are continuous on D. A function $\mathbf{f}$ defined on D with range in $\mathbb{R}^m$, $m > 1$, is said to be of *class* $C^{(k)}$ on D if each of its component functions $f_i, i = 1, \ldots, m$, is of class $C^{(k)}$ on D.

We saw that if ϕ is a real-valued function defined on an open interval $D = \{t : \alpha < t < \beta\}$ in $\mathbb{R}^1$, then $\phi'(t_0)h$ is a "good" approximation to $\phi(t_0+h)-\phi(t_0)$ for small h at any point t_0 in D at which ϕ is differentiable. If ϕ is of class $C^{(k)}$ in D, then a better approximation can be obtained using Taylor's theorem from elementary calculus, one form of which states that for t_0 in D and h such that $t_0 + h$ is in D

$$\phi(t_0 + h) - \phi(t_0) = \sum_{i=1}^{k-1} \frac{\phi^{(i)}(t_0)}{i!} h^i + \frac{\phi^{(k)}(\bar{t})h^k}{k!}, \tag{6}$$

where $\phi^{(i)}$ denotes the ith derivative of ϕ and $\bar{t}$ is a point lying between t_0 and $t_0 + h$. If $k = 1$, then the summation term in (6) is absent and we have a restatement of the mean-value theorem. Since $\phi^{(k)}$ is continuous,

$$\phi^{(k)}(\bar{t}) = \phi^{(k)}(t_0) + \varepsilon(h),$$

where $\varepsilon(h) \to 0$ as $h \to 0$. Substituting this relation into (6) gives

$$\phi(t_0 + h) - \phi(t_0) = \sum_{i=1}^{k} \frac{\phi^{(i)}(t_0)h^i}{i!} + \eta(h), \tag{7}$$

where $\eta(h) = \varepsilon(h)h^k/k!$. Thus $\eta(h)/h^k \to 0$ as $h \to 0$. For small h, the polynomial in (7) is thus a "good" approximation to $\phi(t_0 + h) - \phi(t_0)$, in the sense that the error committed in using the approximation tends to zero faster than h^k.

We now generalize (7) to the case in which f is a real-valued function of class $C^{(1)}$ or $C^{(2)}$ on an open set D in $\mathbb{R}^n$. We restrict our attention to functions of class $C^{(1)}$ or $C^{(2)}$ because for functions of class $C^{(k)}$ with $k > 2$ the statement of the result is very cumbersome, and in this text we shall only need $k = 1, 2$.

THEOREM 11.2. *Let D be an open set in $\mathbb{R}^n$, let $\mathbf{x}_0$ be a point in D, and let f be a real-valued function defined on D. Let $\mathbf{h}$ be a vector such that $\mathbf{x}_0 + \mathbf{h}$ is in D.*

(i) *If f is of class $C^{(1)}$ on D, then*

$$f(\mathbf{x}_0 + \mathbf{h}) - f(\mathbf{x}_0) = \langle \nabla f(\mathbf{x}_0), \mathbf{h} \rangle + \eta_1(\mathbf{h}), \tag{8}$$

where $\eta_1(\mathbf{h})/\|\mathbf{h}\| \to 0$ as $\mathbf{h} \to \mathbf{0}$.

(ii) *If f is of class $C^{(2)}$ on D, then*

$$f(\mathbf{x}_0 + \mathbf{h}) - f(\mathbf{x}_0) = \langle \nabla f(\mathbf{x}_0), \mathbf{h} \rangle + \tfrac{1}{2}\langle \mathbf{h}, H(\mathbf{x}_0)\mathbf{h} \rangle + \eta_2(\mathbf{h}), \tag{9}$$

where $\eta_2(\mathbf{h})/\|\mathbf{h}\|^2 \to 0$ *as* $\mathbf{h} \to \mathbf{0}$ *and*

$$H(\mathbf{x}_0) = \left(\frac{\partial^2 f}{\partial x_j \partial x_i}(\mathbf{x}_0)\right). \tag{10}$$

Proof. Let $\phi(t) = f(\mathbf{x}_0 + t\mathbf{h})$. Then

$$\phi(0) = f(\mathbf{x}_0) \quad \text{and} \quad \phi(1) = f(\mathbf{x}_0 + \mathbf{h}). \tag{11}$$

It follows from the chain rule that ϕ is a function of class $C^{(k)}$ on an open interval containing the closed interval $[0, 1] = \{t : 0 \leqslant t \leqslant 1\}$ in its interior whenever f is of class $C^{(k)}$. If f is of class $C^{(1)}$, then using the chain rule, we get

$$\phi'(t) = \sum_{i=1}^{n} \frac{\partial f}{\partial x_i}(\mathbf{x}_0 + t\mathbf{h})h_i = \langle \nabla f(\mathbf{x}_0 + t\mathbf{h}), \mathbf{h} \rangle. \tag{12}$$

If f is of class $C^{(2)}$, we get

$$\phi''(t) = \sum_{i=1}^{n} \sum_{j=1}^{n} \frac{\partial^2 f}{\partial x_j \partial x_i}(\mathbf{x}_0 + t\mathbf{h})h_i h_j = \langle \mathbf{h}, H(\mathbf{x}_0 + t\mathbf{h})\mathbf{h} \rangle, \tag{13}$$

where H is the matrix defined in (10).

Let f be of class $C^{(1)}$. If we take $k = 1$, $t_0 = 0$, and $h = 1$ in (6), recall that the summation term in (6) is absent when $k = 1$, and use (11) and (12), we get

$$f(\mathbf{x}_0 + \mathbf{h}) - f(\mathbf{x}_0) = \phi(1) - \phi(0) = \langle \nabla f(\mathbf{x}_0 + \bar{t}\mathbf{h}), \mathbf{h} \rangle, \tag{14}$$

where $0 < \bar{t} < 1$. Since f is of class $C^{(1)}$, the function ∇f is continuous and thus

$$\nabla f(\mathbf{x}_0 + \bar{t}\mathbf{h}) = \nabla f(\mathbf{x}_0) + \varepsilon_1(\mathbf{h}),$$

where $\boldsymbol{\varepsilon}_1(\mathbf{h}) \to \mathbf{0}$ as $\mathbf{h} \to \mathbf{0}$. Substituting this relation into (14) and using the Cauchy–Schwarz inequality

$$|\langle \boldsymbol{\varepsilon}_1(\mathbf{h}), \mathbf{h} \rangle| \leqslant \|\boldsymbol{\varepsilon}_1(\mathbf{h})\| \, \|\mathbf{h}\|$$

give (8), with $\eta_1(\mathbf{h}) = \langle \boldsymbol{\varepsilon}_1(\mathbf{h}), \mathbf{h} \rangle$.

If f is of class $C^{(2)}$, we proceed in similar fashion. We take $k = 2$, $t_0 = 0$, and $h = 1$ in (6) and use (11), (12), and (13) to get

$$f(\mathbf{x}_0 + \mathbf{h}) - f(\mathbf{x}_0) = \langle \nabla f(\mathbf{x}_0), \mathbf{h} \rangle + \tfrac{1}{2}\langle \mathbf{h}, H(\mathbf{x}_0 + \bar{t}\mathbf{h})\mathbf{h} \rangle, \tag{15}$$

where $0 < \bar{t} < 1$. Since f is of class $C^{(2)}$, each entry of H is continuous. Hence

$$H(\mathbf{x}_0 + \bar{t}\mathbf{h}) = H(\mathbf{x}_0) + M(\mathbf{h}),$$

where $M(\mathbf{h})$ is a matrix with entries $m_{ij}(\mathbf{h})$ such that $m_{ij}(\mathbf{h}) \to 0$ as $\mathbf{h} \to \mathbf{0}$. Substituting this relation into (15) and using the relations

$$|\langle \mathbf{h}, M(\mathbf{h})\mathbf{h}\rangle| \leqslant \|\mathbf{h}\| \, \|M(\mathbf{h})\mathbf{h}\| \leqslant m(\mathbf{h})\|\mathbf{h}\|^2,$$

where $m(\mathbf{h}) \to 0$ as $\mathbf{h} \to \mathbf{0}$, give (9) with $\eta_2(\mathbf{h}) = \langle \mathbf{h}, M(\mathbf{h})\mathbf{h}\rangle$.

COROLLARY 1 (TAYLOR'S THEOREM). *Let D be an open set in $\mathbb{R}^n$ and let f be a real-valued function defined in D. If f is of class $C^{(1)}$ in D, then for each $\mathbf{x}_0$ in D and $\mathbf{h}$ such that $\mathbf{x}_0 + \mathbf{h}$ is in D*

$$f(\mathbf{x}_0 + \mathbf{h}) - f(\mathbf{x}_0) = \langle \nabla f(\mathbf{x}_0 + \bar{t}\mathbf{h}), \mathbf{h}\rangle, \tag{16}$$

where $0 < \bar{t} < 1$. If f is of class $C^{(2)}$, then

$$f(\mathbf{x}_0 + \mathbf{h}) - f(\mathbf{x}_0) = \langle \nabla f(\mathbf{x}_0), \mathbf{h}\rangle + \tfrac{1}{2}\langle \mathbf{h}, H(\mathbf{x}_0 + \bar{t}\mathbf{h})\mathbf{h}\rangle, \tag{17}$$

where $0 < \bar{t} < 1$.

Proof. Equation (16) follows from (14) and equation (17) from (15).

We now give a condition under which the existence of partial derivatives does imply differentiability.

THEOREM 11.3. *Let D be an open set in $\mathbb{R}^n$, let $\mathbf{f}$ be a function defined on D with range in $\mathbb{R}^m$, and let $\mathbf{x}_0$ be a point in D. If $\mathbf{f}$ is of class $C^{(1)}$ on some open ball $B(\mathbf{x}_0, r)$ centered at $\mathbf{x}_0$, then $\mathbf{f}$ is differentiable at $\mathbf{x}_0$.*

Proof. Since each component function f_i, $i = 1, \ldots, m$, is of class $C^{(1)}$, from (8) of Theorem 11.2 we have that

$$f_i(\mathbf{x}_0 + \mathbf{h}) - f_i(\mathbf{x}_0) = \langle \nabla f_i(\mathbf{x}_0), \mathbf{h}\rangle + \eta_i(\mathbf{h}),$$

where $\eta_i(\mathbf{h})/\|\mathbf{h}\| \to 0$ as $\mathbf{h} \to \mathbf{0}$.

Let $\boldsymbol{\eta}(\mathbf{h}) = (\eta_1(\mathbf{h}), \ldots, \eta_m(\mathbf{h}))$. Then

$$\mathbf{f}(\mathbf{x}_0 + \mathbf{h}) - \mathbf{f}(\mathbf{x}_0) = \nabla\mathbf{f}(\mathbf{x}_0)\mathbf{h} + \boldsymbol{\eta}(\mathbf{h}),$$

where $\boldsymbol{\eta}(\mathbf{h})/\|\mathbf{h}\| \to \mathbf{0}$ as $\|\mathbf{h}\| \to \mathbf{0}$. Therefore $\mathbf{f}$ is differentiable at $\mathbf{x}_0$.

II

CONVEX SETS IN $\mathbb{R}^n$

1. LINES AND HYPERPLANES IN $\mathbb{R}^n$

Convex sets and convex functions play an important role in many optimization problems. The study of convex sets in $\mathbb{R}^n$, in turn, involves the notions of line segments, lines, and hyperplanes in $\mathbb{R}^n$. We therefore begin with a discussion of these objects.

In $\mathbb{R}^2$ and $\mathbb{R}^3$ the vector equation of a line and in $\mathbb{R}^3$ the vector equation of a plane are obtained from geometric considerations. In $\mathbb{R}^n$ we shall reverse this process. We shall define a line to be the set of points satisfying an equation which in vector notation is the same as that of a line in $\mathbb{R}^2$ or $\mathbb{R}^3$. We shall define a hyperplane in $\mathbb{R}^n$ in similar fashion.

In $\mathbb{R}^2$ and $\mathbb{R}^3$ a vector equation of the line through two points $\mathbf{x}_1$ and $\mathbf{x}_2$ is given by (see Figure 2.1)

$$\mathbf{x} = \mathbf{x}_1 + t(\mathbf{x}_2 - \mathbf{x}_1), \qquad -\infty < t < \infty. \tag{1}$$

The closed oriented line segment $[\mathbf{x}_1, \mathbf{x}_2]$ with orientation from initial point $\mathbf{x}_1$ to terminal point $\mathbf{x}_2$ corresponds to values of t in $[0, 1]$. The positive ray from $\mathbf{x}_1$ corresponds to values of $t \geqslant 0$, and the negative ray from $\mathbf{x}_1$ corresponds to values of $t \leqslant 0$.

In $\mathbb{R}^n$ we define the line through two points $\mathbf{x}_1$ and $\mathbf{x}_2$ to be the set of points $\mathbf{x}$ in $\mathbb{R}^n$ that satisfy (1). Thus:

Definition 1.1. The *line in* $\mathbb{R}^n$ through two points $\mathbf{x}_1$ and $\mathbf{x}_2$ is defined to be the set of points $\mathbf{x}$ such that $\mathbf{x} = \mathbf{x}_1 + t(\mathbf{x}_2 - \mathbf{x}_1)$, where t is any real number, or in set notation

$$\{\mathbf{x} : \mathbf{x} = \mathbf{x}_1 + t(\mathbf{x}_2 - \mathbf{x}_1), \ -\infty < t < \infty\}. \tag{2}$$

The set in (2) can also be written as

$$\{\mathbf{x} : \mathbf{x} = (1 - t)\mathbf{x}_1 + t\mathbf{x}_2, \ -\infty < t < \infty\}, \tag{3}$$

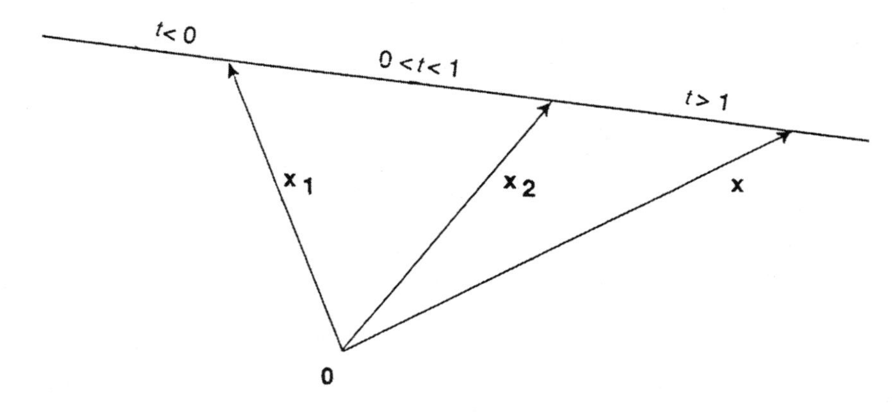

Figure 2.1.

and, on setting $\alpha = (1 - t)$, $\beta = t$,

$$\{\mathbf{x} : \mathbf{x} = \alpha\mathbf{x}_1 + \beta\mathbf{x}_2, \alpha + \beta = 1\}. \tag{4}$$

Definition 1.2. The *closed line segment* joining the points $\mathbf{x}_1$ and $\mathbf{x}_2$ is denoted by $[\mathbf{x}_1, \mathbf{x}_2]$ and is defined by

$$[\mathbf{x}_1, \mathbf{x}_2] = \{\mathbf{x} : \mathbf{x} = (1 - t)\mathbf{x}_1 + t\mathbf{x}_2, 0 \leqslant t \leqslant 1\},$$

or equivalently,

$$[\mathbf{x}_1, \mathbf{x}_2] = \{\mathbf{x} : \mathbf{x} = \alpha\mathbf{x}_1 + \beta\mathbf{x}_2, \alpha \geqslant 0, \beta \geqslant 0, \alpha + \beta = 1\}.$$

Definition 1.3. The *open line segment* joining the points $\mathbf{x}_1$ and $\mathbf{x}_2$ is denoted by $(\mathbf{x}_1, \mathbf{x}_2)$ and is defined by

$$(\mathbf{x}_1, \mathbf{x}_2) = \{\mathbf{x} : \mathbf{x} = (1 - t)\mathbf{x}_1 + t\mathbf{x}_2, 0 < t < 1\},$$

or equivalently,

$$(\mathbf{x}_1, \mathbf{x}_2) = \{\mathbf{x} : \mathbf{x} = \alpha\mathbf{x}_1 + \beta\mathbf{x}_2, \alpha > 0, \beta > 0, \alpha + \beta = 1\}.$$

The *half-open segment* which includes $\mathbf{x}_1$ but not $\mathbf{x}_2$ will be denoted by $[\mathbf{x}_1, \mathbf{x}_2)$ and is obtained by restricting t to the half-open interval $0 \leqslant t < 1$, or equivalently by restricting α and β to satisfy $\alpha > 0$, $\beta \geqslant 0$, $\alpha + \beta = 1$. Similarly, the half-open line segment which includes $\mathbf{x}_2$ but not $\mathbf{x}_1$ will be denoted by $(\mathbf{x}_1, \mathbf{x}_2]$ and is obtained by restricting t to the half-open interval $0 < t \leqslant 1$, or equivalently by restricting α and β to satisfy $\alpha \geqslant 0$, $\beta > 0$, $\alpha + \beta = 1$.

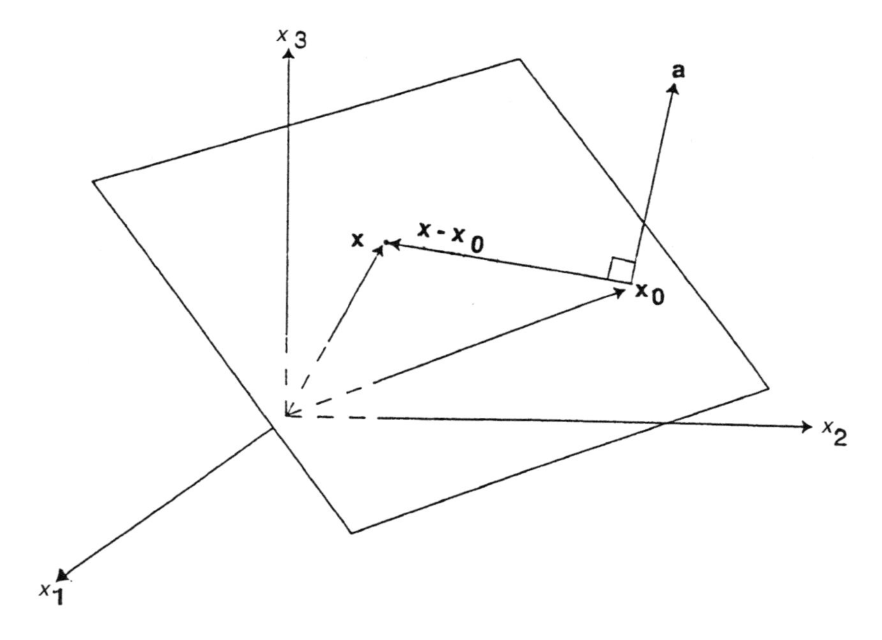

Figure 2.2.

Let $\mathbf{y}$ be a point in the open line segment $(\mathbf{x}_1, \mathbf{x}_2)$. Then $\mathbf{y} = \alpha\mathbf{x}_1 + \beta\mathbf{x}_2$, $\alpha > 0$, $\beta > 0$, $\alpha + \beta = 1$, and $\mathbf{y} - \mathbf{x}_1 = \beta\mathbf{x}_2 - (1 - \alpha)\mathbf{x}_1 = \beta(\mathbf{x}_2 - \mathbf{x}_1)$. Similarly $\mathbf{y} - \mathbf{x}_2 = \alpha(\mathbf{x}_1 - \mathbf{x}_2)$. Hence

$$\frac{\|\mathbf{y} - \mathbf{x}_1\|}{\|\mathbf{y} - \mathbf{x}_2\|} = \frac{\beta}{\alpha}. \tag{5}$$

This observation will be useful in the sequel.

In $\mathbb{R}^3$ a plane through the point $\mathbf{x}_0$ with normal $\mathbf{a}$ consists of all points $\mathbf{x}$ such that $\mathbf{x} - \mathbf{x}_0$ is orthogonal to $\mathbf{a}$. Thus, the plane is the set defined by (see Figure 2.2)

$$\{\mathbf{x} : \langle \mathbf{a}, \mathbf{x} - \mathbf{x}_0 \rangle = 0\}.$$

If $\mathbf{a} = (a_1, a_2, a_3)$ and $\mathbf{x}_0 = (x_{01}, x_{02}, x_{03})$, we can describe the plane as the set of $\mathbf{x} = (x_1, x_2, x_3)$ satisfying

$$a_1(x_1 - x_{01}) + a_2(x_2 - x_{02}) + a_3(x_3 - x_{03}) = 0$$

or

$$a_1x_1 + a_2x_2 + a_3x_3 = \gamma,$$

where $\gamma = a_1 x_{01} + a_2 x_{02} + a_3 x_{03}$. These are the familiar forms of the equation of a plane in $\mathbb{R}^3$. In vector notation these equations can be written as

$$\langle \mathbf{a}, \mathbf{x} - \mathbf{x}_0 \rangle = 0 \quad \text{or} \quad \langle \mathbf{a}, \mathbf{x} \rangle = \gamma, \tag{6}$$

where $\gamma = \langle \mathbf{a}, \mathbf{x}_0 \rangle$. Conversely, every equation of the form (6) is the equation of a plane with normal vector $\mathbf{a}$. To see this, let $\mathbf{x}_0$ be a point that satisfies (6). Then $\langle \mathbf{a}, \mathbf{x}_0 \rangle = \gamma$, so if $\mathbf{x}$ is any other point satisfying (6), $\langle \mathbf{a}, \mathbf{x} \rangle = \langle \mathbf{a}, \mathbf{x}_0 \rangle$. Hence $\langle \mathbf{a}, \mathbf{x} - \mathbf{x}_0 \rangle = 0$ for all $\mathbf{x}$ satisfying (6). But $\langle \mathbf{a}, \mathbf{x} - \mathbf{x}_0 \rangle = 0$ is an equation of the plane through $\mathbf{x}_0$ with normal $\mathbf{a}$.

In $\mathbb{R}^n$ we take (6) to define a hyperplane.

Definition 1.4. A *hyperplane* $H_{\mathbf{a}}^{\alpha}$ in $\mathbb{R}^n$ is defined to be the set of points that satisfy the equation $\langle \mathbf{a}, \mathbf{x} \rangle = \alpha$. Thus

$$H_{\mathbf{a}}^{\alpha} = \{\mathbf{x} : \langle \mathbf{a}, \mathbf{x} \rangle = \alpha\}.$$

The vector $\mathbf{a}$ is said to be a *normal* to the hyperplane.

As was the case in $\mathbb{R}^3$, $H_{\mathbf{a}}^{\alpha} = \{\mathbf{x} : \langle \mathbf{a}, \mathbf{x} \rangle = \alpha\}$ can be written as

$$\{\mathbf{x} : \langle \mathbf{a}, \mathbf{x} - \mathbf{x}_0 \rangle = 0\}$$

for any $\mathbf{x}_0$ satisfying $\langle \mathbf{a}, \mathbf{x}_0 \rangle = \alpha$, or equivalently for any $\mathbf{x}_0$ in $H_{\mathbf{a}}^{\alpha}$. Thus we may say that an equation of the hyperplane in $\mathbb{R}^n$ through the point $\mathbf{x}_0$ with normal $\mathbf{a}$ is

$$\langle \mathbf{a}, \mathbf{x} - \mathbf{x}_0 \rangle = 0. \tag{7}$$

Note that in $\mathbb{R}^2$ a hyperplane is a line.

Recall that an $(n - 1)$-dimensional subspace of $\mathbb{R}^n$ can be represented as the set of vectors $\mathbf{x} = (x_1, \ldots, x_n)$ whose coordinates satisfy $a_1 x_1 + \cdots + a_n x_n = 0$ for some nonzero vector $\mathbf{a} = (a_1, \ldots, a_n)$. Thus, if $\alpha = 0$ and $\mathbf{a} \neq \mathbf{0}$, then the hyperplane $H_{\mathbf{a}}^{0}$ passes through the origin and is an $(n - 1)$-dimensional subspace of $\mathbb{R}^n$.

In Section 10 of Chapter I we saw that for a given linear functional $\mathbf{L}$ on $\mathbb{R}^n$ there exists a unique vector $\mathbf{a}$ such that $\mathbf{L}(\mathbf{x}) = \langle \mathbf{a}, \mathbf{x} \rangle$ for all $\mathbf{x}$ in $\mathbb{R}^n$. This representation of linear functionals and the definition of the hyperplane $H_{\mathbf{a}}^{\alpha}$ show that hyperplanes are level surfaces of linear functionals. This interpretation will prove to be very useful.

Let A be any set in $\mathbb{R}^n$ and let $\mathbf{x}_0$ be any vector in $\mathbb{R}^n$. The set $A + \mathbf{x}_0$ is defined to be

$$A + \mathbf{x}_0 = \{\mathbf{y} : \mathbf{y} = \mathbf{a} + \mathbf{x}_0, \mathbf{a} \in A\}$$

and is called the *translate* of A by $\mathbf{x}_0$.

LEMMA 1.1 *For any* $\mathbf{a}$ *in* $\mathbb{R}^n$ *and scalar* α, *the hyperplane* $H_{\mathbf{a}}^{\alpha}$ *is the translate of* $H_{\mathbf{a}}^{0}$, *the hyperplane through the origin with normal* $\mathbf{a}$, *by any vector* $\mathbf{x}_0$ *in* $H_{\mathbf{a}}^{\alpha}$.

Proof. Fix an $\mathbf{x}_0$ in $H_{\mathbf{a}}^{\alpha}$. Let $\mathbf{x}$ be an arbitrary point in $H_{\mathbf{a}}^{\alpha}$ and let $\mathbf{u} = \mathbf{x} - \mathbf{x}_0$. Then

$$\alpha = \langle \mathbf{a}, \mathbf{x} \rangle = \langle \mathbf{a}, \mathbf{u} + \mathbf{x}_0 \rangle = \langle \mathbf{a}, \mathbf{u} \rangle + \alpha.$$

Thus $\langle \mathbf{a}, \mathbf{u} \rangle = 0$ and so $\mathbf{u} \in H_{\mathbf{a}}^{0}$. Since $\mathbf{x} = \mathbf{x}_0 + \mathbf{u}$, we have shown that

$$H_{\mathbf{a}}^{\alpha} \subseteq H_{\mathbf{a}}^{0} + \mathbf{x}_0, \qquad \mathbf{x}_0 \in H_{\mathbf{a}}^{\alpha}. \tag{8}$$

Now let $\mathbf{u}$ be an arbitrary vector in $H_{\mathbf{a}}^{0}$. Then $\mathbf{u} + \mathbf{x}_0 \in H_{\mathbf{a}}^{0} + \mathbf{x}_0$ and

$$\langle \mathbf{a}, \mathbf{u} + \mathbf{x}_0 \rangle = \langle \mathbf{a}, \mathbf{x}_0 \rangle = \alpha.$$

Thus we get the inclusion opposite to the one in (8), and the lemma is proved.

Two hyperplanes $H_{\mathbf{a}_1}^{\alpha_1}$ and $H_{\mathbf{a}_2}^{\alpha_2}$ are said to be *parallel* if their normals are scalar multiples of each other. Thus two hyperplanes are parallel if and only if they are translates of the same hyperplane through the origin. The parallel hyperplane through the origin is called the *parallel subspace.*

Exercise 1.1. (a) Draw a sketch in $\mathbb{R}^2$ and $\mathbb{R}^3$ that illustrates Lemma 1.1.
(b) In $\mathbb{R}^2$ sketch the hyperplane $x + 2y = 1$ and the parallel subspace.

Exercise 1.2. (a) In $\mathbb{R}^4$ find an equation of the hyperplane through the points $(1, 1, 1, 1), (2, 0, 1, 0), (0, 2, 0, 1)$, and $(1, 1, -1, 0)$.
(b) Let $\mathbf{x}_1, \ldots, \mathbf{x}_n$ be n points in $\mathbb{R}^n$. Find a sufficient condition for the existence of a unique hyperplane containing these points.

Exercise 1.3. Show that a hyperplane is a closed set.

Exercise 1.4. Given a hyperplane $H_{\mathbf{a}}^{\alpha}$ show that $\mathbf{a}$ is indeed normal to $H_{\mathbf{a}}^{\alpha}$ in the sense that if $\mathbf{x}_1$ and $\mathbf{x}_2$ are any two points in $H_{\mathbf{a}}^{\alpha}$, then $\mathbf{a}$ is orthogonal to $\mathbf{x}_2 - \mathbf{x}_1$.

Exercise 1.5 (Linear Algebra Review). Let V be an $(n - 1)$-dimensional subspace of $\mathbb{R}^n$. Let $\mathbf{y}$ be a vector in $\mathbb{R}^n$ not in V. Show that every $\mathbf{x}$ in $\mathbb{R}^n$ has a unique representation $\mathbf{x} = \mathbf{v} + \alpha\mathbf{y}$ where $\mathbf{v} \in V$ and $\alpha \in \mathbb{R}$.

2. PROPERTIES OF CONVEX SETS

A subset C of $\mathbb{R}^n$ is said to be *convex* if given any two points $\mathbf{x}_1$, $\mathbf{x}_2$ in C the line segment $[\mathbf{x}_1, \mathbf{x}_2]$ joining these points is contained in C. We now formulate this definition analytically.

Definition 2.1. A subset C of $\mathbb{R}^n$ is *convex* if for every pair of points $\mathbf{x}_1$, $\mathbf{x}_2$ in C the line segment

$$[\mathbf{x}_1, \mathbf{x}_2] = \{\mathbf{x} : \mathbf{x} = \alpha\mathbf{x}_1 + \beta\mathbf{x}_2, \alpha \geqslant 0, \beta \geqslant 0, \alpha + \beta = 1\}$$

belongs to C.

At this point the reader may find it helpful to do the following exercise.

Exercise 2.1. Sketch the following sets in $\mathbb{R}^2$ and determine from your figure which sets are convex and which are not:

(a) $\{(x, y) : x^2 + y^2 \leqslant 1\}$,
(b) $\{(x, y) : 0 < x^2 + y^2 \leqslant 1\}$,
(c) $\{(x, y) : y \geqslant x^2\}$,
(d) $\{(x, y) : |x| + |y| \leqslant 1\}$, and
(e) $\{(x, y) : y \geqslant 1/(1 + x^2)\}$.

In $\mathbb{R}^3$ a plane determines two sets, one on each side of the plane. We extend this notion to $\mathbb{R}^n$. The *positive half space determined by* $H_{\mathbf{a}}^{\alpha}$ and denoted by $H_{\mathbf{a}}^{\alpha+}$ is defined to be

$$H_{\mathbf{a}}^{\alpha+} = \{\mathbf{x} : \langle \mathbf{a}, \mathbf{x} \rangle > \alpha\}. \tag{1}$$

The *negative half space* $H_{\mathbf{a}}^{\alpha-}$ is defined to be

$$H_{\mathbf{a}}^{\alpha-} = \{\mathbf{x} : \langle \mathbf{a}, \mathbf{x} \rangle < \alpha\}. \tag{2}$$

The *closed positive half space*, denoted by $\bar{H}_{\mathbf{a}}^{\alpha+}$, is defined to be the closure of $H_{\mathbf{a}}^{\alpha+}$. The *closed negative half space* $\bar{H}_{\mathbf{a}}^{\alpha-}$ is defined to be the closure of $H_{\mathbf{a}}^{\alpha-}$.

Note that a half space is not a subspace of $\mathbb{R}^n$, since there exist points $\mathbf{x}$ in a half space such that not all scalar multiples of $\mathbf{x}$ are in the half space.

Exercise 2.2. Show that

$$\bar{H}_{\mathbf{a}}^{\alpha+} = \{\mathbf{x} : \langle \mathbf{a}, \mathbf{x} \rangle \geqslant \alpha\}, \qquad \bar{H}_{\mathbf{a}}^{\alpha-} = \{\mathbf{x} : \langle \mathbf{a}, \mathbf{x} \rangle \leqslant \alpha\}.$$

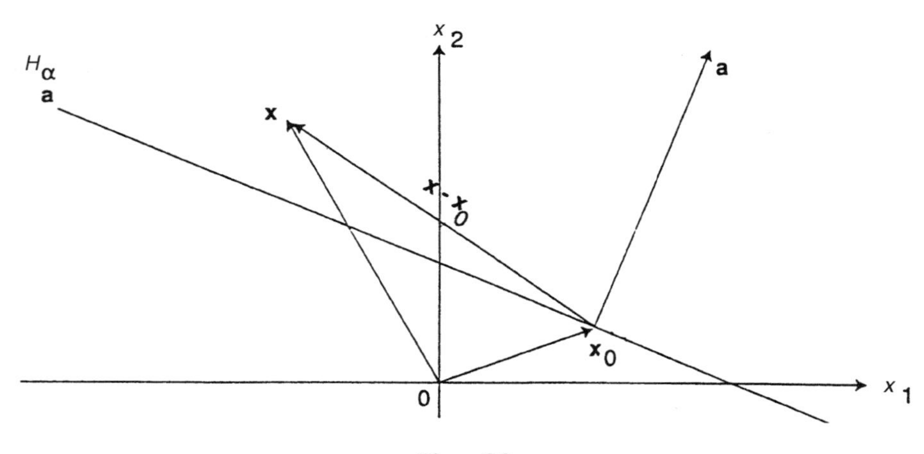

Figure 2.3.

If $\mathbf{x}_0 \in H^\alpha_\mathbf{a}$, then $\langle \mathbf{a}, \mathbf{x}_0 \rangle = \alpha$. Hence, for $\mathbf{x} \in H^{\alpha +}_\mathbf{a}$, we have $\langle \mathbf{a}, \mathbf{x} \rangle > \alpha = \langle \mathbf{a}, \mathbf{x}_0 \rangle$. Therefore

$$\langle \mathbf{a}, \mathbf{x} - \mathbf{x}_0 \rangle > 0 \quad \text{for all } \mathbf{x} \in H^{\alpha +}_\mathbf{a}.$$

Thus, vectors $\mathbf{x}$ in $H^{\alpha +}_\mathbf{a}$ are such that the angle between $\mathbf{a}$, the normal to $H^\alpha_\mathbf{a}$, and $\mathbf{x} - \mathbf{x}_0$ lies between 0 and $\pi/2$. In $\mathbb{R}^2$ and $\mathbb{R}^3$ this says that $\mathbf{a}$ and $\mathbf{x} - \mathbf{x}_0$ point in the same direction or have the same orientation. See Figure 2.3.

Half spaces are convex sets. Although this and other properties of convex sets are "geometrically obvious" in $\mathbb{R}^2$ and $\mathbb{R}^3$, they require proof in $\mathbb{R}^n$. To show that $H^{\alpha +}_\mathbf{a}$ is convex, we must show that if $\mathbf{x}_1$ and $\mathbf{x}_2$ are in $H^{\alpha +}_\mathbf{a}$, then so is $[\mathbf{x}_1, \mathbf{x}_2]$. Let $\mathbf{x} \in [\mathbf{x}_1, \mathbf{x}_2]$. Then there exist scalars $\lambda \geqslant 0$, $\mu \geqslant 0$ with $\lambda + \mu = 1$ such that $\mathbf{x} = \lambda \mathbf{x}_1 + \mu \mathbf{x}_2$. Hence

$$\langle \mathbf{a}, \mathbf{x} \rangle = \langle \mathbf{a}, \lambda \mathbf{x}_1 + \mu \mathbf{x}_2 \rangle = \lambda \langle \mathbf{a}, \mathbf{x}_1 \rangle + \mu \langle \mathbf{a}, \mathbf{x}_2 \rangle > \lambda\alpha + \mu\alpha = \alpha,$$

and so $\mathbf{x} \in H^{\alpha +}_\mathbf{a}$. Similar arguments show that $H^{\alpha -}_\mathbf{a}$, $\bar{H}^{\alpha +}_\mathbf{a}$, $\bar{H}^{\alpha -}_\mathbf{a}$ and the hyperplane $H^\alpha_\mathbf{a}$ are convex sets.

In the series of lemmas that follow we list some elementary properties of convex sets. We shall provide proofs to show the reader how one "proves the obvious." When speaking of a convex set, we shall always assume that the *convex set is not empty.*

LEMMA 2.1. *Let $\{C_\alpha\}$ be a collection of convex sets such that $C = \bigcap_\alpha C_\alpha$ is not empty. Then C is convex.*

Proof. Let $\mathbf{x}_1$ and $\mathbf{x}_2$ belong to C. Then for each α the points $\mathbf{x}_1$ and $\mathbf{x}_2$ belong to C_α. Since for each α the set C_α is convex, for each α the line segment $[\mathbf{x}_1, \mathbf{x}_2]$ belongs to C_α. Hence $[\mathbf{x}_1, \mathbf{x}_2] \subseteq C$, so C is convex.

Let A be an $m \times n$ matrix with ith row $\mathbf{a}_i = (a_{i1}, \dots, a_{in})$ and let $\mathbf{b} = (b_1, \dots, b_m)$ be a vector in $\mathbb{R}^m$. Let S denote the solution set of $A\mathbf{x} \leqslant \mathbf{b}$. Then S is the intersection of half spaces $\{\mathbf{x} : \langle \mathbf{a}_i, \mathbf{x} \rangle \leqslant b_i\}$, so if S is nonempty, then S is convex.

Note that if A and B are convex sets, then $A \cup B$ need not be convex.

If A and B are two sets in $\mathbb{R}^n$ and if λ and μ are scalars, we define

$$\lambda A + \mu B = \{\mathbf{x} : \mathbf{x} = \lambda \mathbf{a} + \mu \mathbf{b},\ \mathbf{a} \in A,\ \mathbf{b} \in B\}.$$

LEMMA 2.2. *If A and B are convex sets in $\mathbb{R}^n$ and λ and μ are scalars, then $\lambda A + \mu B$ is convex.*

Proof. Let $\mathbf{x}_1$ and $\mathbf{x}_2$ belong to $\lambda A + \mu B$. We must show that $[\mathbf{x}_1, \mathbf{x}_2]$ belongs to $\lambda A + \mu B$. We have $\mathbf{x}_1 = \lambda \mathbf{a}_1 + \mu \mathbf{b}_1$ for some $\mathbf{a}_1 \in A$ and $\mathbf{b}_1 \in B$ and $\mathbf{x}_2 = \lambda \mathbf{a}_2 + \mu \mathbf{b}_2$ for some $\mathbf{a}_2 \in A$ and $\mathbf{b}_2 \in B$. If $\mathbf{x} \in [\mathbf{x}_1, \mathbf{x}_2]$, then there exist scalars $\alpha \geqslant 0$, $\beta \geqslant 0$ with $\alpha + \beta = 1$ such that $\mathbf{x} = \alpha \mathbf{x}_1 + \beta \mathbf{x}_2$. Hence

$$\begin{aligned} \mathbf{x} = \alpha \mathbf{x}_1 + \beta \mathbf{x}_2 &= \alpha[\lambda \mathbf{a}_1 + \mu \mathbf{b}_1] + \beta[\lambda \mathbf{a}_2 + \mu \mathbf{b}_2] \\ &= \lambda(\alpha \mathbf{a}_1 + \beta \mathbf{a}_2) + \mu(\alpha \mathbf{b}_1 + \beta \mathbf{b}_2). \end{aligned}$$

Since A and B are convex, the first term in parentheses on the right is an element $\mathbf{a}_3$ in A and the second term in parentheses on the right is an element $\mathbf{b}_3$ in B. Hence for arbitrary $\mathbf{x} \in [\mathbf{x}_1, \mathbf{x}_2]$,

$$\mathbf{x} = \lambda \mathbf{a}_3 + \mu \mathbf{b}_3,$$

so $[\mathbf{x}_1, \mathbf{x}_2] \subseteq \lambda A + \mu B$ as required.

LEMMA 2.3. *Let $A_1 \subseteq \mathbb{R}^{n_1}, A_2 \subseteq \mathbb{R}^{n_2}, \dots, A_k \subseteq \mathbb{R}^{n_k}$ and let $A_1, \dots, A_k$ be convex. Then $A_1 \times \cdots \times A_k$ is a convex set in $\mathbb{R}^{n_1 + \cdots + n_k}$.*

We leave the proof as an exercise for the reader.

LEMMA 2.4. *If A is convex, then so is $\bar{A}$, the closure of A.*

Proof. Let $\mathbf{x}$ and $\mathbf{y}$ be points of $\bar{A}$. Let $\mathbf{z} \in [\mathbf{x}, \mathbf{y}]$, so $\mathbf{z} = \alpha \mathbf{x} + \beta \mathbf{y}$ for some $\alpha \geqslant 0$, $\beta \geqslant 0$, $\alpha + \beta = 1$. By Remark I.5.2, there exist sequences $\{\mathbf{x}_k\}$ and $\{\mathbf{y}_k\}$ of points in A such that $\lim \mathbf{x}_k = \mathbf{x}$ and $\lim \mathbf{y}_k = \mathbf{y}$. Since A is convex, $\mathbf{z}_k = \alpha \mathbf{x}_k + \beta \mathbf{y}_k \in A$ for each k. Letting $k \to \infty$, we get that $\lim \mathbf{z}_k = \mathbf{z}$. From Remark I.5.1 we have that $\mathbf{z} \in \bar{A}$.

The next lemma and its corollaries are fundamental results.

LEMMA 2.5. *Let C be a convex set with nonempty interior. Let $\mathbf{x}_1$ and $\mathbf{x}_2$ be two points with $\mathbf{x}_1 \in \operatorname{int}(C)$ and $\mathbf{x}_2 \in \bar{C}$. Then the line segment $[\mathbf{x}_1, \mathbf{x}_2)$ is contained in the interior of C.*

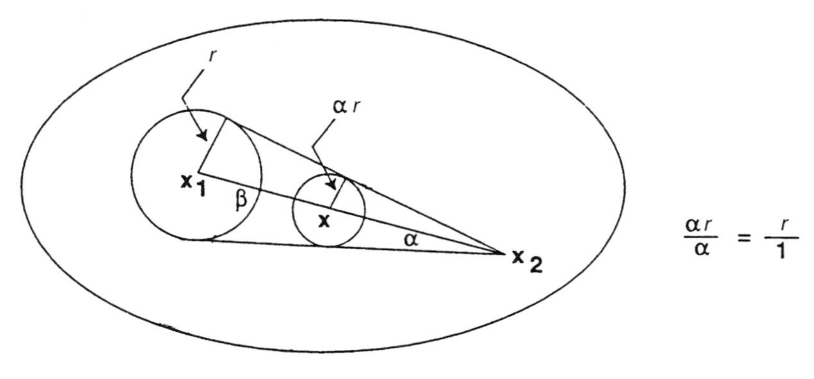

Figure 2.4.

We first assume that $\mathbf{x}_2 \in C$. The proof of the lemma in this case is based on Figure 2.4, which illustrates how one uses geometry to motivate analytical proofs. In the figure we have normalized the dimensions so that $\|\mathbf{x}_2 - \mathbf{x}_2\| = 1$. Let $\mathbf{x}$ be an arbitrary point in $(\mathbf{x}_1, \mathbf{x}_2)$. Then $\mathbf{x} = \alpha\mathbf{x}_1 + \beta\mathbf{x}_2$ with $\alpha > 0$, $\beta > 0$, and $\alpha + \beta = 1$. We must show that $\mathbf{x} \in \text{int}(C)$. Since the point $\mathbf{x}_1$ is in $\text{int}(C)$, there exists a circle of radius $r > 0$ centered at $\mathbf{x}_1$ that is contained entirely within C. From $\mathbf{x}_2$ draw tangents to the circle. Since $\|\mathbf{x}_2 - \mathbf{x}\| = \alpha\|\mathbf{x}_2 - \mathbf{x}_1\|$, by similar triangles we would expect the circle with center at $\mathbf{x}$ and radius αr to lie in C. This would show that $\mathbf{x} \in \text{int}(C)$.

We now verify analytically that the suggested argument is indeed valid. To prove the lemma, we must show that any point $\mathbf{x} \in [\mathbf{x}_1, \mathbf{x}_2)$ is in $\text{int}(C)$. By hypothesis, $\mathbf{x}_1 \in \text{int}(C)$, so we need only consider $\mathbf{x} \in (\mathbf{x}_1, \mathbf{x}_2)$.

Since $\mathbf{x}_1 \in \text{int}(C)$, there exists an $r > 0$ such that $B(\mathbf{x}_1, r) \subset C$. If $\mathbf{x} \in (\mathbf{x}_1, \mathbf{x}_2)$, then $\mathbf{x} = \alpha\mathbf{x}_1 + \beta\mathbf{x}_2$ for some $\alpha > 0$, $\beta > 0$, $\alpha + \beta = 1$. We shall show that $B(\mathbf{x}, \alpha r) \subset C$, and so $\mathbf{x} \in \text{int}(C)$. Let $\mathbf{y} \in B(\mathbf{x}, \alpha r)$ and let

$$\mathbf{z} = \mathbf{x}_1 + \frac{\mathbf{y} - \mathbf{x}}{\alpha}.$$

Then,

$$\|\mathbf{z} - \mathbf{x}_1\| = \frac{\|\mathbf{y} - \mathbf{x}\|}{\alpha} < \frac{\alpha r}{\alpha} = r.$$

Hence $\mathbf{z} \in B(\mathbf{x}_1, r)$ and therefore in C. Now,

$$\mathbf{y} = \alpha(\mathbf{z} - \mathbf{x}_1) + \mathbf{x} = \alpha(\mathbf{z} - \mathbf{x}_1) + \alpha\mathbf{x}_1 + \beta\mathbf{x}_2 = \alpha\mathbf{z} + \beta\mathbf{x}_2.$$

Since C is convex, $\mathbf{y} \in C$.

We now suppose that $\mathbf{x}_2 \in \bar{C}$ and $\mathbf{x}_2 \notin C$. Figure 2.5 will illustrate the idea of the proof. In the figure we again take $\|\mathbf{x}_2 - \mathbf{x}_1\| = 1$.

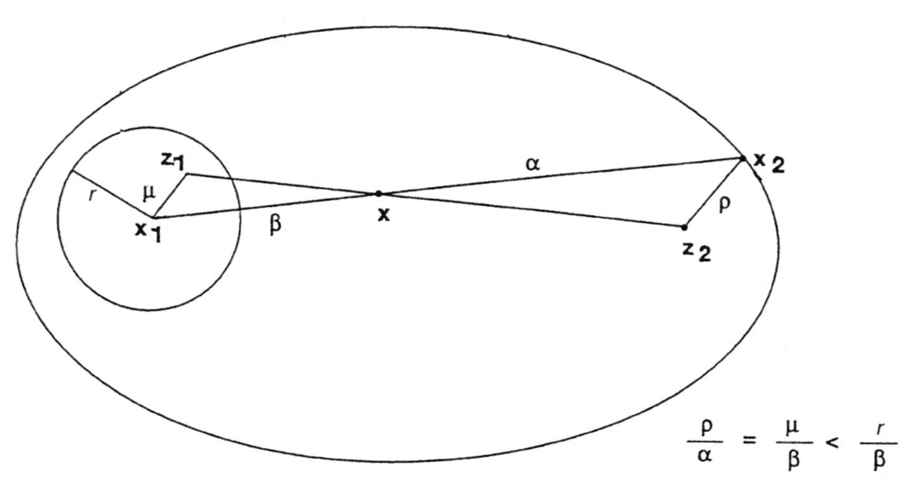

Figure 2.5.

Since $\mathbf{x}_1 \in \text{int}(C)$, there exists an $r > 0$ such that $B(\mathbf{x}_1, r) \subset C$. To prove the lemma, we must show that if $\mathbf{x} \in (\mathbf{x}_1, \mathbf{x}_2)$, then $\mathbf{x} \in \text{int}(C)$. If $\mathbf{x} \in (\mathbf{x}_1, \mathbf{x}_2)$, there exist $\alpha > 0$, $\beta > 0$, $\alpha + \beta = 1$ such that $\mathbf{x} = \alpha\mathbf{x}_1 + \beta\mathbf{x}_2$. Also by relation (5) in Section 1, $\|\mathbf{x} - \mathbf{x}_1\|/\|\mathbf{x} - \mathbf{x}_2\| = \beta/\alpha$. Since $\mathbf{x}_2 \in \bar{C}$, there exists a $\mathbf{z}_2$ in C such that $\|\mathbf{z}_2 - \mathbf{x}_2\| < r(\alpha/\beta)$. Define

$$\mathbf{z}_1 = \mathbf{x}_1 - \frac{\beta}{\alpha}(\mathbf{z}_2 - \mathbf{x}_2).$$

Hence $\|\mathbf{z}_1 - \mathbf{x}_1\| < r$, and so $\mathbf{z}_1 \in \text{int}(C)$. Also,

$$\begin{aligned} \mathbf{x} = \alpha\mathbf{x}_1 + \beta\mathbf{x}_2 &= \alpha\left(\mathbf{z}_1 + \frac{\beta}{\alpha}(\mathbf{z}_2 - \mathbf{x}_2)\right) + \beta\mathbf{x}_2 \\ &= \alpha\mathbf{z}_1 + \beta\mathbf{z}_2. \end{aligned}$$

Since $\mathbf{z}_1 \in \text{int}(C)$ and $\mathbf{z}_2 \in (C)$, by the case already proved, $\mathbf{x} \in \text{int}(C)$.

COROLLARY 1. *If C is convex, then* int(C) *is either empty or convex.*

COROLLARY 2. *Let C be convex and let* int$(C) \neq \phi$. *Then (i)* $\overline{\text{int}(C)} = \bar{C}$ *and (ii)* int(C) = int$(\bar{C})$.

Before proving Corollary 2, we note that its conclusion need not be true if C is not convex. To see this, take $C = \{\text{rational points in } [0, 1]\} \cup [1, 2]$.

Proof. Since for any set C $\text{int}(C) \subseteq C$, it follows that $\overline{\text{int}(C)} \subseteq \bar{C}$. Let $\mathbf{x} \in \bar{C}$. Then for any $\mathbf{y}$ in $\text{int}(C)$, the line segment $[\mathbf{y}, \mathbf{x})$ belongs to $\text{int}(C)$. Hence $\mathbf{x} \in \overline{\text{int}(C)}$, and so $\bar{C} \subseteq \overline{\text{int}(C)}$.

Since, for any set C, $C \subseteq \bar{C}$, it follows that $\text{int}(C) \subseteq \text{int}(\bar{C})$. Let $\mathbf{x} \in \text{int}(\bar{C})$. Then there exists an $r > 0$ such that $B(\mathbf{x}, r) \subset \bar{C}$. Now let $\mathbf{y} \in \text{int}(C)$ and consider the line segment $[\mathbf{y}, \mathbf{z})$, which is the extension of the segment $[\mathbf{y}, \mathbf{x}]$ a distance less than r beyond $\mathbf{x}$. Then $\mathbf{z} \in \bar{C}$, and since $\mathbf{y} \in \text{int}(C)$, we get that $[\mathbf{y}, \mathbf{z}) \in \text{int}(C)$. In particular, $\mathbf{x} \in \text{int}(C)$. Thus (ii) is proved.

For each positive integer n, define

$$P_n = \left\{ \mathbf{p} = (p_1, \ldots, p_n) : p_i \geqslant 0, \ \sum_{i=1}^{n} p_i = 1 \right\}.$$

For $n = 1$, P_1 is the point 1. For $n = 2$, P_2 is the closed line segment joining $(0, 1)$ and $(1, 0)$. For $n = 3$, P_3 is the closed triangle with vertices $(1, 0, 0)$, $(0, 1, 0)$, and $(0, 0, 1)$. It is easy to verify that, for each n, P_n is a closed convex set.

Definition 2.2. A point $\mathbf{x}$ in $\mathbb{R}^n$ is a *convex combination* of points $\mathbf{x}_1, \ldots, \mathbf{x}_k$ if there exists a $\mathbf{p} = (p_1, \ldots, p_k)$ in P_k such that

$$\mathbf{x} = p_1 \mathbf{x}_1 + p_2 \mathbf{x}_2 + \cdots + p_k \mathbf{x}_k.$$

LEMMA 2.6. *A set C in $\mathbb{R}^n$ is convex if and only if every convex combination of points in C is also in C.*

Proof. If every convex combination of points in C is in C, then every convex combination of every two points in C is in C. But for any fixed pair of points $\mathbf{x}_1$ and $\mathbf{x}_2$ in C, the set of convex combinations of $\mathbf{x}_1$ and $\mathbf{x}_2$ is just $[\mathbf{x}_1, \mathbf{x}_2]$. In other words, for every pair of points $\mathbf{x}_1$ and $\mathbf{x}_2$ in C, the line segment $[\mathbf{x}_1, \mathbf{x}_2]$ belongs to C, and so C is convex.

We shall prove the "only if" statement by induction on k, the number of points in the convex combination. If C is convex, then the definition of convexity says that every convex combination of two points is in C. Suppose that we have shown that if C is convex, then every convex combination of k points is in C. We shall show that every convex combination of $k + 1$ points is in C. Let

$$\mathbf{x} = p_1 \mathbf{x}_1 + \cdots + p_k \mathbf{x}_k + p_{k+1} \mathbf{x}_{k+1}, \qquad \mathbf{p} \in P_{k+1}.$$

If $p_{k+1} = 1$, then there is nothing to prove. If $p_{k+1} < 1$, then $\Sigma_{i=1}^{k} p_i > 0$. Hence,

$$\mathbf{x} = \left(\sum_{i=1}^{k} p_i \right) \left[\frac{p_1}{\Sigma_{i=1}^{k} p_i} \mathbf{x}_1 + \cdots + \frac{p_k}{\Sigma_{i=1}^{k} p_i} \mathbf{x}_k \right] + p_{k+1} \mathbf{x}_{k+1}.$$

The term in square brackets is a convex combination of k points in C and so by the inductive hyotheses is in C. Therefore $\mathbf{x}$ is a convex combination of two points in C, and so since C is convex, $\mathbf{x}$ belongs to C. This proves the lemma.

For a given set A, let $K(A)$ denote the set of all convex combinations of points in A. It is easy to verify that $K(A)$ is convex. Clearly, $K(A) \supseteq A$.

Definition 2.3. The *convex hull* of a set A, denoted by co(A), is the intersection of all convex sets containing A.

Since $\mathbb{R}^n$ is convex, co(A) is nonempty, and since the intersection of convex sets is convex, co(A) is convex. Also, since we have co$(A) \subseteq C$ for any convex set C containing A, it follows that co(A) is the *smallest* convex set containing A.

THEOREM 2.1. *The convex hull of a set A is the set of all convex combinations of points in A; that is,* co$(A) = K(A)$.

Proof. Let $\{C_\alpha\}$ denote the family of convex sets containing A. Since co$(A) = \bigcap C_\alpha$ and $K(A)$ is a convex set containing A, co$(A) \subseteq K(A)$. To prove the opposite inclusion, note that $A \subseteq C_\alpha$ for each α and Lemma 2.6 imply that for each α all convex combinations of points in A belong to C_α. That is, for each α, $K(A) \subseteq C_\alpha$. Hence $K(A) \subseteq \bigcap C_\alpha = $ co(A).

Lemma 2.6 states that a set C is convex if and only if $K(C) \subseteq C$. Since $C \subseteq K(C)$ for any set C, Lemma 2.6 can be restated in the equivalent form that a set C is convex if and only if $K(C) = C$. Since, for any set A, $K(A) = $ co(A), we have the following corollary to Theorem 2.1.

COROLLARY 3. *A set A is convex if and only if $A = $* co$(A)$.

THEOREM 2.2 (CARATHÉODORY). *Let A be a subset of $\mathbb{R}^n$ and let $\mathbf{x} \in$* co(A). *Then there exist $n + 1$ points $\mathbf{x}_1, \ldots, \mathbf{x}_{n+1}$ in A and a point $\mathbf{p}$ in P_{n+1} such that*

$$\mathbf{x} = p_1\mathbf{x}_1 + \cdots + p_{n+1}\mathbf{x}_{n+1}.$$

Note that since some of the p_i's may be zero, the result is often stated that a point $\mathbf{x} \in$ co(A) can be written as a convex combination of at most $n + 1$ points of A.

Proof. If $\mathbf{x} \in$ co(A), then by Theorem 2.1,

$$\mathbf{x} = \sum_{i=1}^{m} q_i\mathbf{x}_i, \qquad \mathbf{x}_i \in A, \qquad i = 1, \ldots, m, \qquad \mathbf{q} = (q_1, \ldots, q_m) \in P_m, \tag{3}$$

for some positive integer m and some $\mathbf{q}$ in P_m. If $m \leqslant n + 1$, then there is nothing to prove. If $m > n + 1$, we shall express $\mathbf{x}$ as a convex combination of at most $m - 1$ points. Repeating this argument, a finite number of times gives the desired result.

Let us then suppose that in (3) $m > n + 1$. Consider the $m - 1 > n$ vectors in $\mathbb{R}^n$, $(\mathbf{x}_1 - \mathbf{x}_m), (\mathbf{x}_2 - \mathbf{x}_m), \ldots, (\mathbf{x}_{m-1} - \mathbf{x}_m)$. Since $m - 1 > n$, these vectors are linearly dependent. Hence there exist scalars $\lambda_1, \ldots, \lambda_{m-1}$, not all zero such that

$$\lambda_1(\mathbf{x}_1 - \mathbf{x}_m) + \lambda_2(\mathbf{x}_2 - \mathbf{x}_m) + \cdots + \lambda_{m-1}(\mathbf{x}_{m-1} - \mathbf{x}_m) = \mathbf{0}.$$

Let $\lambda_m = -(\lambda_1 + \cdots + \lambda_{m-1})$. Then

$$\lambda_1\mathbf{x}_1 + \cdots + \lambda_{m-1}\mathbf{x}_{m-1} + \lambda_m\mathbf{x}_m = \mathbf{0} \tag{4}$$

and

$$\sum_{i=1}^{m} \lambda_i = 0. \tag{5}$$

From (3) and (4) it follows that for any t

$$\mathbf{x} = \sum_{i=1}^{m} (q_i - t\lambda_i)\mathbf{x}_i. \tag{6}$$

We shall show that there is a value of t such that the coefficients of the $\mathbf{x}_i$ are nonnegative, have sum equal to 1, and at least one of the coefficients is zero. This will prove the theorem.

Let $I = \{i : i = 1, \ldots, m, \lambda_i > 0\}$. Since $\lambda_1, \ldots, \lambda_m$ are not all zero, it follows from (5) that $I \neq \varnothing$. Let i_0 denote an index in I such that

$$\frac{q_{i0}}{\lambda_{i0}} = \min\{i \in I : q_i/\lambda_i\}.$$

Now take $t = q_{i0}/\lambda_{i0}$. Then $t \geqslant 0$, and for $i \in I$

$$(q_i - t\lambda_i) = \lambda_i\left(\frac{q_i}{\lambda_i} - \frac{q_{i0}}{\lambda_{i0}}\right) \geqslant 0,$$

with equality holding for $i = i_0$. If $i \notin I$, then $\lambda_i \leqslant 0$, so $q_i - t\lambda_i \geqslant 0$ whenever $t \geqslant 0$. Thus, if $t = q_{i0}/\lambda_{i0}$, then

$$q_i - t\lambda_i \geqslant 0, \qquad i = 1, \ldots, m \quad \text{with } q_{i0} - t\lambda_{i0} = 0.$$

Upon comparing this statement with (6), we see that we have written $\mathbf{x}$ as a nonnegative linear combination of at most $m - 1$ points. This combination is also a convex combination since, using (5), we have

$$\sum_{i=1}^{m} (q_i - t\lambda_i) = \sum_{i=1}^{m} q_i - t \sum_{i=1}^{m} \lambda_i = \sum_{i=1}^{m} q_i = 1.$$

LEMMA 2.7. *If A is a compact subset of R^n, then so is* co(A).

If A is closed, then co(A) need not be closed, as the following example shows. Let

$$A_1 = \{(x_1, x_2) : x_1 > 0,\ x_2 > 0,\ x_1 x_2 \geqslant 1\}$$

and let

$$A_2 = \{(x_1, x_2) : x_1 > 0,\ x_2 < 0,\ x_1 x_2 \leqslant -1\}.$$

Let $A = A_1 \cup A_2$. Then A is closed, but $\text{co}(A) = \{(x_1, x_2) : x_1 > 0\}$ is open.

We now prove the lemma. By Theorem 2.2

$$\text{co}(A) = \left\{ \mathbf{x} : \mathbf{x} = \sum_{i=1}^{n+1} p_i \mathbf{x}_i \quad \mathbf{x}_i \in A \quad \mathbf{p} = (p_1, \ldots, p_{n+1}) \in P_{n+1} \right\}.$$

Let

$$\pi = P_{n+1} \times \underbrace{A \times \cdots \times A}_{n+1 \text{ times}}$$

Since A is compact and P_{n+1} is compact, by Theorem I.8.1 the set π is compact in $\mathbb{R}^m$, where $m = (n+1) + n(n+1) = (n+1)^2$. The mapping $\varphi : \mathbb{R}^m \to \mathbb{R}^n$ defined by

$$\varphi(\mathbf{p}, \mathbf{x}_1, \ldots, \mathbf{x}_n) = \sum_{i=1}^{n+1} p_i \mathbf{x}_i$$

is continuous. Since π is compact, it follows from Theorem I.9.2 that $\varphi(\pi)$ is compact. By Theorem 2.2, $\varphi(\pi) = \text{co}(A)$, so co($A$) is compact.

LEMMA 2.8. *If O is an open subset of $\mathbb{R}^n$, then* co(O) *is also open.*

Proof. Since $O \subseteq \text{co}(O)$, the set co(O) has nonempty interior. Therefore, by Corollary 1 of Lemma 2.5 int(co(O)) is convex and $O \subseteq \text{int}(\text{co}(O))$. (Why?) Since co($O$) is the intersection of all convex sets containing O, we have $\text{co}(O) \subseteq \text{int}(\text{co}(O))$. Since we always have the reverse inclusion, we have that $\text{co}(O) = \text{int}(\text{co}(O))$.

Exercise 2.3. Using the definition of a convex set, show that (a) the nonnegative orthant in $\mathbb{R}^n = \{\mathbf{x}:\mathbf{x} = (x_1, \ldots, x_n), x_i \geqslant 0\ i = 1, \ldots, n\}$ is convex and (b) a hyperplane $H_{\mathbf{a}}^{\alpha}$ is convex.

Exercise 2.4. A mapping S from $\mathbb{R}^n$ to $\mathbb{R}^m$ is said to be affine if $S\mathbf{x} = T\mathbf{x} + \mathbf{b}$, where T is a linear map from $\mathbb{R}^n \to \mathbb{R}^m$ and $\mathbf{b}$ is a fixed vector. (If $\mathbf{b} = 0$, then S is linear.) Show that if C is a convex set in $\mathbb{R}^n$, then $S(C) = \{\mathbf{y}:\mathbf{y} = T\mathbf{x} + \mathbf{b}, \mathbf{x} \in C\}$ is a convex set in $\mathbb{R}^m$. In mathematical jargon, you are about to show that under an affine map convex sets are mapped onto convex sets or under an affine map the image of a convex set is convex.

Exercise 2.5. Show that the sets P_n are compact and convex.

Exercise 2.6. Show that for any set A the set $K(A)$ of all convex combinations of points in A is convex.

Exercise 2.7. Show that the open ball $B(\mathbf{0}, r)$ is convex.

Exercise 2.8. Consider the linear programming (LP) problem: Minimize $\langle \mathbf{c}, \mathbf{x} \rangle$ subject to $A\mathbf{x} = \mathbf{b}$, $\mathbf{x} \geqslant \mathbf{0}$. Let $S = \{\mathbf{x}:\mathbf{x}$ is a solution of the problem LP$\}$. Show that if S is not empty, then it is convex.

Exercise 2.9. Let $C \subseteq \mathbb{R}^n$. Show that C is convex if and only if $\lambda C + \mu C = (\lambda + \mu)C$ for all $\lambda \geqslant 0$, $\mu \geqslant 0$.

Exercise 2.10. A set C is said to be a *cone with vertex at the origin*, or simply *a cone*, if whenever $\mathbf{x} \in C$, all vectors $\lambda\mathbf{x}$, $\lambda \geqslant 0$, belong to C. If C is also convex, C is said to be a *convex cone*.

(a) Give an example of a cone that is not convex.
(b) Give an example of a cone that is convex.
(c) Let C be a nonempty set in $\mathbb{R}^n$. Show that C is a convex cone if and only if $\mathbf{x}_1$ and $\mathbf{x}_2 \in C$ implies that $\lambda_1\mathbf{x}_1 + \lambda_2\mathbf{x}_2 \in C$ for all $\lambda_1 \geqslant 0$, $\lambda_2 \geqslant 0$.

Exercise 2.11. Show that if C_1 and C_2 are convex cones, then so is $C_1 + C_2$ and that $C_1 + C_2 = \mathrm{co}(C_1 \cup C_2)$.

Exercise 2.12. Show that if A is a bounded set in $\mathbb{R}^n$, then so is $\mathrm{co}(A)$.

Exercise 2.13. Let X be a convex set in $\mathbb{R}^n$ and let $\mathbf{y}$ be any vector not in X. Let $[\mathbf{y}, X]$ denote the union of all line segments $[\mathbf{y}, \mathbf{x}]$ with $\mathbf{x}$ in X.

(a) Sketch a set $[\mathbf{y}, X]$ in $\mathbb{R}^3$.
(b) Show that $[\mathbf{y}, X]$ is convex.

Exercise 2.14. Let C be a proper subset of $\mathbb{R}^n$. For each $\varepsilon > 0$ define a set $\eta_\varepsilon(C)$, the ε-neighborhood of C, by $\eta_\varepsilon(C) = \bigcup_{\mathbf{x}\in C} B(\mathbf{x}, \varepsilon)$. Show that

$$\eta_\varepsilon(C) = \{\mathbf{y} : \|\mathbf{x} - \mathbf{y}\| < \varepsilon \text{ for some } \mathbf{x} \in C\}$$
$$= \bigcup_{\mathbf{x}\in C} (B(\mathbf{0}, \varepsilon) + \mathbf{x}).$$

Show that $\eta_\varepsilon(C)$ is open. Show that if C is convex, then so is $\eta_\varepsilon(C)$.

Exercise 2.15. Show that if a vector $\mathbf{x}$ in $\mathbb{R}^n$ has distinct representations as a convex combination of a set of vectors $\mathbf{x}_0, \mathbf{x}_1, \ldots, \mathbf{x}_r$, then the vectors $\mathbf{x}_1 - \mathbf{x}_0, \ldots, \mathbf{x}_r - \mathbf{x}_0$ are linearly dependent.

Exercise 2.16. Let X be a set contained in the closed negative half space determined by the hyperplane $H^\alpha_\mathbf{a}$. Show that if $\mathbf{x} \in \text{int}(X)$, then $\langle \mathbf{a}, \mathbf{x} \rangle < \alpha$.

Exercise 2.17. Let X be a convex set and let $H^\alpha_\mathbf{a}$ be a hyperplane such that $X \cap H^\alpha_\mathbf{a} = \varnothing$. Show that X is contained in one of the open half spaces determined by $H^\alpha_\mathbf{a}$.

3. SEPARATION THEOREMS

This section is devoted to the establishment of *separation theorems*. In some sense, these theorems are the fundamental theorems of optimization theory. The validity of this assertion will be made evident in subsequent chapters of this book.

We have seen that a hyperplane $H^\alpha_\mathbf{a}$ divides $\mathbb{R}^n$ into two half spaces, one on each side of $H^\alpha_\mathbf{a}$. It therefore seems natural to say that two sets X and Y are separated by a hyperplane $H^\alpha_\mathbf{a}$ if they are contained in different half spaces determined by $H^\alpha_\mathbf{a}$. There are various types of separation, which we will now define precisely and discuss. The reader should draw sketches in $\mathbb{R}^2$. (Recall that in $\mathbb{R}^2$ a hyperplane is a line.)

Definition 3.1. Two sets X and Y are *separated* by the hyperplane $H^\alpha_\mathbf{a}$ if, for every $\mathbf{x} \in X$, $\langle \mathbf{a}, \mathbf{x} \rangle \geqslant \alpha$ and, for every $\mathbf{y} \in Y$, $\langle \mathbf{a}, \mathbf{y} \rangle \leqslant \alpha$.

This type of separation does not always correspond to what one intuitively considers a separation. For example, in $\mathbb{R}^2$, let

$$X_1 = \{(x_1, x_2) : x_2 = 0, 0 \leqslant x_1 \leqslant 2\}$$

and let

$$Y_1 = \{(x_1\, x_2) : x_2 = 0, 1 \leqslant x_1 \leqslant 3\}.$$

According to Definition 3.1, these sets are separated by the hyperplane $x_2 = 0$. On the other hand, one would not consider these sets as being separated in any reasonable way. This example illustrates that two sets which intuitively should not be considered as being separated can be separated by a hyperplane $H_{\mathbf{a}}^{\alpha}$ according to Definition 3.1 if the sets both lie in $H_{\mathbf{a}}^{\alpha}$. To rule out the possibility just discussed, the notion of proper separation is introduced.

Definition 3.2. Two sets X and Y are *properly separated* by a hyperplane $H_{\mathbf{a}}^{\alpha}$ if, for every $\mathbf{x} \in X$, $\langle \mathbf{a}, \mathbf{x} \rangle \geqslant \alpha$, for every $\mathbf{y}$ in Y, $\langle \mathbf{a}, \mathbf{y} \rangle \leqslant \alpha$, and at least one of the sets is not contained in $H_{\mathbf{a}}^{\alpha}$.

Geometrically, the definition requires that X and Y be in opposite closed half spaces and at least one of the sets not be contained in $H_{\mathbf{a}}^{\alpha}$. Note that the sets X_1 and Y_1 are not properly separated by the hyperplane $x_2 = 0$ and that they cannot be properly separated by any hyperplane.

Proper separation of two sets X and Y does not require that the sets be disjoint. The sets

$$X_2 = \{(x_1, x_2): 0 \leqslant x_1 \leqslant 1, 0 \leqslant x_2 \leqslant 1\}$$

and

$$Y_2 = \{(x_1, x_2): 0 \leqslant x_1 \leqslant 1, \ -1 \leqslant x_2 \leqslant 0\}$$

are not disjoint but are properly separated by the hyperplane $x_2 = 0$. A set may be a subset of another set, yet the two sets can be properly separated. To see this, let $X_2^* = X_2$ and let $Y_2^* = \{(x_1, x_2): 0 \leqslant x_1 \leqslant 1, x_2 = 0\}$. Then $Y_2^* \subseteq X_2^*$ and the sets are properly separated by the hyperplane $x_2 = 0$. To rule out the possibilities just described, the notion of strict separation is introduced.

Definition 3.3. Two sets X and Y are *strictly separated* by a hyperplane $H_{\mathbf{a}}^{\alpha}$ if, for every $\mathbf{x}$ in X, $\langle \mathbf{a}, \mathbf{x} \rangle > \alpha$ and, for every $\mathbf{y}$ in Y, $\langle \mathbf{a}, \mathbf{y} \rangle < \alpha$.

Geometrically, the definition requires that X and Y be in opposite open half spaces determined by $H_{\mathbf{a}}^{\alpha}$. Note that the sets X_2 and Y_2 cannot be strictly separated by any hyperplane. The sets

$$X_3 = \{(x_1, x_2): 0 \leqslant x_1 \leqslant 1, 0 < x_2 \leqslant 1\}$$

and

$$Y_3 = \{(x_1, x_2): 0 \leqslant x_1 \leqslant 1, \ -1 < x_2 < 0\}$$

are strictly separated by the hyperplane $x_2 = 0$. Note that $\bar{X}_3$ and $\bar{Y}_3$ are properly separated but are not strictly separated by the hyperplane $x_2 = 0$.

Finally, note that the sets X_2 and Y_2 are not strictly separated by the hyperplane $x_2 = 0$.

Definition 3.4. Two sets X and Y are *strongly separated* by a hyperplane $H_{\mathbf{a}}^{\alpha}$ if there exists an $\varepsilon > 0$ such that the sets $X + \varepsilon B(\mathbf{0}, 1)$ and $Y + \varepsilon B(\mathbf{0}, 1)$ are strictly separated by $H_{\mathbf{a}}^{\alpha}$.

The sets X_3 and Y_3 are not strongly separated by the hyperplane $x_2 = 0$ and cannot be strongly separated by any hyperplane. Let $0 < \eta < 1$ be fixed. Then for each such η the sets

$$X_{4\eta} = \{(x_1, x_2) : 0 \leqslant x_1 \leqslant 1, \eta \leqslant x_2 \leqslant 1\}$$

and

$$Y_{4\eta} = \{(x_1, x_2) : 0 \leqslant x_1 \leqslant 1, -1 \leqslant x_2 \leqslant -\eta\}$$

are strongly separated by the hyperplane $x_2 = 0$.

It is clear from the definitions that strong separation implies strict separation, which implies proper separation.

Exercise 3.1. Sketch the pairs of sets (X_1, Y_1), (X_2, Y_2), (X_2^*, Y_2^*), (X_3, Y_3), and $(X_{4\eta}, Y_{4\eta})$.

The next lemma gives two conditions, each of which is equivalent to strong separation as given in Definition 3.4.

LEMMA 3.1. *The following statements are equivalent:*

(i) *Two sets X and Y are strongly separated by a hyperplane $H_{\mathbf{a}}^{\alpha}$.*

(ii) *There exists an $\eta > 0$ such that $\langle \mathbf{a}, \mathbf{x} \rangle > \alpha + \eta$ for all $\mathbf{x}$ in X and $\langle \mathbf{a}, \mathbf{y} \rangle < \alpha - \eta$ for all $\mathbf{y}$ in Y.*

(iii) *There exists an $\eta' > 0$ such that*

$$\inf\{\langle \mathbf{a}, \mathbf{x} \rangle : \mathbf{x} \in X\} \geqslant \alpha + \eta'$$

and

$$\sup\{\langle \mathbf{a}, \mathbf{y} \rangle : \mathbf{y} \in Y\} \leqslant \alpha - \eta'.$$

Proof. The equivalence of (ii) and (iii) is clear, so we need only prove the equivalence of (i) and (ii). We note that $\mathbf{z} \in B(\mathbf{0}, 1)$ if and only if $-\mathbf{z} \in B(\mathbf{0}, 1)$. Suppose X and Y are strongly separated by $H_{\mathbf{a}}^{\alpha}$. Then there exists an $\varepsilon > 0$ such that, for all $\mathbf{x}$ in X and all $\mathbf{z}$ in $B(0, 1)$, $\langle \mathbf{a}, \mathbf{x} - \varepsilon \mathbf{z} \rangle > \alpha$. Thus for all $\mathbf{x}$ in X and all $\mathbf{z} \in B(\mathbf{0}, 1)$,

$$-\varepsilon\langle \mathbf{a}, \mathbf{z}\rangle > \alpha - \langle \mathbf{a}, \mathbf{x}\rangle.$$

If we take $\mathbf{z} = (\theta\mathbf{a})/\|\mathbf{a}\|$, $0 < \theta < 1$, and then let $\theta \to 1$, we get $-\varepsilon\|\mathbf{a}\| \geqslant \alpha - \langle \mathbf{a}, \mathbf{x}\rangle$, or

$$\langle \mathbf{a}, \mathbf{x}\rangle \geqslant \alpha + \varepsilon\|\mathbf{a}\|,$$

for all $\mathbf{x}$ in X. A similar argument shows that

$$\langle \mathbf{a}, \mathbf{y}\rangle \leqslant \alpha - \varepsilon\|\mathbf{a}\|$$

for all $\mathbf{y}$ in Y. Hence (ii) holds for any positive number η strictly less than $\varepsilon\|\mathbf{a}\|$.

We now suppose that there exists an $\eta > 0$ such that $\langle \mathbf{a}, \mathbf{x}\rangle > \alpha + \eta$ for all $\mathbf{x}$ in X. Let $\mathbf{z}$ be an arbitrary element of $B(\mathbf{0}, 1)$. Since $\|\mathbf{z}\| < 1$, from the Cauchy–Schwarz inequality we get that $\langle \mathbf{a}, \mathbf{z}\rangle < \|\mathbf{a}\|$, and consequently

$$\langle \mathbf{a}, \mathbf{x} - \varepsilon\mathbf{z}\rangle = \langle \mathbf{a}, \mathbf{x}\rangle - \varepsilon\langle \mathbf{a}, \mathbf{z}\rangle > \alpha + \eta - \varepsilon\|\mathbf{a}\|.$$

Similarly, for all $\mathbf{y}$ in Y and $\mathbf{z} \in B(0, 1)$,

$$\langle \mathbf{a}, \mathbf{y} + \varepsilon\mathbf{z}\rangle < \alpha - \eta + \varepsilon\|\mathbf{a}\|.$$

Hence if we take $\varepsilon < \eta/\|\mathbf{a}\|$, we have that $\langle \mathbf{a}, \mathbf{u}\rangle > \alpha$ for all $\mathbf{u}$ in $X - \varepsilon B(\mathbf{0}, 1) = X + \varepsilon B(\mathbf{0}, 1)$ and that $\langle \mathbf{a}, \mathbf{v}\rangle < \alpha$ for all $\mathbf{v}$ in $Y + \varepsilon B(0, 1)$. This proves (i).

Remark 3.1. From the preceding lemma it should be clear that a necessary and sufficient condition for two sets X and Y to be strongly separated is that there exists a vector $\mathbf{a}$ such that

$$\inf\{\langle \mathbf{a}, \mathbf{x}\rangle : \mathbf{x} \in X\} > \sup\{\langle \mathbf{a}, \mathbf{y}\rangle : \mathbf{y} \in Y\}.$$

Our principal objective in this section is to prove Theorem 3.4. Although this theorem is not the best possible separation theorem in $\mathbb{R}^n$, it is the one that is valid in infinite-dimensional spaces and it does cover the situations that occur in optimization problems. The best possible separation theorem in $\mathbb{R}^n$ will be discussed in Section 7. Our proof of Theorem 3.4 will not be the most economical one but will be one that is suggested by rather obvious geometric considerations.

The principal step in the proof is to show that if C is a convex set and $\mathbf{y}$ is a point not in C, then $\mathbf{y}$ and C can be properly separated. If C is not convex, then it may not be possible to separate $\mathbf{y}$ and C, as can be seen by taking, in $\mathbb{R}^2$, $\mathbf{y} = \mathbf{0}$ and C to be the circumference of any circle with center at the origin.

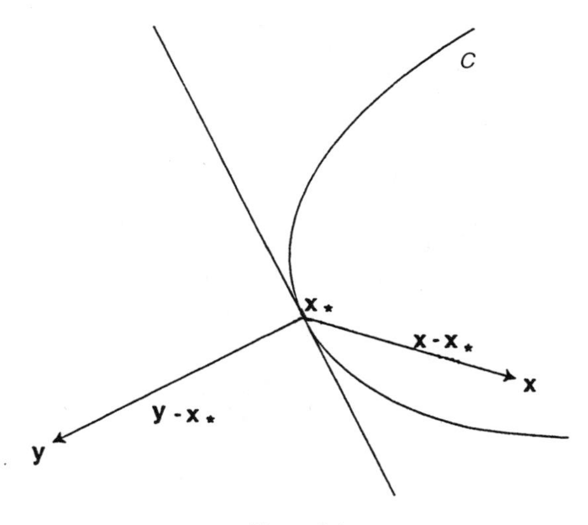

Figure 2.6.

We first show that if C is closed and convex, then strong separation is possible.

THEOREM 3.1. *Let C be a closed convex set and $\mathbf{y}$ a vector such that $\mathbf{y} \notin C$. Then there exists a hyperplane $H_{\mathbf{a}}^{\alpha}$ that strongly separates $\mathbf{y}$ and C.*

To motivate the proof, we argue heuristically from Figure 2.6, in which C is assumed to have a tangent at each boundary point.

Draw a line from $\mathbf{y}$ to $\mathbf{x}_*$, the point of C that is closest to $\mathbf{y}$. The vector $\mathbf{y} - \mathbf{x}_*$ will be perpendicular to C in the sense that $\mathbf{y} - \mathbf{x}_*$ is the normal to the tangent line at $\mathbf{x}_*$. The tangent line, whose equation is $\langle \mathbf{y} - \mathbf{x}_*, \mathbf{x} - \mathbf{x}_* \rangle = 0$, properly separates $\mathbf{y}$ and C. The point $\mathbf{x}_*$ is characterized by the fact that $\langle \mathbf{y} - \mathbf{x}_*, \mathbf{x} - \mathbf{x}_* \rangle \leqslant 0$ for all $\mathbf{x} \in C$. To obtain strong separation, we merely move the line parallel to itself so as to pass through a point $\mathbf{x}_0 \in (\mathbf{x}_*, \mathbf{y})$. We now justify these steps in a series of lemmas, some of which are of interest in their own right.

If X is a set and $\mathbf{y}$ is a vector, then we define the *distance from* $\mathbf{y}$ *to* X, denoted by $d(\mathbf{y}, X)$, to be

$$d(\mathbf{y}, X) = \inf\{\|\mathbf{y} - \mathbf{x}\| : \mathbf{x} \in X\}.$$

A point $\mathbf{x}_*$ in X is said to *attain the distance* from $\mathbf{y}$ to X, or to be *closest to* $\mathbf{y}$, if $d(\mathbf{y}, X) = \|\mathbf{y} - \mathbf{x}_*\|$.

LEMMA 3.2. *Let C be a convex subset of $\mathbb{R}^n$ and let $\mathbf{y} \notin C$. If there exists a point in C that is closest to $\mathbf{y}$, then it is unique.*

Proof. Suppose that there were two points $\mathbf{x}_1$ and $\mathbf{x}_2$ of C that were closest to $\mathbf{y}$. Then since C is convex, $(\mathbf{x}_1 + \mathbf{x}_2)/2$ belongs to C, and so

$$\begin{aligned} d(\mathbf{y}, C) &\leqslant \|(\mathbf{x}_1 + \mathbf{x}_2)/2 - \mathbf{y}\| = \tfrac{1}{2}\|(\mathbf{x}_1 - \mathbf{y}) + (\mathbf{x}_1 - \mathbf{y})\| \\ &\leqslant \tfrac{1}{2}(\|\mathbf{x}_1 - \mathbf{y}\| + \|\mathbf{x}_2 - \mathbf{y}\|) = d(\mathbf{y}, C). \end{aligned}$$

Hence equality holds in the application of the triangle inequality, and so by (5) of Section 2 in Chapter I, we have that, for some $\kappa \geqslant 0$, $(\mathbf{x}_1 - \mathbf{y}) = \kappa(\mathbf{x}_2 - \mathbf{y})$. Suppose $\kappa > 0$. Since $\|\mathbf{x}_1 - \mathbf{y}\| = \|\mathbf{x}_2 - \mathbf{y}\| = d(\mathbf{y}, C)$, we get that $\kappa = 1$, and so $\mathbf{x}_1 = \mathbf{x}_2$. If $\kappa = 0$, we get $\mathbf{x}_1 = \mathbf{y}$, contradicting the assumption that $\mathbf{y} \notin C$.

LEMMA 3.3. *Let C be a closed subset of $\mathbb{R}^n$ and let $\mathbf{y} \notin C$. Then there exists a point $\mathbf{x}_*$ in C that is closest to $\mathbf{y}$.*

Proof. Let $\mathbf{x}_0 \in C$ and let $r > \|\mathbf{x}_0 - \mathbf{y}\|$. Then $C_1 = \overline{B(\mathbf{y}, r)} \cap C$ is nonempty, closed, and bounded and hence is compact. The function $\mathbf{x} \to \|\mathbf{x} - \mathbf{y}\|$ is continuous on C_1 and so attains its minimum at some point $\mathbf{x}_*$ in C_1. Thus, for all $\mathbf{x} \in C_1$, $\|\mathbf{x} - \mathbf{y}\| \geqslant \|\mathbf{x}_* - \mathbf{y}\|$. For $\mathbf{x} \in C$ and $\mathbf{x} \notin C_1$ we have

$$\|\mathbf{x} - \mathbf{y}\| > r > \|\mathbf{x}_0 - \mathbf{y}\| \geqslant \|\mathbf{x}_* - \mathbf{y}\|,$$

since $\mathbf{x}_0 \in \overline{B(\mathbf{y}, r)}$.

Lemmas 3.2 and 3.3 show that if *C is a closed convex set and $\mathbf{y} \notin C$, then there is a unique closest point in C to $\mathbf{y}$.*

The next lemma characterizes closest points.

LEMMA 3.4. *Let C be a convex set and let $\mathbf{y} \notin C$. Then $\mathbf{x}_* \in C$ is a closest point in C to $\mathbf{y}$ if and only if*

$$\langle \mathbf{y} - \mathbf{x}_*, \mathbf{x} - \mathbf{x}_* \rangle \leqslant 0 \quad \textit{for all } \mathbf{x} \in C. \tag{1}$$

Note that if C is not convex, then the above characterization of closest point need not hold. To see this, let $\mathbf{y}$ be the origin in $\mathbb{R}^2$ and let C be the circumference of the unit circle. Let $\mathbf{x}_*$ be a fixed point in C. Then it is not true that $\langle \mathbf{0} - \mathbf{x}_*, \mathbf{x} - \mathbf{x}_* \rangle \leqslant 0$ for all $\mathbf{x}$ in C.

Proof. Let $\mathbf{x}_*$ be a closest point to $\mathbf{y}$ and let $\mathbf{x}$ be any point in C. Since C is convex, the line segment $[\mathbf{x}_*, \mathbf{x}] = \{\mathbf{z}(t) : \mathbf{z}(t) = \mathbf{x}_* + t(\mathbf{x} - \mathbf{x}_*), 0 \leqslant t \leqslant 1\}$ belongs to C. Let

$$\varphi(t) = \|\mathbf{z}(t) - \mathbf{y}\|^2 = \langle \mathbf{x}_* + t(\mathbf{x} - \mathbf{x}_*) - \mathbf{y}, \mathbf{x}_* + t(\mathbf{x} - \mathbf{x}_*) - \mathbf{y} \rangle. \tag{2}$$

For $0 \leqslant t \leqslant 1$, $\varphi(t)$ is the square of the distance between the point $\mathbf{z}(t) \in [\mathbf{x}_*, \mathbf{x}]$ and $\mathbf{y}$. If $t = 0$, then $\mathbf{z} = \mathbf{x}_*$. Since φ is continuously differentiable on $(0, 1]$ and

$\mathbf{x}_*$ is a point in C closest to $\mathbf{y}$, we have $\varphi'(0+) \geqslant 0$. Calculating $\varphi'(t)$ from (2) gives

$$\varphi'(t) = 2[-\langle \mathbf{y} - \mathbf{x}_*, \mathbf{x} - \mathbf{x}_* \rangle + t\,\|\mathbf{x} - \mathbf{x}_*\|^2]. \tag{3}$$

If we now let $t \to 0+$ and use $\varphi'(0+) \geqslant 0$, we get (1).

We now suppose that (1) holds. Let $\mathbf{x}$ be any other point in C. It follows from (3) that $\varphi'(t) > 0$ for $0 < t \leqslant 1$. This φ is strictly increasing function on $[0, 1]$, and so for any point $\mathbf{z}(t)$ in $(\mathbf{x}_*, \mathbf{x}]$. We have $\|\mathbf{z}(t) - \mathbf{y}\| > \|\mathbf{x}_* - \mathbf{y}\|$. In particular, this is true for $\mathbf{z} = \mathbf{x}$, and so $\mathbf{x}_*$ is a closest point to $\mathbf{y}$.

We can now complete the proof of Theorem 3.1. Let $\mathbf{x}_*$ be the closest point in C to $\mathbf{y}$ and let $\mathbf{a} = \mathbf{y} - \mathbf{x}_*$. Then for all $\mathbf{x} \in C$, $\langle \mathbf{a}, \mathbf{x} - \mathbf{x}_* \rangle \leqslant 0$, and so $\langle \mathbf{a}, \mathbf{x} \rangle \leqslant \langle \mathbf{a}, \mathbf{x}_* \rangle$, with equality occurring when $\mathbf{x} = \mathbf{x}_*$. Therefore $\sup\{\langle \mathbf{a}, \mathbf{x} \rangle : \mathbf{x} \in C\} = \langle \mathbf{a}, \mathbf{x}_* \rangle$. On the other hand, $\langle \mathbf{a}, \mathbf{y} - \mathbf{x}_* \rangle = \|\mathbf{a}\|^2 > 0$, so

$$\langle \mathbf{a}, \mathbf{y} \rangle = \langle \mathbf{a}, \mathbf{x}_* \rangle + \|\mathbf{a}\|^2 > \sup\{\langle \mathbf{a}, \mathbf{x} \rangle : \mathbf{x} \in C\}.$$

The conclusion of the theorem now follows from Remark 3.1 with $X = \{\mathbf{y}\}$ and $Y = C$.

We now take up the separation of a point $\mathbf{y}$ and an arbitrary convex set; that is, we shall no longer assume that C is closed. We shall obtain separation, but shall not be able to guarantee proper separation unless we assume that C has nonempty interior.

THEOREM 3.2. *Let C be a convex set and let $\mathbf{y} \notin C$. Then there exists a hyperplane $H^{\alpha}_{\mathbf{a}}$ such that for all $\mathbf{x}$ in C*

$$\langle \mathbf{a}, \mathbf{x} \rangle \leqslant \alpha \quad \textit{and} \quad \langle \mathbf{a}, \mathbf{y} \rangle = \alpha. \tag{i}$$

If $\operatorname{int}(C) \neq \varnothing$, *then for all* $\mathbf{x}$ *in* $\operatorname{int}(C)$

$$\langle \mathbf{a}, \mathbf{x} \rangle < \alpha. \tag{ii}$$

Proof. We first suppose that $\mathbf{y} = \mathbf{0}$. Then to establish (i), we must find an $\mathbf{a} \neq \mathbf{0}$ such that

$$\langle \mathbf{a}, \mathbf{x} \rangle \leqslant 0 \quad \text{for all } \mathbf{x} \text{ in } C. \tag{4}$$

For each $\mathbf{x}$ in C define

$$N_{\mathbf{x}} = \{\mathbf{z} : \|\mathbf{z}\| = 1, \langle \mathbf{z}, \mathbf{x} \rangle \leqslant 0\}.$$

The set $N_{\mathbf{x}}$ is nonempty since it contains the element $-\mathbf{x}/\|\mathbf{x}\|$.

To establish (4), it suffices to show that

$$\bigcap_{\mathbf{x} \in C} N_{\mathbf{x}} \neq \varnothing.$$

Each $N_{\mathbf{x}}$ is closed and is contained in $S(\mathbf{0}, 1) = \{\mathbf{u}: \|\mathbf{u}\| = 1\}$. Thus we may write the last relation as

$$\bigcap_{\mathbf{x} \in C} N_{\mathbf{x}} = \left(\bigcap_{\mathbf{x} \in C} N_{\mathbf{x}}\right) \cap S(\mathbf{0}, 1) \neq \varnothing. \tag{5}$$

Since $S(\mathbf{0}, 1)$ is compact, it has the finite-intersection property. Thus, if the intersection in (5) were empty, there would exist a finite subcollection $N_{\mathbf{x}_1}, \ldots, N_{\mathbf{x}_r}$ that is empty. Therefore, we may establish (5) by showing that every finite subcollection $N_{\mathbf{x}_1}, \ldots, N_{\mathbf{x}_k}$ has nonempty intersection.

Now consider any finite collection of sets $N_{\mathbf{x}_1}, \ldots, N_{\mathbf{x}_k}$ and the corresponding k points $\mathbf{x}_1, \ldots, \mathbf{x}_k$ in C. Let $\text{co}[\mathbf{x}_1, \ldots, \mathbf{x}_k]$ denote the convex hull of $\mathbf{x}_1, \ldots, \mathbf{x}_k$. Then $\text{co}[\mathbf{x}_1, \ldots, \mathbf{x}_k]$ is compact and convex. Since C is convex, $\text{co}[\mathbf{x}_1, \ldots, \mathbf{x}_k] \subseteq C$. Thus, since $\mathbf{0} \notin C$, we have $\mathbf{0} \notin \text{co}[\mathbf{x}_1, \ldots, \mathbf{x}_k]$. Hence, by Theorem 3.1, there exists a vector $\mathbf{w} \neq \mathbf{0}$ such that

$$0 = \langle \mathbf{w}, \mathbf{0} \rangle > \langle \mathbf{w}, \mathbf{x} \rangle, \qquad \mathbf{x} \in \text{co}[\mathbf{x}_1, \ldots, \mathbf{x}_k]. \tag{6}$$

We may divide through by $\|\mathbf{w}\| \neq 0$ in (6) and so assume that $\|\mathbf{w}\| = 1$. Thus, (6) says that

$$\mathbf{w} \in \left(\bigcap_{i=1}^{k} N_{\mathbf{x}_i}\right) \cap S(\mathbf{0}, 1),$$

and we have shown that the intersection of any finite subcollection $N_{\mathbf{x}_1}, \ldots, N_{\mathbf{x}_k}$ of the closed sets $\{N_{\mathbf{x}}\}_{\mathbf{x} \in C}$ has nonempty intersection with $S(\mathbf{0}, 1)$. Hence (5) holds.

We now remove the restriction that $\mathbf{y} = \mathbf{0}$. We have

$$\mathbf{y} \notin C \quad \text{if and only if } \mathbf{0} \notin C - \mathbf{y} = \{\mathbf{x}' : \mathbf{x}' = \mathbf{x} - \mathbf{y}, \mathbf{x} \in C\}.$$

Therefore, there exists an $\mathbf{a} \neq \mathbf{0}$ such that $\langle \mathbf{a}, \mathbf{x}' \rangle \leqslant 0$ for all $\mathbf{x}'$ in $C - \mathbf{y}$. Hence for all $\mathbf{x}$ in C,

$$\langle \mathbf{a}, \mathbf{x} - \mathbf{y} \rangle \leqslant 0,$$

and so

$$\langle \mathbf{a}, \mathbf{x} \rangle \leqslant \langle \mathbf{a}, \mathbf{y} \rangle \quad \text{for all } \mathbf{x} \text{ in } C.$$

If we now let $\alpha = \langle \mathbf{a}, \mathbf{y} \rangle$, we get the first conclusion of the theorem. The second follows from Exercise 2.16.

Our first separation result for convex sets will follow from Theorem 3.2.

THEOREM 3.3. *Let X_0 and Y be two disjoint convex sets. Then there exists a hyperplane $H_{\mathbf{a}}^{\alpha}$ that separates them.*

Note that the theorem does not assert that proper separation can be achieved. Theorem 7.2 in the sequel will allow us to conclude that the separation is proper. Our proof of Theorem 3.3 does not yield this fact.

To prove the theorem, we first note that

$$X_0 \cap Y = \varnothing \quad \text{if and only if } \mathbf{0} \notin X_0 - Y.$$

Let

$$A = X_0 - Y = X_0 + (-1)Y.$$

Since X_0 and Y are convex, by Lemma 2.2, so is A. Since X_0 and Y are disjoint, $\mathbf{0} \notin A$. Hence by Theorem 3.2, there exists an $\mathbf{a} \neq \mathbf{0}$ such that

$$\langle \mathbf{a}, \mathbf{z} \rangle \leqslant 0 \quad \text{all } \mathbf{z} \text{ in } A.$$

Let $\mathbf{x}$ be an arbitrary element of X_0 and let $\mathbf{y}$ be an arbitrary element of Y. Then $\mathbf{z} = \mathbf{x} - \mathbf{y}$ is in A and

$$\langle \mathbf{a}, \mathbf{x} \rangle \leqslant \langle \mathbf{a}, \mathbf{y} \rangle. \tag{7}$$

Let $\beta = \sup\{\langle \mathbf{a}, \mathbf{x} \rangle : \mathbf{x} \in X_0\}$ and let $\gamma = \inf\{\langle \mathbf{a}, \mathbf{y} \rangle : \mathbf{y} \in Y\}$. Then from (7) we get that β and γ are finite, and for any number α such that $\beta \leqslant \alpha \leqslant \gamma$,

$$\langle \mathbf{a}, \mathbf{x} \rangle \leqslant \alpha \leqslant \langle \mathbf{a}, \mathbf{y} \rangle$$

for all $\mathbf{x}$ in X_0 and all $\mathbf{y}$ in Y. Thus the hyperplane $H_{\mathbf{a}}^{\alpha}$ separates X_0 and Y.

THEOREM 3.4. *Let X and Y be two convex sets such that* int(X) *is not empty and* int(X) *is disjoint from Y. Then there exists a hyperplane $H_{\mathbf{a}}^{\alpha}$ that properly separates $\bar{X}$ and $\bar{Y}$.*

To illustrate the theorem, let $X = \{(x_1, x_2) : x_1 \leqslant 0, \ -1 < x_2 \leqslant 1\}$ and let $Y = \{(x_1, x_2) : x_1 = 0, \ -1 \leqslant x_2 \leqslant 1\}$. The hypotheses of the theorem are fulfilled and $x_1 = 0$ properly separates $\bar{X}$ and $\bar{Y}$ and hence X and Y. Note that strict separation of X and Y is not possible.

Proof. Since int(X) is convex and disjoint from Y, we can apply Theorem 3.3 with $X_0 = \text{int}(X)$ and obtain the existence of an $\mathbf{a} \neq \mathbf{0}$ and an α such that

$$\langle \mathbf{a}, \mathbf{x} \rangle \leqslant \alpha \leqslant \langle \mathbf{a}, \mathbf{y} \rangle \tag{8}$$

for all $\mathbf{x}$ in int(X) and all $\mathbf{y}$ in Y.

Let $\mathbf{x} \in \bar{X}$. By Corollary 2 to Lemma 2.5, $\bar{X} = \overline{\text{int}(X)}$. Hence $\mathbf{x} \in \overline{\text{int}(X)}$. By Remark I.5.2 there exists a sequence of points $\{\mathbf{x}_k\}$ in int(X) such that $\mathbf{x}_k \to \mathbf{x}$. It then follows from (8) and the continuity of the inner product that (8) holds for all $\mathbf{x}$ in $\bar{X}$ and all $\mathbf{y} \in \bar{Y}$. Thus, the hyperplane $H_{\mathbf{a}}^{\alpha}$ separates $\bar{X}$ and $\bar{Y}$. By Corollary 2 to Lemma 2.5, $\text{int}(\bar{X}) = \text{int}(X)$. By Exercise 2.16, for $\mathbf{x} \in \text{int}(\bar{X})$, $\langle \mathbf{a}, \mathbf{x} \rangle < \alpha$, so the separation is proper.

THEOREM 3.5. *Let K be a compact convex set and let C be a closed convex set such that K and C are disjoint. Then K and C can be strongly separated.*

Proof. Let

$$K_\varepsilon = K + B(\mathbf{0}, \varepsilon) = \{\mathbf{u} : \mathbf{u} = \mathbf{x} + \mathbf{z},\ \mathbf{x} \in K,\ \|\mathbf{z}\| < \varepsilon\},$$
$$C_\varepsilon = C + B(\mathbf{0}, \varepsilon) = \{\mathbf{v} : \mathbf{v} = \mathbf{y} + \mathbf{z},\ \mathbf{y} \in C,\ \|\mathbf{z}\| < \varepsilon\}.$$

Since $K_\varepsilon = \bigcup_{\mathbf{x} \in K} (B(\mathbf{0}, \varepsilon) + \mathbf{x})$, it is a union of open sets and hence is open. It is also readily verified that K_ε is convex. Similarly, the set C_ε is open and convex. (See Exercise 2.14.)

We now show that there exists an $\varepsilon > 0$ such that $K_\varepsilon \cap C_\varepsilon = \varnothing$. If the assertion were false, there would exist a sequence $\{\varepsilon_k\}$ with $\varepsilon_k > 0$ and $\varepsilon_k \to 0$ and a sequence $\{\mathbf{w}_k\}$ such that, for each k, $\mathbf{w}_k \in K_{\varepsilon_k} \cap C_{\varepsilon_k}$. Since $\mathbf{w}_k = \mathbf{u}_k + \mathbf{z}_k$ with $\mathbf{u}_k \in K$, $\|\mathbf{z}_k\| < \varepsilon_k$ and $\mathbf{w}_k = \mathbf{v}_k + \mathbf{z}'_k$ with $\mathbf{v}_k \in C$ and $\|\mathbf{z}'_k\| < \varepsilon_k$, we have a sequence $\{\mathbf{u}_k\}$ in K and a sequence $\{\mathbf{v}_k\}$ in C such that

$$\|\mathbf{w}_k - \mathbf{u}_k\| < \varepsilon_k, \qquad \|\mathbf{w}_k - \mathbf{v}_k\| < \varepsilon_k.$$

Hence

$$\|\mathbf{u}_k - \mathbf{v}_k\| = \|(\mathbf{u}_k - \mathbf{w}_k) + (\mathbf{w}_k - \mathbf{v}_k)\| \leqslant \|\mathbf{u}_k - \mathbf{w}_k\| + \|\mathbf{v}_k - \mathbf{w}_k\| < 2\varepsilon_k. \tag{9}$$

Since K is compact, there exists a subsequence $\{\mathbf{u}_{kj}\}$ that converges to an element $\mathbf{u}_0$ in K. It follows from (9) that $\mathbf{v}_{kj} \to \mathbf{u}_0$. Since C is closed, $\mathbf{u}_0 \in C$. This contradicts the assumption that C and K are disjoint, and so the assertion is true.

We have shown that there exists an $\varepsilon > 0$ such that K_ε and C_ε are disjoint open convex sets. Hence by Theorem 3.4 there is a hyperplane that properly separates K_ε and C_ε. Since both K_ε and C_ε are open sets, it follows from Exercise 2.16 that K_ε and C_ε are strictly separated. According to Definition 3.4, this says that K and C are strongly separated.

We now apply the separation theorems to obtain an additional characterization of convex sets.

THEOREM 3.6. *Let A be a set contained in some half space. Then the closure of the convex hull of A, $\overline{\text{co}(A)}$, is the intersection of all closed half spaces containing A.*

Proof. We orient the hyperplanes that determine the closed half spaces containing A to be the negative closed half spaces. If $\bar{H}_{\mathbf{a}}^{\alpha-}$ is a closed half space containing A, then $\overline{\text{co}(A)}$ is contained in $\bar{H}_{\mathbf{a}}^{\alpha-}$ and hence in the intersection of all such half spaces. Now let $\mathbf{x}$ be a point in the intersection. If $\mathbf{x}$ were not in $\overline{\text{co}(A)}$, then by Theorem 3.1, there would exist a hyperplane $H_{\mathbf{a}}^{\alpha}$ such that

$$\langle \mathbf{a}, \mathbf{z} \rangle < \alpha < \langle \mathbf{a}, \mathbf{x} \rangle, \qquad \mathbf{z} \in \overline{\text{co}(A)}.$$

Hence $\mathbf{x}$ would not be in the closed half space by $\bar{H}_{\mathbf{a}}^{\alpha-}$ which contains A. Therefore $\mathbf{x}$ is not in the intersection of all negative closed half spaces containing A, which contradicts the assumption that $\mathbf{x}$ is in the intersection. Thus the intersection is contained in $\overline{\text{co}(A)}$, and the theorem is proved.

The following corollary follows from the theorem and the fact that a closed convex set not equal to $\mathbb{R}^n$ is contained in some closed half space.

COROLLARY 1. *If A is a closed convex set not equal to $\mathbb{R}^n$, then A is the intersection of all closed half spaces containing A.*

Note that the theorem is false if we replace $\overline{\text{co}(A)}$ by $\text{co}(A)$. To see this, consider the set A in $\mathbb{R}^2$ defined by

$$A = \{(x_1, x_2) : x_1 > 0, \ x_1 x_2 \geqslant 1\} \cup \{(x_1, x_2) : x_1 > 0, \ x_1 x_2 \leqslant -1\}.$$

Then $\text{co}(A) = \{(x_1, x_2) : x_1 > 0\}$, and the intersection of all closed half spaces is $\{(x_1 \ x_2) : x_1 \geqslant 0\}$.

Exercise 3.2. Let V be a linear subspace and let $\mathbf{y} \notin V$. Show that $\mathbf{x}_*$ in V is the closest point in V to $\mathbf{y}$ if and only if $\mathbf{y} - \mathbf{x}_*$ is orthogonal to V; that is, for every $\mathbf{w}$ in V, $\mathbf{y} - \mathbf{x}_*$ is orthogonal to $\mathbf{w}$.

Exercise 3.3. Let C be a closed convex set and let $\mathbf{y} \notin C$. Show that $\mathbf{x}_*$ in C is closest to $\mathbf{y}$ if and only if $\langle \mathbf{x} - \mathbf{y}, \mathbf{x}_* - \mathbf{y} \rangle \geqslant \|\mathbf{x}_* - \mathbf{y}\|^2$ for all $\mathbf{x}$ in C.

Exercise 3.4. In reference to Theorem 3.5, show that if K is assumed to be convex and closed, then it may not be possible to separate K and C strictly, let alone strongly.

Exercise 3.5. Show that if F is closed and K is compact, then $K + F$ is closed. Give an example in which F and K are closed yet $F + K$ is not.

Exercise 3.6. Let A and B be two compact sets. Show that A and B can be strongly separated if and only if $\text{co}(A) \cap \text{co}(B) = \varnothing$.

Exercise 3.7. Prove Theorem 3.5 using Theorem 3.1 and Exercise 3.5.

Exercise 3.8. Let A be a bounded set. Show that $\text{co}(\bar{A})$ is the intersection of all closed half spaces containing A. Show that the statement is false if A is not bounded and a proper subset of $\mathbb{R}^n$.

Exercise 3.9. Let A be a closed convex set such that cA (the complement of A) is convex. Show that A is a closed half space.

Exercise 3.10. Let C_1 and C_2 be two convex subsets of $\mathbb{R}^n$. Show that there exists a hyperplane that strongly separates C_1 and C_2 if and only if

$$\inf\{\|\mathbf{x} - \mathbf{y}\| : \mathbf{x} \in C_1, \mathbf{y} \in C_2\} > 0.$$

4. SUPPORTING HYPERPLANES: EXTREME POINTS

In studying convex sets we do not wish to restrict ourselves to sets with smooth boundaries, that is, sets such that every boundary point has a tangent plane. The concept that replaces that of tangent plane is that of supporting hyperplane. It is a concept that has many important consequences, as we shall see.

We first define what is meant by a boundary point of a set. A point $\mathbf{z}$ is said to be a *boundary point of* a set S if for every $\varepsilon > 0$ the ball $B(\mathbf{z}, \varepsilon)$ contains a point $\mathbf{x} \in S$ and a point $\mathbf{y} \notin S$. Note that the definition allows us to have $\mathbf{y} = \mathbf{z}$ or $\mathbf{x} = \mathbf{z}$. For example, in $\mathbb{R}^2$ let $S = \{\mathbf{x} = (x_1, x_2) : 0 < \|\mathbf{x}\| < 1\}$. Then $\mathbf{0}$ is a boundary point of S, and for every $0 < \varepsilon < 1$ the only point in $B(\mathbf{0}, \varepsilon)$ that is not in S is $\mathbf{0}$. The other boundary points of S are the points $\mathbf{x}$ with $\|\mathbf{x}\| = 1$. If there exists an $\varepsilon > 0$ such that the only point in $B(\mathbf{z}, \varepsilon)$ that is in S is $\mathbf{z}$ itself, then we say that $\mathbf{z}$ is an *isolated point* of S.

Definition 4.1. A hyperplane $H_{\mathbf{a}}^{\alpha}$ is said to be a *supporting hyperplane* to a set S if, for every $\mathbf{x} \in S$, $\langle \mathbf{a}, \mathbf{x} \rangle \leqslant \alpha$ and there exists at least one point $\mathbf{x}_0 \in \bar{S}$ such that $\langle \mathbf{a}, \mathbf{x}_0 \rangle = \alpha$. The hyperplane $H_{\mathbf{a}}^{\alpha}$ is said to *support* S at $\mathbf{x}_0$. The hyperplane $H_{\mathbf{a}}^{\alpha}$ is a *nontrivial supporting hyperplane* if there exists a point $\mathbf{x}_1$ in S such that $\langle \mathbf{a}, \mathbf{x}_1 \rangle < \alpha$.

To illustrate this definition, consider the set S in $\mathbb{R}^3$ defined by

$$S = \{\mathbf{x} : 0 \leqslant x_1^2 + x_2^2 < 1, x_3 = 0\}.$$

As a set in $\mathbb{R}^3$, every point of S is a boundary point. The only supporting hyperplane at a point of S is the trivial one, $x_3 = 0$. Every point in the set $E = \{\mathbf{x}: x_1^2 + x_2^2 = 1, x_3 = 0\}$ is also a boundary point of S. At every point of E there is a nontrivial supporting hyperplane with equation $ax_1 + bx_2 = 1$ for appropriate a and b.

Remark 4.1. Let K be a compact set in $\mathbb{R}^n$. Then for each $\mathbf{a}$ in R^n there exists an α such that the hyperplane $H_{\mathbf{a}}^{\alpha}$ with equation $\langle \mathbf{a}, \mathbf{x} \rangle = \alpha$ is a supporting hyperplane to K. To see this, note that the linear functional L defined by $L(\mathbf{x}) = \langle \mathbf{a}, \mathbf{x} \rangle$ is continuous on K and so attains its maximum at a point $\mathbf{x}_*$ in K. Then $\langle \mathbf{a}, \mathbf{x} \rangle \leqslant \langle \mathbf{a}, \mathbf{x}_* \rangle$ for all $\mathbf{x} \in K$. The desired hyperplane is obtained by taking $\alpha = \langle \mathbf{a}, \mathbf{x}_* \rangle$.

THEOREM 4.1. *Let C be a convex set and let $\mathbf{z}$ be a boundary point of C. Then there exists a supporting hyperplane $H_{\mathbf{a}}^{\alpha}$ to C such that $\mathbf{z} \in H_{\mathbf{a}}^{\alpha}$. If* $\mathrm{int}(C) \neq \varnothing$, *the supporting hyperplane is nontrivial.*

Proof. If $\mathbf{z} \notin C$, then by Theorem 3.2 with $\mathbf{y} = \mathbf{z}$ there exists a hyperplane $H_{\mathbf{a}}^{\alpha}$ such that for all $\mathbf{x}$ in C

$$\langle \mathbf{a}, \mathbf{x} \rangle \leqslant \alpha \quad \text{and} \quad \langle \mathbf{a}, \mathbf{z} \rangle = \alpha.$$

Since $\mathbf{z} \notin C$ and $\mathbf{z}$ is a boundary point of C, it follows that $\mathbf{z} \in \bar{C}$. Thus, $H_{\mathbf{a}}^{\alpha}$ is a supporting hyperplane to C with $\mathbf{z} \in H_{\mathbf{a}}^{\alpha}$.

If $\mathbf{z} \in C$, then from the definition of boundary point it follows that there exists a sequence of points $\{\mathbf{y}_k\}$ with $\mathbf{y}_k \notin C$ such that $\mathbf{y}_k \to \mathbf{z}$. It follows from Theorem 3.2 that for each positive integer k there exists a vector $\mathbf{a}_k \neq \mathbf{0}$ such that

$$\langle \mathbf{a}_k, \mathbf{x} \rangle \leqslant \langle \mathbf{a}_k, \mathbf{y}_k \rangle \quad \text{for all } \mathbf{x} \text{ in } C. \tag{1}$$

If we divide through by $\|\mathbf{a}_k\| \neq 0$ in (1), we see that we may assume that $\|\mathbf{a}_k\| = 1$. Since $S(0, 1)$ is compact, there exists a subsequence of $\{\mathbf{a}_k\}$ that we again label as $\{\mathbf{a}_k\}$ and a vector $\mathbf{a}$ with $\|\mathbf{a}\| = 1$ such that $\mathbf{a}_k \to \mathbf{a}$. If we now let $k \to \infty$ in (1), we get that

$$\langle \mathbf{a}, \mathbf{x} \rangle \leqslant \langle \mathbf{a}, \mathbf{z} \rangle \quad \text{for all } \mathbf{x} \text{ in } C.$$

If we set $\alpha = \langle \mathbf{a}, \mathbf{z} \rangle$, we see that the hyperplane $H_{\mathbf{a}}^{\alpha}$ contains $\mathbf{z}$ and is a supporting hyperplane to C.

It follows from Exercise 2.16 that the supporting hyperplane is nontrivial if $\mathrm{int}(C) \neq \varnothing$.

Definition 4.2. A point $\mathbf{x}_0$ belonging to a convex set C is an *extreme point* of C if, for no pair of distinct points $\mathbf{x}_1$, $\mathbf{x}_2$ in C and for no $\alpha > 0$, $\beta > 0$, $\alpha + \beta = 1$, is it true that $\mathbf{x}_0 = \alpha\mathbf{x}_1 + \beta\mathbf{x}_2$.

In other words, a point $\mathbf{x}_0$ in a convex set C is an extreme point of C if $\mathbf{x}_0$ is interior to no closed interval $[\mathbf{x}_1, \mathbf{x}_2]$ in C. We leave it as an exercise for the reader to show that an extreme point is a boundary point.

Let

$$C_1 = \{(x_1, x_2) : x_1^2 + x_2^2 \leqslant 1\}.$$

The set of extreme points of C_1 is $\{(x_1, x_2) : x_1^2 + x_2^2 = 1\}$.

Let

$$C_2 = \{(x_1, x_2) : x_1^2 + x_2^2 < 1\}, \qquad C_3 = \{(x_1, x_2) : x_1^2 + x_2^2 = 1,\ x_2 \geqslant 0\},$$

and let $C_4 = C_2 \cup C_3$. Then the set of extreme points of C_4 is C_3.

A convex set need not have any extreme points, as evidenced by C_2. Another example of a convex set without any extreme points is $\{(x_1, x_2) : x_2 = 0\}$.

THEOREM 4.2. *Let C be a compact convex set. Then the set of extreme points of C is not empty, and every supporting hyperplane of C contains at least one extreme point of C.*

The following observation will be used in the proof of the theorem.

LEMMA 4.1. *Let C be a convex set and let H be a supporting hyperplane to C. A point $\mathbf{x}_0$ in $H \cap C$ is an extreme point of C if and only if $\mathbf{x}_0$ is an extreme point of $H \cap C$.*

Proof. Let $\mathbf{x}_0$ in $H \cap C$ be an extreme point of C. If $\mathbf{x}_0$ were not an extreme point of $H \cap C$, then $\mathbf{x}_0$ would be an interior point of some closed interval $[\mathbf{x}_1, \mathbf{x}_2]$ contained in $H \cap C$. But then $[\mathbf{x}_1, \mathbf{x}_2]$ would be contained in C, and $\mathbf{x}_0$ would not be an extreme point of C. Thus, if $\mathbf{x}_0$ is an extreme point of C, then it is also an extreme point of $H \cap C$.

Let $\mathbf{x}_0$ be an extreme point of $H \cap C$. If $\mathbf{x}_0$ were not an extreme point of C, then there would exist points $\mathbf{x}_1$ and $\mathbf{x}_2$ in C with at least one of $\mathbf{x}_1$ and $\mathbf{x}_2$ not in H such that for some $0 < t < 1$

$$\mathbf{x}_0 = t\mathbf{x}_1 + (1 - t)\mathbf{x}_2.$$

Suppose that $H = H_{\mathbf{a}}^{\alpha}$, that $\langle \mathbf{a}, \mathbf{x} \rangle \leqslant \alpha$ for $\mathbf{x}$ in C, and that $\mathbf{x}_1 \notin H_{\mathbf{a}}^{\alpha}$. Then $\langle \mathbf{a}, \mathbf{x}_1 \rangle < \alpha$ and

$$\alpha = \langle \mathbf{a}, \mathbf{x}_0 \rangle = t\langle \mathbf{a}, \mathbf{x}_1 \rangle + (1 - t)\langle \mathbf{a}, \mathbf{x}_2 \rangle < t\alpha + (1 - t)\alpha = \alpha.$$

This contradiction shows that $\mathbf{x}_0$ must be an extreme point of C.

We now show that the set of extreme points of a compact convex set is not empty.

LEMMA 4.2. *The set of extreme points of a compact convex set C is not empty.*

Proof. If C is a singleton, then there is nothing to prove. Therefore we may suppose that C is not a singleton. Since C is compact and the norm is a continuous function, there exists a point $\mathbf{x}^*$ in C of maximum norm. Thus $\|\mathbf{x}\| \leqslant \|\mathbf{x}^*\|$ for all $\mathbf{x}$ in C and $\|\mathbf{x}^*\| > 0$.

We assert that $\mathbf{x}^*$ is an extreme point of C. If $\mathbf{x}^*$ were not an extreme point, there would exist distinct points $\mathbf{x}$, and $\mathbf{x}_2$ in C different from $\mathbf{x}^*$ and scalars $\alpha > 0$, $\beta > 0$, $\alpha + \beta = 1$ such that

$$\mathbf{x}^* = \alpha\mathbf{x}_1 + \beta\mathbf{x}_2.$$

Since

$$\|\mathbf{x}^*\| = \|\alpha\mathbf{x}_1 + \beta\mathbf{x}_2\| \leqslant \alpha\|\mathbf{x}_1\| + \beta\|\mathbf{x}_2\|,$$

we see that if either $\|\mathbf{x}_1\|$ or $\|\mathbf{x}_2\|$ were strictly less than $\|\mathbf{x}^*\|$, we would have

$$\|\mathbf{x}^*\| < (\alpha + \beta)\|\mathbf{x}^*\| = \|\mathbf{x}^*\|.$$

This contradiction implies that

$$\|\mathbf{x}_1\| = \|\mathbf{x}_2\| = \|\mathbf{x}^*\|. \tag{2}$$

But then,

$$\begin{aligned}
\|\mathbf{x}^*\|^2 &= \|\alpha\mathbf{x}_1 + \beta\mathbf{x}_2\|^2 = \alpha^2\|\mathbf{x}_1\|^2 + 2\alpha\beta\langle\mathbf{x}_1, \mathbf{x}_2\rangle + \beta^2\|\mathbf{x}_2\|^2 \\
&\leqslant \alpha^2\|\mathbf{x}_1\|^2 + 2\alpha\beta\|\mathbf{x}_1\|\,\|\mathbf{x}_2\| + \beta^2\|\mathbf{x}_2\|^2 \\
&= (\alpha + \beta)^2\|\mathbf{x}^*\|^2 = \|\mathbf{x}^*\|^2.
\end{aligned}$$

Hence we must have

$$2\alpha\beta\langle\mathbf{x}_1, \mathbf{x}_2\rangle = 2\alpha\beta\|\mathbf{x}_1\|\,\|\mathbf{x}_2\|.$$

Since $\alpha > 0$, $\beta > 0$, we have $\langle\mathbf{x}_1, \mathbf{x}_2\rangle = \|\mathbf{x}_1\|\,\|\mathbf{x}_2\|$. Therefore $\mathbf{x}_2 = \kappa\mathbf{x}_1$ for some $\kappa \geqslant 0$. The relation (2) implies that $\kappa = 1$, and so $\mathbf{x}_2 = \mathbf{x}_1$. This contradicts the assumption that $\mathbf{x}_1$ and $\mathbf{x}_2$ are distinct. Thus $\mathbf{x}^*$ is an extreme point.

We now complete the proof of the theorem. Let $H_{\mathbf{a}}^{\alpha}$ be a supporting hyperplane to C. Since C is compact, the point $\mathbf{x}_0$ in $\bar{C}$ such that $\langle\mathbf{a}, \mathbf{x}_0\rangle = \alpha$ is also in C. Hence $H_{\mathbf{a}}^{\alpha} \cap C$ is nonempty and is compact and convex. By Lemma 4.2, $H_{\mathbf{a}}^{\alpha} \cap C$ contains an extreme point $\mathbf{x}_e$. By Lemma 4.1, $\mathbf{x}_e$ is also an extreme point of C.

THEOREM 4.3. *Let C be a compact convex set. Let C_e denote the set of extreme points of C. Then $C = \overline{\text{co}}(C_e)$.*

We first note that by Theorem 4.2 the set C_e is not empty.

Since $C_e \subseteq C$ and C is convex, we have that $\text{co}(C_e) \subseteq \text{co}(C) = C$. Since C is compact, we have that $\overline{\text{co}}(C_e) \subseteq \bar{C} = C$.

We now assert that $C \subseteq \overline{\text{co}}(C_e)$. If the last assertion were not true, then there would exist an $\mathbf{x}_0 \notin \overline{\text{co}}(C_e)$ and in C. Since $\overline{\text{co}}(C_e)$ is closed and convex, it follows from Theorem 3.1 that there exists a hyperplane $H_{\mathbf{a}}^{\alpha}$ such that for all $\mathbf{x} \in \overline{\text{co}}(C_e)$

$$\langle \mathbf{a}, \mathbf{x} \rangle < \alpha < \langle \mathbf{a}, \mathbf{x}_0 \rangle. \tag{3}$$

Let

$$\beta = \max\{\langle \mathbf{a}, \mathbf{x} \rangle : \mathbf{x} \in C\}. \tag{4}$$

Since C is compact, the maximum is attained at some point $\mathbf{x}_*$ in C, and thus $H_{\mathbf{a}}^{\beta}$ is a supporting hyperplane to C at $\mathbf{x}_*$. By Theorem 4.2, $H_{\mathbf{a}}^{\beta}$ contains an extreme point $\mathbf{x}_1$ of C. Thus, $\mathbf{x}_1 \in H_{\mathbf{a}}^{\beta} \cap \overline{\text{co}}(C_e)$, and so by (3),

$$\beta = \langle \mathbf{a}, \mathbf{x}_1 \rangle < \alpha.$$

Since $\mathbf{x}_0 \in C$, it follows from (3) and (4) that

$$\alpha < \langle \mathbf{a}, \mathbf{x}_0 \rangle \leqslant \beta.$$

Combining the last two inequalities gives $\beta < \beta$. This contradiction shows that $C \subseteq \overline{\text{co}}(C_e)$.

In more general contexts Theorem 4.3 is known as the *Krein–Milman theorem.* We shall see in Section 7 that in $\mathbb{R}^n$ the conclusion of Theorem 4.3 can be strengthened to $C = \text{co}(C_e)$.

We conclude this section with an example of a compact convex set whose set of extreme points is not closed. In $\mathbb{R}^3$ let

$$C_1 = \{\mathbf{x} : (x_1 - 1)^2 + x_2^2 \leqslant 1,\ x_3 = 0\},$$
$$C_2 = \{\mathbf{x} : x_1 = 0,\ x_2 = 0,\ -1 \leqslant x_3 \leqslant 1\},$$

and let $C = \text{co}(C_1 \cup C_2)$. Then C_e is the union of the points $(0, 0, 1)$, $(0, 0, -1)$ and all points other than the origin that are on the circumference of the circle in the plane $x_3 = 0$ whose center is at $(1, 0, 0)$ and whose radius is 1. The origin is a limit point of this set yet is not in the set.

Exercise 4.1. Let S be a convex set. Show that a point $\mathbf{x}_0$ in S is an extreme point of S if and only if the set $S \sim \{\mathbf{x}_0\} \equiv \{\mathbf{x}:\mathbf{x} \in S, \mathbf{x} \neq \mathbf{x}_0\}$ is convex.

Exercise 4.2. Show by examples that Theorem 4.3 fails if the convex set C has a nonempty set of extreme points and is merely assumed to be closed or is merely assumed to be bounded rather than closed and bounded.

Exercise 4.3. Let S be a set in $\mathbb{R}^n$. Show that every point of S is either an interior point or a boundary point. Show that if S is compact, then the set of boundary points of S is not empty. Show that if $\mathbf{x}_0$ is an extreme point of S, then it is a boundary point of S.

Exercise 4.4. Let C be a convex subset of $\mathbb{R}^n$ with nonempty interior. Show that $C_e \cap \text{int}(C)$ is empty.

Exercise 4.5. Show directly, without using Lemma 4.2, that the set of extreme points of a closed ball is the set of its boundary points. *Hint:* To simplify notation, take the center of the ball to be the origin.

Exercise 4.6. Let C be a closed convex set in $\mathbb{R}^n$ not equal to $\mathbb{R}^n$. Show that C is the intersection of all closed half spaces containing C that are determined by the supporting hyperplanes of C.

Exercise 4.7. Let $\mathbf{x}_0$ be a point in $\mathbb{R}^n$. Let C_δ denote the closed cube with center at $\mathbf{x}_0$ and length of side equal to 2δ. Thus

$$C_\delta = \{\mathbf{x}: |x_i - x_{0i}| \leqslant \delta, i = 1, \ldots, n\}.$$

The vertices of the cube are the 2^n points of the form $(\varepsilon_1, \varepsilon_2, \ldots, \varepsilon_n)$ where the ε_i take on the values ± 1.

(a) Show that the set of 2^n vertices is the set of extreme points of C_δ.

(b) Show directly, without using Theorem 4.3, that C_δ is the convex hull of the set of vertices. (*Hint:* Use induction on n.)

5. SYSTEMS OF LINEAR INEQUALITIES: THEOREMS OF THE ALTERNATIVE

In this section we shall use the separation theorems to obtain so-called theorems of the alternative. Theorems of the alternative have the following form. Given two systems of inequalities, I and II, either system I has a solution or system II has a solution but never both. Necessary conditions for many optimization problems follow from these theorems.

The following definitions and lemma are needed in our discussion. A vector $\mathbf{x} = (x_1, \ldots, x_n)$ in $\mathbb{R}^n$ is said to be *nonnegative, or* $\mathbf{x} \geqslant \mathbf{0}$, *if, for every* $i = 1, 2, \ldots, n$, $x_i \geqslant 0$. A vector $\mathbf{x}$ in $\mathbb{R}^n$ is said to be *positive*, or $\mathbf{x} > \mathbf{0}$, if, for every $i = 1, 2, \ldots, n$, $x_i > 0$.

LEMMA 5.1. *Let A be an $m \times n$ matrix and let*

$$C = \{\mathbf{w} : \mathbf{w} = A\mathbf{x},\ \mathbf{w} \in \mathbb{R}^m,\ \mathbf{x} \in \mathbb{R}^n,\ \mathbf{x} \geqslant 0\}.$$

Then C is a closed convex cone.

Proof. A straightforward calculation verifies that C is a convex cone.

To show that C is closed, we first consider the case in which the columns of A are linearly independent. Let $\mathbf{w}_0$ be a limit point of C, and let $\{\mathbf{w}_k\}$ be a sequence of points in C converging to $\mathbf{w}_0$. Then there exists a sequence of points $\{\mathbf{x}_k\}$ in $\mathbb{R}^n$ with $\mathbf{x}_k \geqslant \mathbf{0}$ such that $A\mathbf{x}_k \to \mathbf{w}_0$.

We show that $\{\mathbf{x}_k\}$ is bounded. If $\{\mathbf{x}_k\}$ were unbounded, there would exist a subsequence, again denoted by $\{\mathbf{x}_k\}$, such that $\|\mathbf{x}_k\| \to \infty$. All points in the sequence $\{\mathbf{x}_k/\|\mathbf{x}_k\|\}$ have norm 1, so there exists a subsequence and a point $\mathbf{x}_*$ of norm 1 such that $\mathbf{x}_k/\|\mathbf{x}_k\| \to \mathbf{x}_*$. From

$$\|\mathbf{x}_k\| A(\mathbf{x}_k/\|\mathbf{x}_k\|) \to \mathbf{w}_0, \qquad \|\mathbf{x}_k\| \to \infty,$$

we see that we must have $A(\mathbf{x}_k/\|\mathbf{x}_k\|) \to \mathbf{0}$. On the other hand, $\mathbf{x}_k/\|\mathbf{x}_k\| \to \mathbf{x}_*$ implies that $A(\mathbf{x}_k/\|\mathbf{x}_k\|) \to A\mathbf{x}_*$. Hence $A\mathbf{x}_* = \mathbf{0}$. Recall that $\|\mathbf{x}_*\| = 1$, so that $\mathbf{x}_* \neq \mathbf{0}$. Therefore the relation $A\mathbf{x}_* = \mathbf{0}$ contradicts the linear independence of the columns of A.

Since $\{\mathbf{x}_k\}$ is bounded, $\{\mathbf{x}_k\}$ has a convergent subsequence, which we denote again by $\{\mathbf{x}_k\}$. Let $\mathbf{x}_0 = \lim \mathbf{x}_k$. Then $\mathbf{x}_0 \geqslant \mathbf{0}$. Also, $A\mathbf{x}_k \to A\mathbf{x}_0$, so that $A\mathbf{x}_0 = \mathbf{w}_0$. Thus $\mathbf{w}_0 \in C$, and C is closed.

If the columns of A are not linearly independent, we first show that any point $\mathbf{z}$ in C can be expressed as a nonnegative linear combination of $p < n$ linearly independent columns of A. Let $\mathbf{A}_j$, $j = 1, \ldots, n$, denote the jth column of A. Then there exists an $\mathbf{x} \geqslant \mathbf{0}$ such that

$$\mathbf{z} = \sum_{j=1}^{n} x_j \mathbf{A}_j, \qquad x_j \geqslant 0.$$

Since the columns of A are linearly dependent, there exists a $\boldsymbol{\mu} = (\mu_1, \ldots, \mu_n) \neq \mathbf{0}$ such that

$$\mu_1 \mathbf{A}_1 + \mu_2 \mathbf{A}_2 + \cdots + \mu_n \mathbf{A}_n = \mathbf{0}.$$

Therefore, for any $\rho \in \mathbb{R}$,

$$\mathbf{z} = \sum_{j=1}^{n} (x_j - \rho\mu_j)\mathbf{A}_j = \sum_{\mu_j \neq 0} (x_j - \rho\mu_j)\mathbf{A}_j + \sum_{\mu_j = 0} x_j \mathbf{A}_j.$$

Since $\mu \neq \mathbf{0}$, the first sum on the right exists. If $\mu_j < 0$, then $x_j - \rho\mu_j \geqslant 0$ whenever $\rho \geqslant x_j/\mu_j$. In particular, since $x_j \geqslant 0$, we have $x_j - \rho\mu_j \geqslant 0$ whenever $\rho \geqslant 0$. If $\mu_j > 0$, then $x_j/\mu_j \geqslant 0$ and $x_j - \rho\mu_j \geqslant 0$ if and only if $\rho \leqslant x_j/\mu_j$. If there exist indices j such that $\mu_j > 0$, set

$$\bar{\rho} = \min\left\{\frac{x_j}{\mu_j} : \mu_j > 0\right\}.$$

If $\mu_j < 0$ for all indices j for which $\mu_j \neq 0$, set

$$\bar{\rho} = \max\left\{\frac{x_j}{\mu_j} : \mu_j \neq 0\right\}.$$

Then $x_j - \bar{\rho}\mu_j \geqslant 0$ for all $j = 1, \ldots, n$ and $x_j - \bar{\rho}\mu_j = 0$ for at least one value of j. We have now expressed $\mathbf{z}$ as a nonnegative linear combination of $q < n$ columns of A. We continue the process until we express $\mathbf{z}$ as a nonnegative linear combination of $p < n$ linearly independent columns of A.

Let σ denote a choice of $p < n$ linearly independent columns of A and let $A(\sigma)$ denote the corresponding $m \times p$ matrix. There are a finite number of such choices. Let

$$C(\sigma) = \{\mathbf{w} : \mathbf{w} = A(\sigma)\mathbf{y},\ \mathbf{w} \in \mathbb{R}^m,\ \mathbf{y} \in \mathbb{R}^p,\ \mathbf{y} \geqslant \mathbf{0}\}.$$

We have shown that each set $C(\sigma)$ is closed and that

$$C \subseteq \bigcup_{\sigma} C(\sigma).$$

We now show the opposite inclusion. Let $\mathbf{w} \in C(\sigma)$ for some σ. By relabeling columns of A, we can assume without loss of generality that σ selects the first p columns of A. Then there exists a $\mathbf{y} \in \mathbb{R}^p$, $\mathbf{y} \geqslant 0$ such that

$$\mathbf{w} = \sum_{j=1}^{p} y_j \mathbf{A}_j = \sum_{j=1}^{p} y_j \mathbf{A}_j + \sum_{j=p+1}^{n} (0)\mathbf{A}_j, \qquad y_j \geqslant 0, \qquad j = 1, \ldots, p.$$

Let $\mathbf{x} = (y_1, \ldots, y_p, 0, \ldots, 0)$. Then $\mathbf{w} = A\mathbf{x}$, and so $\mathbf{w} \in C$.

We have expressed C as the union of a finite number of closed sets. Hence C is closed and Lemma 5.1 is proved.

THEOREM 5.1 (FARKAS'S LEMMA). *Let A be an $m \times n$ matrix and let $\mathbf{b}$ be a vector in $\mathbb{R}^n$. Then one and only one of the systems*

$$\text{I:} \quad A\mathbf{x} \leqslant \mathbf{0}, \qquad \langle \mathbf{b}, \mathbf{x} \rangle > 0, \qquad \mathbf{x} \in \mathbb{R}^n,$$

$$\text{II:} \quad A^t\mathbf{y} = \mathbf{b}, \qquad \mathbf{y} \geqslant \mathbf{0}, \qquad \mathbf{y} \in \mathbb{R}^m,$$

has a solution.

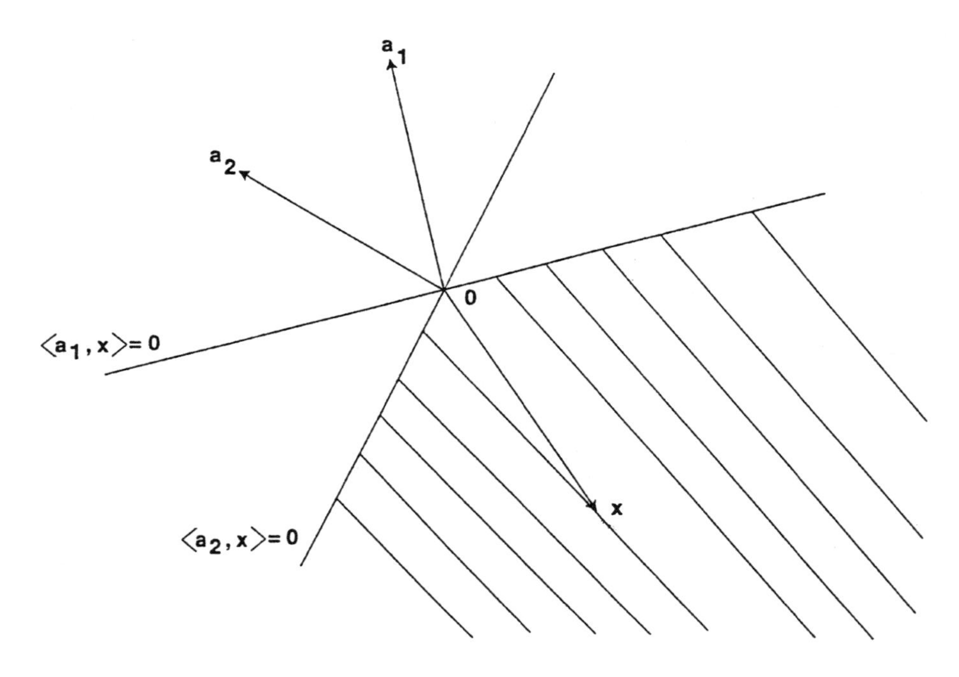

Figure 2.7.

The conclusion of the theorem is sometimes stated in the logically equivalent form: Either I has a solution or II has a solution but never both. It can also be stated in the following form: System I has a solution if and only if system II has no solution.

Figure 2.7 illustrates Farkas's lemma when A is a 2×2 matrix with rows $\mathbf{a}_1$ and $\mathbf{a}_2$. Any vector $\mathbf{x}$ in the cross-hatched region satisfies $A\mathbf{x} \leqslant 0$. Thus, if $\mathbf{b}$ does not lie in the acute angle determined by $\mathbf{a}_1$ and $\mathbf{a}_2$, then I has a solution. (Draw the hyperplane $\langle \mathbf{b}, \mathbf{x} \rangle = 0$ for such a $\mathbf{b}$.) If we write II in the equivalent form $\mathbf{y}^t A = \mathbf{b}^t$, $\mathbf{y}^t \geqslant \mathbf{0}$, we see that for II to have a solution $\mathbf{b}$ must lie in the angle determined by $\mathbf{a}_1, \mathbf{a}_2$. The vector $\mathbf{b}$ is given, so its position is fixed. It either lies in the angle or does not. Thus, precisely one of the systems I and II has a solution.

We now prove the theorem. The reader will get an appreciation of the efficiency and esthetic appeal of convexity arguments by looking up the original proof [Farkas, 1901].

We first suppose that II has a solution $\mathbf{y}_0$. We shall show that I has no solution. If $A\mathbf{x} > \mathbf{0}$ for all $\mathbf{x}$ in $\mathbb{R}^n$, then I has no solution. Suppose there exists an $\mathbf{x}_0$ in $\mathbb{R}^n$ such that $A\mathbf{x}_0 \leqslant \mathbf{0}$. Then

$$\langle \mathbf{b}, \mathbf{x}_0 \rangle = \mathbf{b}^t \mathbf{x}_0 = (A^t \mathbf{y}_0)^t \mathbf{x}_0 = \mathbf{y}_0^t A \mathbf{x}_0 = \langle \mathbf{y}_0, A\mathbf{x}_0 \rangle.$$

Since $\mathbf{y}_0 \geqslant \mathbf{0}$ and $A\mathbf{x}_0 \leqslant \mathbf{0}$, it follows that $\langle \mathbf{b}, \mathbf{x}_0 \rangle \leqslant 0$. Hence I has no solution.

We now suppose that II has no solution. Let

$$C = \{\mathbf{z} : \mathbf{z} = A^t\mathbf{y},\ \mathbf{y} \in \mathbb{R}^m,\ \mathbf{y} \geqslant 0,\ \mathbf{z} \in \mathbb{R}^n\}.$$

The statement that II has no solution is equivalent to the statement that $\mathbf{b} \notin C$. By Lemma 5.1, C is closed and convex. Therefore, by Theorem 3.1 there exists a hyperplane $H^{\alpha}_{\mathbf{x}_0}$ that strongly separates $\mathbf{b}$ and C. Thus $\mathbf{x}_0 \neq \mathbf{0}$ and

$$\langle \mathbf{x}_0, \mathbf{b} \rangle > \alpha > \langle \mathbf{x}_0, \mathbf{z} \rangle$$

for all $\mathbf{z}$ in C. Since $\mathbf{0} \in C$, we get that $\alpha > \mathbf{0}$. Thus

$$\langle \mathbf{x}_0, \mathbf{b} \rangle > 0. \tag{1}$$

For all $\mathbf{z}$ in C we have

$$\langle \mathbf{x}_0, \mathbf{z} \rangle = \langle \mathbf{x}_0, A^t\mathbf{y} \rangle = \mathbf{y}^t A\mathbf{x}_0 < \alpha,$$

the last inequality now holding for all $\mathbf{y}$ in $\mathbb{R}^m$ satisfying $\mathbf{y} \geqslant \mathbf{0}$.

We claim that the last inequality implies that

$$A\mathbf{x}_0 \leqslant \mathbf{0}. \tag{2}$$

If (2) were false, there would exist at least one component of $A\mathbf{x}_0$, say the jth component, that is positive. Let ξ_j denote this component. Let $\mathbf{y}_j = \lambda\mathbf{e}_j = \lambda(0, \dots, 0, 1, 0, \dots, 0)$, where $\lambda > 0$. Then $\langle \mathbf{y}_j, A\mathbf{x}_0 \rangle = \lambda\xi_j < \alpha$. Since $\xi_j > 0$, if we take λ to be sufficiently large, we get a contradiction. Thus (2) is true. But (1) and (2) say that $\mathbf{x}_0$ is a solution of I.

THEOREM 5.2 [GORDAN 1873]. *Let A be an $m \times n$ matrix. Then one and only one of the following systems has a solution:*

$$\text{I:}\quad A\mathbf{x} < \mathbf{0}, \qquad \mathbf{x} \in \mathbb{R}^n,$$

$$\text{II:}\quad A^t\mathbf{y} = \mathbf{0}, \qquad \mathbf{y} \in \mathbb{R}^m, \qquad \mathbf{y} \neq \mathbf{0}, \qquad \mathbf{y} \geqslant \mathbf{0}.$$

The reader should draw a figure which illustrates this theorem when A is a 3×2 matrix with rows $\mathbf{a}_1$, $\mathbf{a}_2$, $\mathbf{a}_3$.

Proof. We shall show that the system I has a solution if and only if the system

$$\text{I}':\quad A\mathbf{x} \leqslant \zeta\mathbf{e}, \qquad \zeta < 0,$$

has a solution, where $\mathbf{e}$ in the vector in $\mathbb{R}^m$ all of whose entries equal 1. Let

$(\mathbf{x}_0, \zeta_0)$ be a solution of I′. Then

$$A\mathbf{x}_0 \leqslant \zeta_0 \mathbf{e} < \mathbf{0},$$

so I has a solution. Now let $\mathbf{x}_0$ be a solution of I and let $\xi = A\mathbf{x}_0$. If $\xi = (\xi_1, \ldots, \xi_m)$, then $\xi_i < 0$ for $i = 1, \ldots, m$. Let $\zeta_0 = \max\{\xi_1, \ldots, \xi_m\}$. Then $(\mathbf{x}_0, \zeta_0)$ is a solution of I′.

The system I′ has a solution if and only if the system

$$\text{I}'': \quad (A, -\mathbf{e})\begin{pmatrix} \mathbf{x} \\ \zeta \end{pmatrix} \leqslant \mathbf{0}, \qquad \langle (\mathbf{0}_n, -1), (\mathbf{x}, \zeta) \rangle > 0,$$

where $\mathbf{0}_n$ is the zero vector in $\mathbb{R}^n$, has a solution. By Farkas's lemma, either I″ (and hence I) has a solution or

$$\text{II}'': \quad \begin{pmatrix} A^t \\ -\mathbf{e}^t \end{pmatrix} \mathbf{y} = \begin{pmatrix} \mathbf{0} \\ -1 \end{pmatrix}, \qquad \mathbf{y} \geqslant \mathbf{0}, \qquad \mathbf{y} \in \mathbb{R}^m,$$

has a solution but never both. System II″ can be written as

$$\text{II}': \quad A^t\mathbf{y} = \mathbf{0}, \qquad \langle \mathbf{e}, \mathbf{y} \rangle = 1, \qquad \mathbf{y} \geqslant 0.$$

From $\langle \mathbf{e}, \mathbf{y} \rangle = 1$ it follows that $\mathbf{y} \neq \mathbf{0}$. Hence if II′ has a solution, so does

$$\text{II}: \quad A^t\mathbf{y} = \mathbf{0}, \qquad \mathbf{y} \geqslant \mathbf{0}, \qquad \mathbf{y} \neq \mathbf{0}.$$

If II has a solution, then $\langle \mathbf{e}, \mathbf{y} \rangle \neq 0$. If we set $\mathbf{y}' = \mathbf{y}/\langle \mathbf{e}, \mathbf{y} \rangle$, then II′ will have a solution. Hence either I has a solution or II has a solution but never both.

THEOREM 5.3 [MOTZKIN 1936]. *Let A, B, C be given matrices with n columns and let $\mathbf{A} \neq \mathbf{O}$. Then one and only one of the following systems has a solution:*

$$\text{I}: \quad A\mathbf{x} > \mathbf{0}, \qquad B\mathbf{x} \geqslant \mathbf{0}, \qquad C\mathbf{x} = \mathbf{0}, \qquad \mathbf{x} \in \mathbb{R}^n,$$

$$\text{II}: \quad A^t\mathbf{y}_1 + B^t\mathbf{y}_2 + C^t\mathbf{y}_3 = \mathbf{0}, \qquad \mathbf{y}_1 \geqslant \mathbf{0}, \qquad \mathbf{y}_1 \neq \mathbf{0}, \qquad \mathbf{y}_2 \geqslant \mathbf{0}.$$

Proof. Let $\mathbf{e} = (1, 1, 1, \ldots, 1)$ be a vector all of whose entries equal 1 and whose dimension is the row dimension of A. We assert that the system I has a solution if and only if the system I′ has a solution, where

$$\text{I}': \quad A\mathbf{x} \geqslant \zeta\mathbf{e}, \qquad B\mathbf{x} \geqslant \mathbf{0}, \qquad C\mathbf{x} = \mathbf{0}, \qquad \zeta > 0.$$

Here ζ is a real scalar. It is clear that if I′ has a solution, then I has a solution. If I has a solution $\mathbf{x}_0$, let $\boldsymbol{\xi} = (\xi_1, \ldots, \xi_m)^t = A\mathbf{x}_0$. Let $\zeta = \min\{\xi_1, \ldots, \xi_m\}$. Then $\zeta > 0$ and $A\mathbf{x}_0 \geqslant \zeta\mathbf{e}$, and so I′ has a solution.

Since I has a solution if and only if I′ has a solution, I has no solution if and only if I′ has no solution. Now I′ has no solution if and only if the system

$$\text{I}'': \quad \begin{pmatrix} -A & \mathbf{e} \\ -B & \mathbf{0} \\ -C & \mathbf{0} \\ C & \mathbf{0} \end{pmatrix} \begin{pmatrix} \mathbf{x} \\ \zeta \end{pmatrix} \leqslant \begin{pmatrix} \mathbf{0} \\ \mathbf{0} \\ \mathbf{0} \\ \mathbf{0} \end{pmatrix}, \qquad \langle (\mathbf{0}_n, 1), (\mathbf{x}, \zeta) \rangle > 0,$$

where $\mathbf{0}_n$ is the zero vector in $\mathbb{R}^n$, has no solution.

By Farkas's lemma I″ has no solution if and only if the system II″ has a solution, where

$$\text{II}'': \quad \begin{pmatrix} -A^t & -B^t & -C^t & C^t \\ \mathbf{e}^t & \mathbf{0} & \mathbf{0} & \mathbf{0} \end{pmatrix} \begin{pmatrix} \mathbf{y}_1 \\ \mathbf{y}_2 \\ \boldsymbol{\eta}_1 \\ \boldsymbol{\eta}_2 \end{pmatrix} = \begin{pmatrix} \mathbf{0}_n \\ 1 \end{pmatrix}, \qquad (\mathbf{y}_1, \mathbf{y}_2, \boldsymbol{\eta}_1, \boldsymbol{\eta}_2) \geqslant \mathbf{0}.$$

But II″ having a solution is equivalent to the system

$$\text{II}': \quad A^t\mathbf{y}_1 + B^t\mathbf{y}_2 + C^t(\boldsymbol{\eta}_1 - \boldsymbol{\eta}_2) = \mathbf{0}, \qquad \langle \mathbf{e}, \mathbf{y}_1 \rangle = 1,$$

having a solution $\mathbf{y}_1 \geqslant \mathbf{0}$, $\mathbf{y}_2 \geqslant \mathbf{0}$, $\boldsymbol{\eta}_1 \geqslant \mathbf{0}$, $\boldsymbol{\eta}_2 \geqslant \mathbf{0}$. If we let $\mathbf{y}_3 = (\boldsymbol{\eta}_1 - \boldsymbol{\eta}_2)$, we see that if II′ has a solution as indicated, then

$$\text{II}: \quad A^t\mathbf{y}_1 + \mathbf{B}^t\mathbf{y}_2 + C^t\mathbf{y}_3 = \mathbf{0}, \qquad \mathbf{y}_1 \geqslant \mathbf{0}, \qquad \mathbf{y}_1 \neq \mathbf{0}, \qquad \mathbf{y}_2 \geqslant \mathbf{0},$$

has a solution. In summary, we have shown that if I has no solution, then II has a solution.

Now suppose that II has a solution $(\mathbf{y}_1, \mathbf{y}_2, \mathbf{y}_3)$. Then $\langle \mathbf{e}, \mathbf{y}_1 \rangle > 0$. Hence if we set

$$\mathbf{y}_1' = \frac{\mathbf{y}_1}{\langle \mathbf{e}, \mathbf{y}_1 \rangle}, \qquad \mathbf{y}_2' = \frac{\mathbf{y}_2}{\langle \mathbf{e}, \mathbf{y}_1 \rangle}, \qquad \mathbf{y}_3' = \frac{\mathbf{y}_3}{\langle \mathbf{e}, \mathbf{y}_1 \rangle},$$

then $(\mathbf{y}_1', \mathbf{y}_2', \mathbf{y}_3')$ is a solution of II with $\langle \mathbf{e}, \mathbf{y}_1' \rangle = 1$. Thus we may assume that II has a solution $(\mathbf{y}_1, \mathbf{y}_2, \mathbf{y}_3)$ with $\langle \mathbf{e}, \mathbf{y}_1 \rangle = 1$.

Let $\boldsymbol{\eta}_1$ be a vector whose dimension is that of $\mathbf{y}_3$ and whose ith component is equal to y_{3i}, the ith component of $\mathbf{y}_3$, if $y_{3i} \geqslant 0$ and zero otherwise. Let $\boldsymbol{\eta}_2$ be a vector whose dimension is that of $\mathbf{y}_3$ and whose ith component is equal to $|y_{3i}|$ if $y_{3i} < 0$ and zero otherwise. Then $\boldsymbol{\eta}_1 \geqslant \mathbf{0}$, $\boldsymbol{\eta}_2 \geqslant \mathbf{0}$, and $\mathbf{y}_3 = (\boldsymbol{\eta}_1 - \boldsymbol{\eta}_2)$. Hence if II has a solution, then so does II′. Therefore, I has no solution.

Remark 5.1. Theorem 5.3 is valid if either B or C or both are zero. In fact, if both B and C are zero, the lemma becomes Gordan's theorem.

THEOREM 5.4 [GALE 1960]. *Let A be an $m \times n$ matrix and let $\mathbf{c} \neq \mathbf{0}$ be a vector in $\mathbb{R}^m$. Then one and only one of the following systems has a solution:*

$$\text{I:} \quad A\mathbf{x} \leqslant \mathbf{c},$$

$$\text{II:} \quad A^t\mathbf{y} = \mathbf{0}, \qquad \langle \mathbf{c}, \mathbf{y} \rangle = -1, \qquad \mathbf{y} \geqslant \mathbf{0}$$

Proof. The system $A\mathbf{x} \leqslant \mathbf{c}$ has a solution if and only if the system $\zeta > 0$, $A\mathbf{x} \leqslant \zeta\mathbf{c}$ has a solution. Thus the system I is equivalent to the system

$$\text{I}': \quad (A, -\mathbf{c})\begin{pmatrix} \mathbf{x} \\ \zeta \end{pmatrix} \leqslant \mathbf{0}, \qquad \langle (\mathbf{0}_n, 1), (\mathbf{x}, \zeta) \rangle > 0,$$

where $\mathbf{0}_n$ is the zero vector in $\mathbb{R}^n$. By Farkas's lemma I′ has a solution if and only if the system II′:

$$\begin{pmatrix} A^t \\ -\mathbf{c} \end{pmatrix} \mathbf{y}^t = \begin{pmatrix} \mathbf{0}_n \\ 1 \end{pmatrix}, \qquad \mathbf{y} \geqslant \mathbf{0},\ \mathbf{y} \in \mathbb{R}^n,$$

has no solution. But system II′ is the same as

$$\langle \mathbf{c}, \mathbf{y} \rangle = -1, \qquad A^t\mathbf{y} = \mathbf{0}, \qquad \mathbf{y} \geqslant 0,$$

which is system (II).

COROLLARY 1. *Let A be an $m \times n$ matrix and let $\mathbf{c} \neq \mathbf{0}$ in $\mathbb{R}^m$ be such that, for all $\mathbf{x} \in \mathbb{R}^n$, $A\mathbf{x} - \mathbf{c} > 0$. Then there exists a vector $\mathbf{y} \in \mathbb{R}^m$ such that $\mathbf{y} \geqslant 0$, $A^t\mathbf{y} = 0$, and $\langle \mathbf{c}, \mathbf{y} \rangle = -1$.*

To prove the corollary, we need only note that if $A\mathbf{x} - \mathbf{c} > \mathbf{0}$ for all $\mathbf{x}$ in $\mathbb{R}^n$, then $A\mathbf{x} \leqslant \mathbf{c}$ has no solution in $\mathbb{R}^n$.

Exercise 5.1. Show that the set C of Lemma 5.1 is a convex cone.

Exercise 5.2. Let A be an $m \times n$ matrix and $\mathbf{b}$ a vector in $\mathbb{R}^n$. Show that one and only one of the following systems has a solution:

$$\text{I:} \quad A\mathbf{x} \geqslant \mathbf{0}, \qquad \mathbf{x} \geqslant \mathbf{0}, \qquad \langle \mathbf{b}, \mathbf{x} \rangle > 0$$

$$\text{II:} \quad A^t\mathbf{y} \geqslant \mathbf{b}, \qquad \mathbf{y} \leqslant \mathbf{0}.$$

Exercise 5.3. Let A be an $m \times n$ matrix and let $\mathbf{c} \neq \mathbf{0}$ be a vector in $\mathbb{R}^m$. Show that one and only one of the following systems has a solution:

$$\text{I:} \quad A\mathbf{x} = \mathbf{c}, \qquad \mathbf{x} \in \mathbb{R}^n,$$

$$\text{II:} \quad A^t\mathbf{y} = \mathbf{0}, \qquad \langle \mathbf{c}, \mathbf{y} \rangle = 1.$$

6. AFFINE GEOMETRY

Consider a plane Π in $\mathbb{R}^3$ and a line Λ in $\mathbb{R}^3$. Let C be a subset of Π and let S be a subset of Λ. Although C and S are subsets of $\mathbb{R}^3$, it is natural to consider C as a two-dimensional object and S as a one-dimensional object. In this section we shall make these ideas precise for subsets of $\mathbb{R}^n$.

Let V_r denote an r-dimensional vector subspace of $\mathbb{R}^n$. When we do not wish to emphasize the dimension of the subspace, we shall write V. The reader will recall from linear algebra that an $(n-1)$-dimensional subspace V_{n-1} of $\mathbb{R}^n$ can be expressed as

$$V_{n-1} = \left\{ \mathbf{x} : \sum_{i=1}^{n} a_i x_i = 0 \right\} = \{\mathbf{x} : \langle \mathbf{a}, \mathbf{x} \rangle = 0\},$$

where $\mathbf{a}$ is a nonzero vector. Thus V_{n-1} is a hyperplane through the origin.

Also, given a subspace V_{n-k}, there exist k linearly independent vectors $\mathbf{a}_1, \ldots, \mathbf{a}_k$ such that

$$V_{n-k} = \{\mathbf{x} : \langle \mathbf{a}_i, \mathbf{x} \rangle = 0, i = 1, \ldots, k\}.$$

Thus V_{n-k}, a subspace of dimension $n-k$, is the intersection of k hyperplanes through the origin.

Definition 6.1. A *linear manifold* M is a subset of $\mathbb{R}^n$ of the form

$$M = V + \mathbf{b} = \{\mathbf{x} : \mathbf{x} = \mathbf{v} + \mathbf{b}, \mathbf{v} \in V\},$$

where V is a subspace and $\mathbf{b}$ is a vector in $\mathbb{R}^n$.

Note that since $\mathbf{b} = \mathbf{0} + \mathbf{b}$ and $\mathbf{0} \in V$, it follows that $\mathbf{b} \in M$.

Thus a linear manifold is a translate of a subspace. Since subspaces are closed sets in $\mathbb{R}^n$, linear manifolds are closed sets. Note that $\mathbb{R}^n$ is a linear manifold, since $\mathbb{R}^n = \mathbb{R}^n + \mathbf{0}$, and that a subspace V is a linear manifold, since $V = V + \mathbf{0}$. In Section 1 we saw that a hyperplane is the translate of a hyperplane through the origin and thus the translate of an $(n-1)$-dimensional vector space. Therefore, hyperplanes are linear manifolds. In $\mathbb{R}^3$ the linear manifolds are planes, lines, and points. (A point is the translate of the zero-dimensional subspace consisting of the origin.)

Linear manifolds are also called *flats* or *affine varieties* or *affine sets.*

LEMMA 6.1. *Let M be a linear manifold, $M = V + \mathbf{b}$. Then the subspace V is unique.*

To prove the lemma, we suppose that M has another representation $M = V' + \mathbf{b}'$ and show that $V' = V$. Since $\mathbf{b} \in M$, from $M = V' + \mathbf{b}'$, we have that $\mathbf{b} = \mathbf{v}' + \mathbf{b}'$ for some $\mathbf{v}' \in V'$. Hence $\mathbf{b}' - \mathbf{b} = -\mathbf{v}' \in V'$. Now let $\mathbf{x} \in V$. Then $\mathbf{x} = \mathbf{m} - \mathbf{b}$ for some $\mathbf{m} \in M$. But $\mathbf{m} = \mathbf{y} + \mathbf{b}'$ for some $\mathbf{y} \in V'$, and so

$$\mathbf{x} = (\mathbf{y} + \mathbf{b}') - \mathbf{b} = \mathbf{y} + (\mathbf{b}' - \mathbf{b}) \in V'.$$

Thus $V \subseteq V'$. A similar argument shows that $V' \subseteq V$.

If $M = V + \mathbf{b}$, then the vector $\mathbf{b}$ is not unique. In fact, $M = V + \mathbf{b}'$ for any $\mathbf{b}' \in M$. To see this, let $\mathbf{b}'$ be any vector in M and let $M' = V + \mathbf{b}'$. Let $\mathbf{m} \in M$, so $\mathbf{m} = \mathbf{v} + \mathbf{b}$ for some $\mathbf{v} \in V$. Since $\mathbf{b}' \in M$, we have $\mathbf{b}' = \mathbf{v}' + \mathbf{b}$ for some $\mathbf{v}' \in V$. Hence $\mathbf{m} = (\mathbf{v} - \mathbf{v}') + \mathbf{b}'$, which is an element of M'. Thus, $M \subseteq M'$. A similar argument shows that $M' \subseteq M$.

On the other hand, if $\mathbf{b}$ is fixed and $M = V + \mathbf{b}$, then the representation of an element $\mathbf{x}$ in M as $\mathbf{x} = \mathbf{v} + \mathbf{b}$ is unique. For if $\mathbf{x} = \mathbf{v}' + \mathbf{b}$, then $\mathbf{v} = \mathbf{x} - \mathbf{b}$ and $\mathbf{v}' = \mathbf{x} - \mathbf{b}$.

If M is a linear manifold, then the unique subspace V such that $M = V + \mathbf{b}$ for some $\mathbf{b}$ is called the *parallel subspace* of M.

Definition 6.2. The *dimension* of a linear manifold M is defined to be the dimension of the parallel subspace V.

If M_1 and M_2 are two linear manifolds with the same parallel subspace, then M_1 and M_2 are said to be *parallel*.

The next lemma characterizes linear manifolds as sets having the property that if $\mathbf{x}_1$ and $\mathbf{x}_2$ belong to the set, then the entire line determined by these points lies in the set.

LEMMA 6.2. *A set M is a linear manifold if and only if, for every $\mathbf{x}_1$, $\mathbf{x}_2$ in M and λ_1, λ_2 real with $\lambda_1 + \lambda_2 = 1$, the point $\lambda_1\mathbf{x}_1 + \lambda_2\mathbf{x}_2$ belongs to M.*

Proof. Let M be a linear manifold, $M = V + \mathbf{b}$. Let $\mathbf{x}_1$ and $\mathbf{x}_2$ be in M. Then $\mathbf{x}_1 = \mathbf{v}_1 + \mathbf{b}$ and $\mathbf{x}_2 = \mathbf{v}_2 + \mathbf{b}$, where $\mathbf{v}_1$ and $\mathbf{v}_2$ are in V. Hence for λ_1, λ_2 with $\lambda_1 + \lambda_2 = 1$,

$$\lambda_1\mathbf{x}_1 + \lambda_2\mathbf{x}_2 = \lambda_1(\mathbf{v}_1 + \mathbf{b}) + \lambda_2(\mathbf{v}_2 + \mathbf{b}) = (\lambda_1\mathbf{v}_1 + \lambda_2\mathbf{v}_2) + \mathbf{b}.$$

Since V is a subspace, $\lambda_1\mathbf{v}_1 + \lambda_2\mathbf{v}_2 \in V$, and so $\lambda_1\mathbf{x}_1 + \lambda_2\mathbf{x}_2 \in M$.

We now prove the reverse implication. Let M be a set such that if $\mathbf{x}_1$ and $\mathbf{x}_2$ are in M and λ_1, λ_2 are scalars such that $\lambda_1 + \lambda_2 = 1$, then $\lambda_1\mathbf{x}_1 + \lambda_2\mathbf{x}_2$ is in M. Let $\mathbf{x}_0 \in M$ and let $W = M - \mathbf{x}_0$. We shall show that W is a subspace, and thus show that M is a linear manifold, since $M = W + \mathbf{x}_0$. Let $\mathbf{y}_1$ and $\mathbf{y}_2$

belong to W and let α and β be arbitrary scalars. Then

$$\alpha \mathbf{y}_1 + \beta \mathbf{y}_2 + \mathbf{x}_0 = [\alpha(\mathbf{y}_1 + \mathbf{x}_0) + (1 - \alpha)\mathbf{x}_0] + [\beta(\mathbf{y}_2 + \mathbf{x}_0) + (1 - \beta)\mathbf{x}_0] - \mathbf{x}_0.$$

Since $\mathbf{y}_1 + \mathbf{x}_0$ and $\mathbf{y}_2 + \mathbf{x}_0$ belong to M, each of the expressions in square brackets belongs to M. If we denote the expression in the first bracket by $\mathbf{m}_1$ and the expression in the second bracket by $\mathbf{m}_2$, we may write

$$\alpha \mathbf{y}_1 + \beta \mathbf{y}_2 + \mathbf{x}_0 = 2\{\tfrac{1}{2}\mathbf{m}_1 + \tfrac{1}{2}\mathbf{m}_2\} - \mathbf{x}_0 = 2\mathbf{m}_3 - \mathbf{x}_0,$$

where $\mathbf{m}_3$ denotes the expression in curly braces. Since $\mathbf{m}_3$ and $\mathbf{x}_0$ are in M and the sum of the coefficients on the extreme right is 1, we get that $\alpha \mathbf{y}_1 + \beta \mathbf{y}_2 + \mathbf{x}_0 \in M$. Thus, $\alpha \mathbf{y}_1 + \beta \mathbf{y}_2 \in W$, and so W is a subspace.

COROLLARY 1. *If M_1 and M_2 are linear manifolds, then so is* $M_1 + M_2 \equiv \{\mathbf{x} : \mathbf{x} = \mathbf{m}_1 + \mathbf{m}_2, \mathbf{m}_1 \in M_1, \mathbf{m}_2 \in M_2\}$.

COROLLARY 2. *If $\{M_\alpha\}$ is a family of linear manifolds, then so is $\bigcap_\alpha M_\alpha$.*

We leave the proofs of these corollaries as exercises for the reader.

Definition 6.3. A point $\mathbf{x}$ is said to be an *affine combination* of points $\mathbf{x}_1, \ldots, \mathbf{x}_k$ if there exists a vector $\mathbf{q} = (q_1, \ldots, q_k)$ in $\mathbb{R}^k$ such that

$$\mathbf{x} = q_1\mathbf{x}_1 + \cdots + q_k\mathbf{x}_k \quad \text{and} \quad \sum_{i=1}^{k} q_i = 1.$$

COROLLARY 3. *A set S is a linear manifold if and only if every affine combination of points in S is in S.*

The proof is similar to that of Lemma 2.6.

Definition 6.4. A finite set of points $\mathbf{x}_1, \ldots, \mathbf{x}_k$ is *affinely dependent* if there exist real numbers $\lambda_1, \ldots, \lambda_k$, not all zero such that $\lambda_1 + \cdots + \lambda_k = 0$ and $\lambda_1\mathbf{x}_1 + \cdots + \lambda_k\mathbf{x}_k = \mathbf{0}$. If the points $\mathbf{x}_1, \ldots, \mathbf{x}_k$ are not affinely dependent, then they are *affinely independent.*

LEMMA 6.3. *A set of points $\mathbf{x}_1, \mathbf{x}_2, \ldots, \mathbf{x}_k$ is affinely dependent if and only if the vectors $(\mathbf{x}_2 - \mathbf{x}_1), (\mathbf{x}_3 - \mathbf{x}_1), \ldots, (\mathbf{x}_k - \mathbf{x}_1)$ are linearly dependent.*

Proof. Let $\mathbf{x}_1, \ldots, \mathbf{x}_k$ be affinely dependent. Then there exist real numbers $\lambda_1, \ldots, \lambda_k$ not all zero such that

$$\lambda_1\mathbf{x}_1 + \lambda_2\mathbf{x}_2 + \cdots + \lambda_k\mathbf{x}_k = \mathbf{0}, \qquad \sum_{i=1}^{k} \lambda_i = 0. \tag{1}$$

Hence

$$\lambda_1 = -(\lambda_2 + \cdots + \lambda_k). \tag{2}$$

Not all of $\lambda_2, \ldots, \lambda_k$ are zero, for if they were, λ_1 would also be zero, contradicting the assumption that not all of $\lambda_1, \ldots, \lambda_k$ are zero. Substituting (2) into the first relation in (1) gives

$$\lambda_2(\mathbf{x}_2 - \mathbf{x}_1) + \lambda_3(\mathbf{x}_3 - \mathbf{x}_1) + \cdots + \lambda_k(\mathbf{x}_k - \mathbf{x}_1) = \mathbf{0},$$

where $\lambda_2, \ldots, \lambda_k$ are not all zero. Thus the vectors $(\mathbf{x}_2 - \mathbf{x}_1), \ldots, (\mathbf{x}_k - \mathbf{x}_1)$ are linearly dependent.

To prove the reverse implication, we reverse the argument.

Remark 6.1. Clearly, an equivalent statement of Lemma 6.3 is the following. *A set of points $\mathbf{x}_1, \ldots, \mathbf{x}_k$ is affinely independent if and only if the vectors $\mathbf{x}_2 - \mathbf{x}_1, \ldots, x_k - \mathbf{x}_1$ are linearly independent.*

LEMMA 6.4. *Let M be a linear manifold. Then the following statements are equivalent:*

(*i*) *The dimension of M equals r.*
(*ii*) *There exist $r + 1$ points $\mathbf{x}_0, \mathbf{x}_1, \ldots, \mathbf{x}_r$ in M that are affinely independent and any set of $r + 2$ points in M is affinely dependent.*
(*iii*) *There exist points $\mathbf{x}_0, \mathbf{x}_1, \ldots, \mathbf{x}_r$ that are affinely independent and such that each point $\mathbf{x}$ in M has a unique representation*

$$\mathbf{x} = \sum_{i=0}^{r} \lambda_i \mathbf{x}_i, \qquad \sum_{i=0}^{r} \lambda_i = 1 \tag{3}$$

in terms of the $\mathbf{x}_i$'s.

Proof. We first show that (i) implies (ii). If (i) holds, then we may write $M = V_r + \mathbf{x}_0$, where V_r is a subspace of dimension r and $\mathbf{x}_0$ is an arbitrary point in M. Let $\mathbf{v}_1, \ldots, \mathbf{v}_r$ be a basis for V_r and let

$$\mathbf{x}_i = \mathbf{v}_i + \mathbf{x}_0, \qquad i = 1, \ldots, r.$$

Then the vectors $\mathbf{x}_i - \mathbf{x}_0$, $i = 1, \ldots, r$, are linearly independent and the $r + 1$ points $\mathbf{x}_0, \mathbf{x}_1, \ldots, \mathbf{x}_r$ are affinely independent.

Let $\mathbf{y}_0, \mathbf{y}_1, \ldots, \mathbf{y}_r, \mathbf{y}_{r+1}$ be any set of $r + 2$ points in M. If we write $M = V_r + \mathbf{y}_0$, then each of the vectors $\mathbf{y}_i - \mathbf{y}_0$, $i = 1, \ldots, r + 1$, is in V_r. Therefore the $r + 1$ vectors in this set are linearly dependent, and hence $\mathbf{y}_0, \mathbf{y}_1, \ldots, \mathbf{y}_r, \mathbf{y}_{r+1}$ are affinely dependent.

Next we show that (ii) implies (iii). Since M is a linear manifold and $\mathbf{x}_0 \in M$, we may write $M = V + \mathbf{x}_0$, where V is a subspace. We assert that V has dimension r. To see this, let

$$\mathbf{v}_i = \mathbf{x}_i - \mathbf{x}_0, \qquad i = 1, \ldots, r.$$

Since the points $\mathbf{x}_0, \mathbf{x}_1, \ldots, \mathbf{x}_r$ are affinely independent, the vectors $\mathbf{v}_1, \ldots, \mathbf{v}_r$ are linearly independent. Now let $\mathbf{w}_1, \ldots, \mathbf{w}_{r+1}$ be a set of $r + 1$ distinct nonzero vectors in V. Then there exist points $\mathbf{y}_1, \ldots, \mathbf{y}_{r+1}$ in M such that

$$\mathbf{w}_i = \mathbf{y}_i - \mathbf{x}_0, \qquad i = 1, \ldots, r + 1.$$

Since the $\mathbf{w}_i$ are distinct and none of the vectors $\mathbf{w}_i$ is zero, the $r + 2$ points $\mathbf{x}_0\,\mathbf{y}_1, \ldots, \mathbf{y}_{r+1}$ are distinct. By hypothesis, they are affinely dependent. Hence the vectors $\mathbf{w}_1, \ldots, \mathbf{w}_{r+1}$ are linearly dependent. Thus, V has dimension r, we may write $V = V_r$, and the vectors $\mathbf{v}_1, \ldots, \mathbf{v}_r$ are a basis for V_r.

For arbitrary $\mathbf{x}$ in M we have $\mathbf{x} = \mathbf{v} + \mathbf{x}_0$ uniquely for some $\mathbf{v}$ in V_r. Hence there exist unique real numbers $\lambda_1, \ldots, \lambda_r$ such that

$$\mathbf{x} = \sum_{i=1}^{r} \lambda_i \mathbf{v}_i + \mathbf{x}_0 = \sum_{i=1}^{r} \lambda_i(\mathbf{x}_i - \mathbf{x}_0) + \mathbf{x}_0 = \left(1 - \sum_{i=1}^{r} \lambda_i\right)\mathbf{x}_0 + \sum_{i=1}^{r} \lambda_i \mathbf{x}_i.$$

We obtain the representation (3) by setting $\lambda_0 = (1 - \lambda_1 - \cdots - \lambda_r)$.

To show that the representation is unique in terms of the $\mathbf{x}_i$'s, we suppose that there were another representation

$$\mathbf{x} = \sum_{i=0}^{r} \lambda_i' \mathbf{x}_i, \qquad \sum_{i=0}^{r} \lambda_i' = 1.$$

Then

$$\mathbf{0} = \sum_{i=0}^{r} (\lambda_i' - \lambda_i)\mathbf{x}_i, \qquad \sum_{i=0}^{r} (\lambda_i' - \lambda_i) = 0,$$

with not all of the coefficients $(\lambda_i' - \lambda_i)$ equal to zero. But then the points $\mathbf{x}_i$, $i = 0, \ldots, r$, would be affinely dependent, contrary to assumption.

We conclude the proof by showing that (iii) implies (i). Let $\mathbf{x}$ be an arbitrary point in M. Then by virtue of (3)

$$\mathbf{x} = \sum_{i=0}^{r} \lambda_i \mathbf{x}_i = \left(1 - \sum_{i=1}^{r} \lambda_i\right)\mathbf{x}_0 + \sum_{i=1}^{r} \lambda_i \mathbf{x}_i = \sum_{i=1}^{r} \lambda_i(\mathbf{x}_i - \mathbf{x}_0) + \mathbf{x}_0. \tag{4}$$

Since $\mathbf{x}_0, \mathbf{x}_1, \ldots, \mathbf{x}_r$ are affinely independent, the vectors

$$\mathbf{x}_1 - \mathbf{x}_0, \mathbf{x}_2 - \mathbf{x}_0, \ldots, \mathbf{x}_r - \mathbf{x}_0$$

are linearly independent.

Let V_r denote the r-dimensional vector space spanned by $\mathbf{x}_1 - \mathbf{x}_0, \ldots, \mathbf{x}_r - \mathbf{x}_0$. It follows from (4) that $\mathbf{x} \in V_r + \mathbf{x}_0$. Thus $M \subseteq V_r + \mathbf{x}_0$.

To prove the opposite inclusion, we let $W = M - \mathbf{x}_0$. We saw in the proof of Lemma 6.2 that W is a subspace. Moreover, since $\mathbf{x}_1 - \mathbf{x}_0, \ldots, \mathbf{x}_r - \mathbf{x}_0$ belong to W, we have that $V_r \subseteq W$. Hence

$$V_r + \mathbf{x}_0 \subseteq W + \mathbf{x}_0 = M$$

and so $M = V_r + \mathbf{x}_0$. This concludes the proof of Lemma 6.4.

Points $\mathbf{x}_0, \mathbf{x}_1, \ldots, \mathbf{x}_r$ as in Lemma 6.4 are said to be an *affine basis* for M. The numbers λ_i, $i = 0, 1, \ldots, r$, are called the *barycentric coordinates* of $\mathbf{x}$ relative to the affine basis $\mathbf{x}_0, \mathbf{x}_1, \ldots, \mathbf{x}_r$.

Definition 6.5. Let A be a subset of $\mathbb{R}^n$. The *affine hull* of A, denoted by $\Lambda[A]$, is defined to be the intersection of all linear manifolds containing A. The *dimension* of A is the dimension of $\Lambda[A]$.

Since $A \subseteq \mathbb{R}^n$, since $\mathbb{R}^n$ is a linear manifold, and since the intersection of linear manifolds is a linear manifold, it follows that, for any nonempty set A, $\Lambda[A]$ is not empty. Thus the dimension of A is well defined.

For any linear manifold M, the relations $M \subseteq \Lambda[M]$ and $\Lambda[M] \subseteq M$ hold, so $M = \Lambda[M]$. Thus, for a linear manifold M, the dimension of M as defined in Definition 6.2 coincides with that of Definition 6.5.

We shall often write dim A to refer to the dimension of A.

LEMMA 6.5. *Let A be a subset of $\mathbb{R}^n$ and let $M(A)$ denote the set of all affine combinations of points in A. Then*

(*i*) $\Lambda[A] = M(A)$ *and*
(*ii*) *if there exist $\rho + 1$ points $\mathbf{x}_0, \mathbf{x}_1, \ldots, \mathbf{x}_\rho$ in A that are affinely independent and if any set of $\rho + 2$ points in A is affinely dependent, then* $\dim A = \rho$.

Remark 6.2. Statement (i) is the analog for affine hulls of Theorem 2.1 for convex hulls.

Proof. It is readily verified that if $\mathbf{x}$ and $\mathbf{y}$ are two points in $M(A)$, then each affine combination of these points is in $M(A)$. Hence, by Lemma 6.2, $M(A)$ is a linear manifold. Since $M(A) \supseteq A$, it follows from the definition of $\Lambda[A]$ that $\Lambda[A] \subseteq M(A)$. On the other hand, it follows from Corollary 3 to Lemma 6.2 that $M(A)$ is contained in every linear manifold containing A. Thus $\Lambda[A] \supseteq M(A)$. Hence $\Lambda[A] = M(A)$.

We now establish (ii). Let $\mathbf{x}_0, \mathbf{x}_1, \ldots, \mathbf{x}_\rho$ be $\rho + 1$ affinely independent points in A and suppose that any set of $\rho + 2$ points is affinely dependent. Since the points $\mathbf{x}_0, \mathbf{x}_1, \ldots, \mathbf{x}_\rho$ are in $\Lambda[A]$, it follows from Lemma 6.4 that $\dim \Lambda[A] \geqslant \rho$.

On the other hand, for arbitrary $\mathbf{x}$ in A the vectors

$$\mathbf{x}_1 - \mathbf{x}_0, \mathbf{x}_2 - \mathbf{x}_0, \ldots, \mathbf{x}_\rho - \mathbf{x}_0, \mathbf{x} - \mathbf{x}_0$$

are linearly dependent, and the first ρ vectors are linearly independent. Hence for arbitrary $\mathbf{x}$ in A there exist unique scalars $\mu_1, \ldots, \mu_\rho$ such that

$$\mathbf{x} - \mathbf{x}_0 = \sum_{i=1}^{\rho} \mu_i(\mathbf{x}_i - \mathbf{x}_0).$$

On setting $\mu_0 = 1 - (\mu_1 + \mu_2 + \cdots + \mu_\rho)$, we get, uniquely,

$$\mathbf{x} = \sum_{i=0}^{\rho} \mu_i \mathbf{x}_i, \qquad \sum_{i=0}^{\rho} \mu_i = 1.$$

Hence A is contained in the linear manifold $\Lambda[\{\mathbf{x}_0, \mathbf{x}_1, \ldots, \mathbf{x}_\rho\}]$ and so $\dim A \leqq \dim \Lambda[\{\mathbf{x}_0, \mathbf{x}_1, \ldots, \mathbf{x}_\rho\}]$.

To complete the proof of (ii), it suffices to show that $\Lambda[\{\mathbf{x}_0, \mathbf{x}_1, \ldots, \mathbf{x}_\rho\}]$ has dimension ρ. For then $\dim A \leqslant \rho$, and since we have already shown that $\dim A \geqslant \rho$, statement (ii) will follow.

We now show that $\dim \Lambda[\{\mathbf{x}_0, \mathbf{x}_1, \ldots, \mathbf{x}_\rho\}] = \rho$. Since $\mathbf{x}_0, \mathbf{x}_1, \ldots, \mathbf{x}_\rho$ are affinely independent to show that $\dim \Lambda[\{\mathbf{x}_0, \mathbf{x}_1, \ldots, \mathbf{x}_\rho\}] = \rho$, we must show that any set $\mathbf{y}_0, \mathbf{y}_1, \ldots, \mathbf{y}_\rho, \mathbf{y}_{\rho+1}$ of $\rho + 2$ vectors in $\Lambda[\{\mathbf{x}_0, \mathbf{x}_1, \ldots, \mathbf{x}_\rho\}]$ is affinely dependent. From (i), we have that $M(\{\mathbf{x}_0, \ldots, \mathbf{x}_\rho\}) = \Lambda[\{\mathbf{x}_0, \ldots, \mathbf{x}_\rho\}]$. Thus for $i = 0, 1, \ldots, \rho + 1$ we have

$$\mathbf{y}_i = \sum_{j=0}^{\rho} q_{ij}\mathbf{x}_j, \qquad \sum_{j=0}^{\rho} q_{ij} = 1.$$

Since $q_{i0} = 1 - q_{i1} - q_{i2} - \cdots - q_{i\rho}$, we have

$$\mathbf{y}_i = \sum_{j=1}^{\rho} q_{ij}(\mathbf{x}_j - \mathbf{x}_0) + \mathbf{x}_0, \qquad i = 0, 1, 2, \ldots, \rho + 1.$$

Hence,

$$\mathbf{y}_i - \mathbf{y}_0 = \sum_{j=1}^{\rho} (q_{ij} - q_{0j})(\mathbf{x}_j - \mathbf{x}_0), \qquad i = 1, \ldots, \rho + 1. \tag{5}$$

Thus, the $\rho + 1$ vectors $\mathbf{y}_i - \mathbf{y}_0$ in (5) are in the vector space spanned by the ρ linearly independent vectors $\mathbf{x}_1 - \mathbf{x}_0, \ldots, \mathbf{x}_\rho - \mathbf{x}_0$. Hence the vectors $\mathbf{y}_i - \mathbf{y}_0$ in (5) are linearly dependent and the corresponding points $\mathbf{y}_0, \mathbf{y}_1, \ldots, \mathbf{y}_{\rho+1}$ are affinely dependent.

COROLLARY 4. *The points* $\mathbf{x}_0, \mathbf{x}_1, \ldots, \mathbf{x}_\rho$ *are an affine basis for* $\Lambda[A]$.

Proof. In the proof of Lemma 6.5 we showed that $A \subseteq \Lambda[\{\mathbf{x}_0, \mathbf{x}_1, \ldots, \mathbf{x}_\rho\}]$. Hence $\Lambda[A] \subseteq \Lambda[\{\mathbf{x}_0, \mathbf{x}_1, \ldots, \mathbf{x}_r\}]$. Since the opposite inclusion clearly holds, we have $\Lambda[A] = \Lambda[\{\mathbf{x}_0, \mathbf{x}_1, \ldots, \mathbf{x}_\rho\}]$. The corollary then follows from the uniqueness of the representation (3) of a point $\mathbf{x}$ in $\Lambda[\{\mathbf{x}_0\, \mathbf{x}_1, \ldots, \mathbf{x}_\rho\}]$.

Let $\mathbf{x}_0, \mathbf{x}_1, \ldots, \mathbf{x}_r$ be affinely independent. We call $\mathrm{co}\{\mathbf{x}_0, \mathbf{x}_1, \ldots, \mathbf{x}_r\}$ the *simplex spanned* by $\mathbf{x}_0, \mathbf{x}_1, \ldots, \mathbf{x}_r$. The points $\mathbf{x}_0, \mathbf{x}_1, \ldots, \mathbf{x}_r$ are called the *vertices* of the simplex.

We leave it as an exercise for the reader to show that any $r + 2$ points in the simplex are affinely dependent. Thus the dimension of the simplex $\mathrm{co}\{\mathbf{x}_0, \mathbf{x}_1, \ldots, \mathbf{x}_r\}$ is r.

COROLLARY 5. *The dimension of a convex set C is the maximum dimension of simplexes in C.*

This corollary is a restatement of Lemma 6.5 when C is a convex set.

Intuitively, in $\mathbb{R}^3$ one would expect that a convex set that is not contained in a plane or in a line to have nonempty interior. In terms of our definition of dimension of a set, we would expect a three-dimensional convex set to have nonempty interior. This is indeed the case, not only in $\mathbb{R}^3$, but also in $\mathbb{R}^n$, as the following theorem shows.

THEOREM 6.1. *A convex set C in* $\mathbb{R}^n$ *has dimension n if and only if C has nonvoid interior.*

Suppose C has nonempty interior. Let $\mathbf{x}_0$ be an interior point of C and let $\mathbf{e}_1, \ldots, \mathbf{e}_n$ denote the standard basis vectors in $\mathbb{R}^n$. Then there exists an $\varepsilon > 0$ such that each of the $n + 1$ points $\mathbf{x}_0$, $\mathbf{x}_0 + \varepsilon\mathbf{e}_1, \ldots, \mathbf{x}_0 + \varepsilon\mathbf{e}_n$ lies in C. These points are affinely independent. Since in $\mathbb{R}^n$ any $n + 2$ points are affinely dependent, it follows from (ii) of Lemma 6.5 that C has dimension n.

Now suppose that C has dimension n. Then $\Lambda[C] = \mathbb{R}^n$, and by Corollary 4 of Lemma 6.5 there exist points $\mathbf{x}_0, \mathbf{x}_1, \ldots, \mathbf{x}_n$ in C that form an affine basis for $\Lambda[C] = \mathbb{R}^n$. Thus each $\mathbf{x}$ in $\mathbb{R}^n$ can be written uniquely as

$$\mathbf{x} = \sum_{i=0}^{n} \lambda_i \mathbf{x}_i, \qquad \sum_{i=0}^{n} \lambda_i = 1. \tag{6}$$

Since C is convex, all points $\mathbf{x}$ in $\mathbb{R}^n$ with $\lambda_i \geqslant 0$, $i = 0, 1, \ldots, n$, lie in C. In particular,

$$\mathbf{x}_* = \frac{1}{n+1}(\mathbf{x}_0 + \mathbf{x}_1 + \cdots + \mathbf{x}_n),$$

the centroid of the simplex $\mathrm{co}\{\mathbf{x}_0, \mathbf{x}_1, \ldots, \mathbf{x}_n\}$, is in C.

We can also write $\mathbf{x}$ uniquely as

$$\mathbf{x} = \sum_{i=1}^{n} \lambda_i(\mathbf{x}_i - \mathbf{x}_0),$$

with the λ_i as in (6). Since the vectors $\mathbf{x}_1 - \mathbf{x}_0, \ldots, \mathbf{x}_n - \mathbf{x}_0$ are linearly independent, they form a basis for $\mathbb{R}^n$, and the linear transformation T from $\mathbb{R}^n$ to $\mathbb{R}^m$ defined by

$$T(\mathbf{x}) = (\lambda_1, \ldots, \lambda_n)$$

is nonsingular and continuous. Let $\mathbf{f}$ be the mapping from $\mathbb{R}^n$ to $\mathbb{R}^{n+1}$ defined by

$$\mathbf{f}(\mathbf{x}) = \mathbf{f}\left(\sum_{i=0}^{n} \lambda_i x_i\right) = (\lambda_0, \lambda_1, \ldots, \lambda_n),$$

where the λ_i are as in (6). The mapping $\mathbf{f}$ can be viewed as a composite map $\mathbf{h} \circ T$, where

$$\mathbf{h}(\lambda_1, \ldots, \lambda_n) = (\lambda_0, \lambda_1, \ldots, \lambda_n), \qquad \lambda_0 = 1 - \sum_{i=1}^{n} \lambda_i.$$

Thus, $\mathbf{f}$ is continuous.

Since $\mathbf{f}$ is continuous and since

$$\mathbf{f}(\mathbf{x}_*) = \left(\frac{1}{n+1}, \frac{1}{n+1}, \ldots, \frac{1}{n+1}\right),$$

there exists a $\delta > 0$ such that for $\mathbf{x} \in B(\mathbf{x}_*, \delta)$ all components λ_i of $f(\mathbf{x})$ will be strictly positive. Thus, all $\mathbf{x} \in B(\mathbf{x}_*, \delta)$ lie in C, and so $\mathbf{x}_*$ is an interior point of C.

By experimenting with a sheet of cardboard, if necessary, the reader should find it plausible that an arbitrary plane in $\mathbb{R}^3$ can be put into coincidence with the plane $x_3 = 0$ by means of an appropriate sequence of rotations followed by a possible translation. The plane $x_3 = 0$ consists of those points in $\mathbb{R}^3$ of the form $(x_1, x_2, 0)$ and thus is essentially $\mathbb{R}^2$. Thus, in some sense a plane in $\mathbb{R}^3$ is equivalent to $\mathbb{R}^2$. Similarly, any line in $\mathbb{R}^3$ can be made to coincide with any one of the coordinate axes by means of an appropriate sequence of rotations followed by a possible translation.

We shall now show that any r-dimensional manifold M in $\mathbb{R}^n$ is equivalent, in a sense to be made precise, to $\mathbb{R}^r$.

Definition 6.6. An *affine transformation* T from $\mathbb{R}^n$ to $\mathbb{R}^m$ is a mapping of the form $T\mathbf{x} = S\mathbf{x} + \mathbf{a}$, where $\mathbf{a}$ is a fixed vector in $\mathbb{R}^m$ and S is a linear transformation from $\mathbb{R}^n$ to $\mathbb{R}^m$.

Since relative to appropriate bases in $\mathbb{R}^n$ and $\mathbb{R}^m$ a linear transformation S from $\mathbb{R}^n$ to $\mathbb{R}^m$ is given by $S\mathbf{x} = A\mathbf{x}$, where A is an $m \times n$ matrix, we have $T\mathbf{x} = A\mathbf{x} + \mathbf{a}$.

LEMMA 6.6. *A mapping T from $\mathbb{R}^n$ to $\mathbb{R}^m$ is affine if and only if for each $\lambda \in \mathbb{R}$ and each pair of points $\mathbf{x}_1$ and $\mathbf{x}_2$ in $\mathbb{R}^n$*

$$T[(1-\lambda)\mathbf{x}_1 + \lambda\mathbf{x}_2] = (1-\lambda)T\mathbf{x}_1 + \lambda T\mathbf{x}_2. \tag{7}$$

It is a straightforward calculation to show that if T is affine, then (7) holds. Now suppose that (7) holds. Let $\mathbf{a} = T(\mathbf{0})$, and let $S\mathbf{x} = T\mathbf{x} - \mathbf{a}$. To show that T is affine, we must show that S is linear.

For any α in $\mathbb{R}$ and $\mathbf{x}$ in $\mathbb{R}^n$,

$$\begin{aligned} S(\alpha\mathbf{x}) &= T[\alpha\mathbf{x} + (1-\alpha)\mathbf{0}] - \mathbf{a} = \alpha T(\mathbf{x}) + (1-\alpha)T(\mathbf{0}) - \mathbf{a} \\ &= \alpha[T(\mathbf{x}) - \mathbf{a}] = \alpha S(\mathbf{x}). \end{aligned}$$

For any $\mathbf{x}_1$ and $\mathbf{x}_2$ in $\mathbb{R}^n$ we have

$$\begin{aligned} S(\mathbf{x}_1 + \mathbf{x}_2) &= S[2(\tfrac{1}{2}\mathbf{x}_1 + \tfrac{1}{2}\mathbf{x}_2)] = 2S(\tfrac{1}{2}\mathbf{x}_1 + \tfrac{1}{2}\mathbf{x}_2) = 2T(\tfrac{1}{2}\mathbf{x}_1 + \tfrac{1}{2}\mathbf{x}_2) - 2\mathbf{a} \\ &= [T(\mathbf{x}_1) - \mathbf{a}] + [T(\mathbf{x}_2) - \mathbf{a}] = S(\mathbf{x}_1) + S(\mathbf{x}_2). \end{aligned}$$

Thus S is linear.

From (7) we see that under an affine map lines are mapped onto lines. Thus under an affine map linear manifolds are mapped onto linear manifolds. By taking $0 \leqslant \lambda \leqslant 1$ in (7), we see that under an affine map convex sets are mapped onto convex sets.

THEOREM 6.2. *Let $\mathbf{x}_0, \mathbf{x}_1, \ldots, \mathbf{x}_r$ and $\mathbf{y}_0, \mathbf{y}_1, \ldots, \mathbf{y}_r$ be two affinely independent sets in $\mathbb{R}^n$. Then there exists a one-to-one affine transformation T from $\mathbb{R}^n$ onto itself such that $T\mathbf{x}_i = \mathbf{y}_i$, $i = 0, 1, \ldots, r$. If $r = n$, then T is unique.*

If $r < n$, obtain two affinely independent sets $\mathbf{x}_0, \ldots, \mathbf{x}_r, \mathbf{x}_{r+1}, \ldots, \mathbf{x}_n$ and $\mathbf{y}_0, \ldots, \mathbf{y}_r, \mathbf{y}_{r+1}, \ldots, \mathbf{y}_n$ of $n + 1$ points by adjoining appropriate points to the original sets. The vectors $\mathbf{x}_1 - \mathbf{x}_0, \ldots, \mathbf{x}_n - \mathbf{x}_0$ constitute a basis for $\mathbb{R}^n$, as do the vectors $\mathbf{y}_1 - \mathbf{y}_0, \ldots, \mathbf{y}_n - \mathbf{y}_0$. Thus, there exists a unique one-to-one linear transformation S of $\mathbb{R}^n$ onto itself such that $S(\mathbf{x}_i - \mathbf{x}_0) = \mathbf{y}_i - \mathbf{y}_0$, $i = 1, \ldots, n$. Let $T\mathbf{x} = S\mathbf{x} + \mathbf{a}$ for a fixed $\mathbf{a}$ to be determined. Then for $i = 0, 1, \ldots, n$,

$$T(\mathbf{x}_i) = S(\mathbf{x}_i) + \mathbf{a} = S(\mathbf{x}_i - \mathbf{x}_0) + S(\mathbf{x}_0) + \mathbf{a} = \mathbf{y}_i - \mathbf{y}_0 + S(\mathbf{x}_0) + \mathbf{a}.$$

Therefore, if we take $\mathbf{a} = \mathbf{y}_0 - S(\mathbf{x}_0)$, we will have $T\mathbf{x}_i = \mathbf{y}_i$, $i = 0, 1, \ldots, n$.

COROLLARY 6. *Let M_1 and M_2 be two linear manifolds of the same dimension. Then there exists a one-to-one affine transformation T of $\mathbb{R}^n$ onto itself such that $TM_1 = M_2$.*

By Lemma 6.4, M_1 has an affine basis $\mathbf{x}_0, \mathbf{x}_1, \ldots, \mathbf{x}_r$ and M_2 has an affine basis $\mathbf{y}_0, \mathbf{y}_1, \ldots, \mathbf{y}_r$. By Theorem 6.2 there exists a one-to-one transformation T from $\mathbb{R}^n$ onto itself such that $T\mathbf{x}_i = \mathbf{y}_i$, $i = 1, \ldots, n$. The corollary now follows from the facts that each $\mathbf{x}$ in M_1 is a unique affine combination of $\mathbf{x}_0, \mathbf{x}_1, \ldots, \mathbf{x}_r$, each $\mathbf{y}$ in M_2 is a unique affine combination of $\mathbf{y}_0, \mathbf{y}_1, \ldots, \mathbf{y}_r$, and T maps affine combinations onto affine combinations.

Remark 6.3. If M is a linear manifold of dimension r, then there exists a one-to-one affine transformation T of $\mathbb{R}^n$ onto itself that maps M onto the manifold

$$V_r^* = \{\mathbf{y} : \mathbf{y} = (y_1, \ldots, y_n),\ y_{r+1} = y_{r+2} = \cdots = y_n = 0\}.$$

The manifold V_r^* can be identified with $\mathbb{R}^r$ by means of the one-to-one mapping. Π from V_r^* onto $\mathbb{R}^r$ given by

$$\Pi : \mathbf{y} = (y_1, \ldots, y_r, 0, \ldots, 0) \to (y_1, \ldots, y_r).$$

Denote the restriction of T to M by T_M, and define a mapping τ from M to $\mathbb{R}^r$ by the formula

$$\mathbf{x} \in M : \tau(\mathbf{x}) = \Pi(T_M \mathbf{x}).$$

Then τ is one-to-one onto, is continuous, and has a continuous inverse, τ^{-1}. Note that τ and τ^{-1} map convex sets onto convex sets and linear manifolds onto linear manifolds of the same dimension.

Exercise 6.1. Let k be a positive integer, $1 \leqslant k \leqslant n - 1$. Show that for a given subspace V_{n-k} of $\mathbb{R}^n$, there exist k linearly independent vectors $\mathbf{a}_1, \ldots, \mathbf{a}_k$ such that

$$V_{n-k} = \{\mathbf{x} \in \mathbb{R}^n,\ \langle \mathbf{a}_i, \mathbf{x} \rangle = 0,\ i = 1, \ldots, k\}.$$

Exercise 6.2. Show that a set of points $\mathbf{x}_1, \ldots, \mathbf{x}_k$ is affinely dependent if and only if one of the points is an affine combination of the other points.

Exercise 6.3. In $\mathbb{R}^3$ consider the plane $2x + y + z = 5$.

(a) Find an affine basis for this plane that does not include the point $\mathbf{x} = (-1, -2, 9)$.

(b) Express $\mathbf{x} = (-1, -2, 9)$ in barycentric coordinates relative to your basis.

Exercise 6.4. Let $\mathbf{x}_0, \mathbf{x}_1, \ldots, \mathbf{x}_r$ be affinely independent. Show that any $r + 2$ points $\mathbf{y}_0, \mathbf{y}_1, \ldots, \mathbf{y}_r, \mathbf{y}_{r+1}$ in $\text{co}\{\mathbf{x}_0, \mathbf{x}_1, \ldots, \mathbf{x}_r\}$ are affinely dependent.

Exercise 6.5. Show that an affine map T maps affine combinations of points onto affine combinations of points.

Exercise 6.6. In Theorem 6.2, why is T unique if $r = n$?

7. MORE ON SEPARATION AND SUPPORT

Although the separation and support theorems of Sections 3 and 4 are adequate for most applications to optimization problems, they are not applicable in some situations where proper separation and nontrivial support of convex sets do occur. To illustrate this, let

$$C_1 = \{\mathbf{x} \in \mathbb{R}^3 : x_1 \geqslant 0,\ x_2 \geqslant 0,\ x_3 = 0\}$$

$$C_2 = \{\mathbf{x} \in \mathbb{R}^3 : x_2 \geqslant 0,\ x_3 \geqslant 0,\ x_1 = 0\}.$$

These two sets are properly separated by the plane (hyperplane in $\mathbb{R}^3$) $x_1 - x_3 = 0$. These sets are convex, are not disjoint, and have empty interiors so that neither Theorem 3.3 nor Theorem 3.4 can be applied to deduce separation.

Let $C = \{\mathbf{x} \in \mathbb{R}^3 : x_1^2 + x_2^2 \leqslant 1,\ x_3 = 0\}$. At each boundary point $\mathbf{x}_0$ with $x_{01}^2 + x_{02}^2 = 1$ there exist a nontrivial supporting hyperplane, namely the plane that is orthogonal to the plane $x_3 = 0$ and whose intersection with this plane is the tangent line to the circle $x_1^2 + x_2^2 = 1$, $x_3 = 0$ at $\mathbf{x}_0$. Again, since the convex set C has empty interior, Theorem 4.1 does not give us the existence of this nontrivial supporting hyperplane.

The sets C_1, C_2, and C, considered as sets in $\mathbb{R}^2$, do have interiors. It turns out that this observation is the key to theorems that are applicable to examples such as those described above. We introduce the notion of relative interior of a convex set C of dimension r. The relative interior is the interior of a convex set C considered as a set in $\Lambda[C]$. In other words we consider the set C as a set in $\mathbb{R}^r$ rather than a set in $\mathbb{R}^n$.

Definition 7.1. Let A be a set of dimension $r \leqslant n$ and let $M = \Lambda[A]$. A point $\mathbf{x}_0$ in A is said to be a *relative interior point* of A if there exists an $\varepsilon > 0$ such that $B(\mathbf{x}_0, \varepsilon) \cap M \subseteq A$. The set of relative interior points of A is called the *relative interior* of A and will be denoted by $\text{ri}(A)$. A set A is said to *relatively open* if $\text{ri}(A) = A$, that is, if every point of A is a relative interior point of A.

Note that if $\dim M = n$, then a relative interior point is an interior point and a relatively open set is an open set.

Since a linear manifold M is a closed set, any limit point of A will also be in M. Thus the closure of A as a subset of M is the same as the closure of A as a subset of $\mathbb{R}^n$.

If A is contained in M and $\dim M < n$, then every point of A is a boundary point of A. We define the *relative boundary* of A to be the set $\bar{A} \sim \text{ri}(A)$, that is, those points of $\bar{A}$ that are not relative interior points of A. If $\dim M = n$, then the relative boundary of A is the boundary of A.

Let M be a linear manifold of dimension $r < n$ and let τ be the mapping defined in Remark 6.3. We noted there that τ and τ^{-1} are one to one and continuous.

LEMMA 7.1. *If O is an open set in $\mathbb{R}^r$, then $\tau^{-1}(O)$ is a relatively open set in M. If O' is a relatively open set in M, then $\tau(O')$ is an open set in $\mathbb{R}^r$.*

Proof. Let O be an open set in $\mathbb{R}^r$ and let $\mathbf{x}_0 \in \tau^{-1}(O)$. Then there is a unique $\mathbf{y}_0$ in O such that $\tau^{-1}(\mathbf{y}_0) = \mathbf{x}_0$, or equivalently, $\mathbf{y}_0 = \tau(\mathbf{x}_0)$. Since $\mathbf{y}_0 \in O$, there exists an $\varepsilon > O$ such that the open ball $B_r(\mathbf{y}_0, \varepsilon)$ in $\mathbb{R}^r$ is contained in O. Since τ is continuous, there exists a $\delta > 0$ such that for $\mathbf{x} \in B(\mathbf{x}_0, \delta) \cap M$ we have $\tau(\mathbf{x}) \in B(\mathbf{y}_0, \varepsilon) \subset 0$. But this says that $B(\mathbf{x}_0, \delta) \cap M \subseteq \tau^{-1}(O)$. Hence $\mathbf{x}_0$ is a relative interior point of $\tau^{-1}(O)$, and so $\tau^{-1}(O)$ is relatively open. Since $\tau = (\tau^{-1})^{-1}$, a similar argument with τ replaced by τ^{-1} shows that if O' is relatively open, then $\tau(O')$ is open in $\mathbb{R}^r$.

Let C be a convex set in $\mathbb{R}^n$ of dimension $r < n$. As a consequence of the one-to-one relationship between $M = \Lambda[C]$ and $\mathbb{R}^r$ that preserves open sets, convexity, dimension of convex sets, and linear manifolds, we can obtain many properties of C by establishing these properties for convex sets of full dimension. These properties then hold for the set $\tau(C)$ considered as a set of full dimension in $\mathbb{R}^r$ and hence for $C = \tau^{-1}(\tau(C))$. In particular, we obtain the following results in this fashion.

THEOREM 7.1. *A convex set of dimension r has nonempty relative interior.*

This theorem follows from Theorem 6.1 and Lemma 7.1.

LEMMA 7.2. *Let C be a convex set and let $\mathbf{x}_1$ and $\mathbf{x}_2$ be two points with $\mathbf{x}_1 \in \text{ri}(C)$ and $\mathbf{x}_2 \in \bar{C}$. Then the line segment $[\mathbf{x}_1, \mathbf{x}_2)$ is contained in* $\text{ri}(C)$.

COROLLARY 1. *If C is convex, then* $\text{ri}(C)$ *is convex.*

COROLLARY 2. *Let C be convex. Then*

(*i*) $\overline{\text{ri}(C)} = \bar{C}$ *and*
(*ii*) $\text{ri}(C) = \text{ri}(\bar{C})$.

Lemma 7.2, Corollary 1 and (i) of Corollary 2 follow from Lemma 2.5 and the corresponding corollaries. In proving (ii) of Corollary 2 to Lemma 2.5, we argued that since $C \subseteq \bar{C}$, it follows that $\text{int}(C) \subseteq \text{int}(\bar{C})$. In general, it is not true that if A and B are convex sets with $A \subseteq B$ that $\text{ri}(A) \subseteq \text{ri}(B)$, as the following example shows. Take B to be a closed unit cube in $\mathbb{R}^3$. Let A be any closed face of B. Then $A \subseteq B$, but it is not true that $\text{ri}(A) \subseteq \text{ri}(B)$. In fact, $\text{ri}(A)$ and $\text{ri}(B)$ are disjoint. Since $\Lambda[C] = \Lambda[\bar{C}]$, it does follow that $\text{ri}(C) \subseteq \text{ri}(\bar{C})$. The proof of the opposite inclusion given in Section 2 is valid if we replace interiors by relative interiors.

We now state and prove the best possible separation theorem in $\mathbb{R}^n$.

THEOREM 7.2. *Let X and Y be two convex subsets of $\mathbb{R}^n$. Then X and Y can be properly separated if and only if* $\text{ri}(X)$ *and* $\text{ri}(Y)$ *are disjoint.*

To prove the "only if" part of the theorem, we shall need to define what is meant by hyperplane cutting a convex set and to develop conditions under which cutting occurs.

Definition 7.2. A hyperplane $H^{\alpha}_{\mathbf{a}}$ is said to *cut* a convex set C if there exist points $\mathbf{x}_1$ and $\mathbf{x}_2$ in C such that $\langle \mathbf{a}, \mathbf{x}_1 \rangle < \alpha$ and $\langle \mathbf{a}, \mathbf{x}_2 \rangle > \alpha$. A hyperplane that does not cut C is said to *bound* C.

LEMMA 7.3. *A hyperplane $H^{\alpha}_{\mathbf{a}}$ cuts a convex set C if and only if the following two conditions hold:*

(*i*) *$H^{\alpha}_{\mathbf{a}}$ does not contain C and*
(*ii*) $H^{\alpha}_{\mathbf{a}} \cap \text{ri}(C) \neq \varnothing$.

Proof. Let $H^{\alpha}_{\mathbf{a}}$ cut C. Then there exist points $\mathbf{x}_1$ and $\mathbf{x}_2$ in C such that $\langle \mathbf{a}, \mathbf{x}_1 \rangle < \alpha$ and $\langle \mathbf{a}, \mathbf{x}_2 \rangle > \alpha$. Since for $\mathbf{y} \in H^{\alpha}_{\mathbf{a}}$ we have $\langle \mathbf{a}, \mathbf{y} \rangle = \alpha$, conclusion (i) follows. Now let $\mathbf{z} \in \text{ri}(C)$. Then by Lemma 7.2 the line segments $(\mathbf{x}_1, \mathbf{z}]$ and $[\mathbf{z}, \mathbf{x}_2)$ belong to $\text{ri}(C)$. Let

$$\mathbf{y}(t) = \begin{cases} \mathbf{z} + t(\mathbf{z} - \mathbf{x}_1), & -1 \leqslant t \leqslant 0, \\ \mathbf{z} + t(\mathbf{x}_2 - \mathbf{z}), & 0 \leqslant t \leqslant 1. \end{cases}$$

Then $\mathbf{y}(t) \in C$ for $-1 \leqslant t \leqslant 1$ and $\mathbf{y}(t) \in \text{ri}(C)$ for $-1 < t < 1$. Let $\varphi(t) = \langle \mathbf{a}, \mathbf{y}(t) \rangle$. Since φ is continuous on $[-1, 1]$ with $\varphi(-1) < \alpha$, and $\varphi(1) > \alpha$, there exists a t_0 with $-1 < t_0 < 1$ such that $\varphi(t_0) = \alpha$. But then $\mathbf{y}(t_0) \in \text{ri}(C)$ and $\langle \mathbf{a}, \mathbf{y}(t_0) \rangle = \alpha$, which establishes (ii).

Now suppose that (i) and (ii) hold. Then there exist a point $\mathbf{x}_0$ in $\mathrm{ri}(C) \cap H_{\mathbf{a}}^{\alpha}$ and a point $\mathbf{x}_1$ in C and not in $H_{\mathbf{a}}^{\alpha}$. Suppose that $\langle \mathbf{a}, \mathbf{x}_1 \rangle < \alpha$. Since $\mathbf{x}_0 \in \mathrm{ri}(C)$, there exists a $\delta > 0$ such that $\mathbf{x}_2 = \mathbf{x}_0 - \delta(\mathbf{x}_1 - \mathbf{x}_0) \in C$. Then

$$\langle \mathbf{a}, \mathbf{x}_2 \rangle = \alpha - \delta(\langle \mathbf{a}, \mathbf{x}_1 \rangle - \alpha) > \alpha.$$

If $\langle \mathbf{a}, \mathbf{x}_1 \rangle > \alpha$, then $\langle \mathbf{a}, \mathbf{x}_2 \rangle < \alpha$.

We now prove Theorem 7.2.

If a hyperplane properly separates the convex sets X and Y, then it certainly properly separates $\mathrm{ri}(X)$ and $\mathrm{ri}(Y)$. Conversely, if a hyperplane properly separates $\mathrm{ri}(X)$ and $\mathrm{ri}(Y)$, then since $\overline{\mathrm{ri}(X)} = \bar{X}$ and $\overline{\mathrm{ri}(Y)} = \bar{Y}$, it follows that the hyperplane properly separates X and Y. Thus a hyperplane properly separates X and Y if and only if it properly separates $\mathrm{ri}(X)$ and $\mathrm{ri}(Y)$. Thus we may assume without loss of generality that X and Y are disjoint relatively open convex sets. The proof of separation will be carried out by a descending induction on $d = \max(\dim X, \dim Y)$.

If $\dim X = n$, then the separation follows from Theorem 3.4. We now assume that proper separation occurs if $d \geqslant k$, where $1 < k \leqslant n$. We now suppose that $\dim X = k - 1$ and $\dim Y \leqslant k - 1$. To simplify the notation, we assume that $\mathbf{0} \in X$. There is no loss of generality in assuming this, since this can always be achieved by a translation of coordinates. Since $\mathbf{0} \in X$, the manifold $\Lambda[X]$ is a subspace of dimension $k \leqslant n - 1$. Let $\mathbf{w}$ be a point in $\mathbb{R}^n$ that is not in $\Lambda[X]$. Let

$$A = \mathrm{co}\{X \cup (X + \mathbf{w})\}, \qquad B = \mathrm{co}\{X \cup (X - \mathbf{w})\}.$$

We assert that Y cannot intersect both A and B. See Figure 2.8.

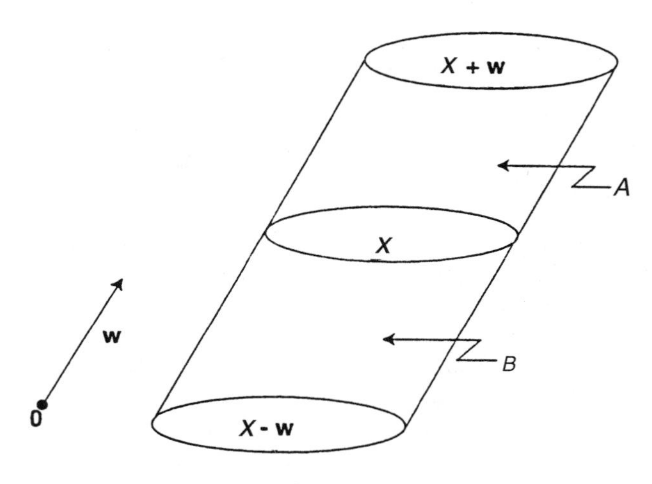

Figure 2.8.

By Theorem 2.2 points $\boldsymbol{\xi}$ in A can be written

$$\boldsymbol{\xi} = \sum_{i=1}^{n+1} \lambda_i(\mathbf{x}_i + \delta_i \mathbf{w}),$$

where $\mathbf{x}_i \in X$ for $i = 1, \ldots, n+1$, the vector $\boldsymbol{\lambda} = (\lambda_1, \ldots, \lambda_{n+1}) \in P_{n+1}$, and δ_i is either zero or 1. Thus, if $\boldsymbol{\xi} \in A$, then there exists an $\mathbf{x} \in X$ such that

$$\boldsymbol{\xi} = \mathbf{x} + \theta \mathbf{w}, \qquad 0 \leqslant \theta \leqslant 1. \tag{1}$$

Since

$$\boldsymbol{\xi} = \mathbf{x} + \theta \mathbf{w} = (1 - \theta)\mathbf{x} + \theta(\mathbf{w} + \mathbf{x}),$$

every point $\boldsymbol{\xi}$ given by (1) is in A. A similar argument shows that

$$B = \{\boldsymbol{\eta} : \boldsymbol{\eta} = \mathbf{x} - \theta \mathbf{w}, \ \mathbf{x} \in X, \ 0 \leqslant \theta \leqslant 1\}.$$

Thus if $A \cap Y \neq \varnothing$ and $B \cap Y \neq \varnothing$, then there exist points $\mathbf{y}_1$ and $\mathbf{y}_2$ in Y such that

$$\mathbf{y}_1 = \mathbf{x}_1 + \theta_1 \mathbf{w}, \qquad \mathbf{y}_2 = \mathbf{x}_2 - \theta_2 \mathbf{w},$$

where $\mathbf{x}_1, \mathbf{x}_2 \in X$, $0 < \theta_1 \leqslant 1$, $0 < \theta_2 \leqslant 1$. The inequalities $\theta_1 > 0$ and $\theta_2 > 0$ follow from $X \cap Y = \varnothing$.

Let $\alpha = \theta_2/(\theta_1 + \theta_2)$ and let $\beta = \theta_1/(\theta_1 + \theta_2)$. Then $\alpha > 0$, $\beta > 0$, $\alpha + \beta = 1$, and

$$\alpha \mathbf{y}_1 + \beta \mathbf{y}_2 = \alpha \mathbf{x}_1 + \beta \mathbf{x}_2. \tag{2}$$

Since X and Y are convex, the left-hand side of (2) belongs to X and the right-hand side belongs to Y. This contradicts $X \cap Y = \varnothing$, and so Y cannot intersect both A and B.

For the sake of definiteness, suppose that $A \cap Y = \varnothing$. Since $\dim A = k$ (why?) and $\dim Y \leqslant k - 1$, by the inductive hypothesis there exists a hyperplane H that properly separates A and Y. Since $X \subseteq A$, the hyperplane H properly separates X and Y.

To prove the "only if" assertion, suppose that there exists a hyperplane H that properly separates X and Y and that there exists a point $\mathbf{z}$ in $\mathrm{ri}(X) \cap \mathrm{ri}(Y)$. Then $\mathbf{z} \in H$, and thus $H \cap \mathrm{ri}(X) \neq \varnothing$ and $H \cap \mathrm{ri}(Y) \neq \varnothing$. Since the separation is proper, at least one of the sets, say X, is not contained in H. Hence by Lemma 7.3, H cuts X, which contradicts the assumption that H separates X and Y.

We conclude this section with results that are stronger than Theorems 4.1 and 4.3.

THEOREM 7.3. *Let C be a convex set and let D be a convex subset of the relative boundary of C. Then there exists a nontrivial supporting hyperplane to C that contains D.*

Proof. Since $\text{ri}(C) \cap D = \varnothing$, we have $\text{ri}(C) \cap \text{ri}(D) = \varnothing$. Hence by Theorem 7.2 there exists a hyperplane $H_{\mathbf{a}}^{\alpha}$ that properly separates C and D. Thus

$$\langle \mathbf{a}, \mathbf{x} \rangle \leqslant \alpha \quad \text{for } \mathbf{x} \in C \quad \text{and} \quad \langle \mathbf{a}, \mathbf{y} \rangle \geqslant \alpha \quad \text{for } \mathbf{y} \in D. \tag{3}$$

Since each $\mathbf{y}$ in D is a relative boundary point of C, for each $\mathbf{y}$ in D there exists a sequence $\{\mathbf{x}_k\}$ of points in C such that $\mathbf{x}_k \to \mathbf{y}$. For this sequence we have $\langle \mathbf{a}, \mathbf{x}_k \rangle \leqslant \alpha$ and $\langle \mathbf{a}, \mathbf{x}_k \rangle \to \langle \mathbf{a}, \mathbf{y} \rangle$. Thus $\langle \mathbf{a}, \mathbf{y} \rangle \leqslant \alpha$. From this and from (3) we get that $\langle \mathbf{a}, \mathbf{y} \rangle = \alpha$. Thus, each $\mathbf{y}$ in D is in $H_{\mathbf{a}}^{\alpha}$. Since $D \subseteq \bar{C}$, it follows that $H_{\mathbf{a}}^{\alpha}$ is a supporting hyperplane to C that contains D. Since the separation of C and D is proper and since $D \subseteq H_{\mathbf{a}}^{\alpha}$, there is an $\mathbf{x}_0$ in C such that $\langle \mathbf{a}, \mathbf{x}_0 \rangle < \alpha$. Thus the supporting hyperplane is nontrivial.

The next theorem is stronger than Theorem 4.3 but is only valid in finite dimensions.

THEOREM 7.4. *Let C be a compact convex subset of $\mathbb{R}^n$. Then C is the convex hull of its extreme points.*

Proof. Let C_e denote the set of extreme points of C. By Theorem 4.2 the set $C_e \neq \varnothing$. Since $C_e \subseteq C$ and $\text{co}(C_e) \subseteq \text{co}(C) = C$, to prove the theorem, we must show that

$$C \subseteq \text{co}(C_e). \tag{4}$$

The proof of (4) proceeds by induction on the dimension k of C.

If $k = 0$, then C is a point, and if $k = 1$, then C is a closed line segment. Thus (4) holds in these cases. Now suppose that (4) holds if $\dim C = k - 1$, where $1 \leqslant k \leqslant n$.

Let $\dim C = k$ and let $\mathbf{x} \in C$. By Theorem 7.1, either $\mathbf{x} \in \text{ri}(C)$ or $\mathbf{x}$ belongs to the relative boundary of C. If $\mathbf{x}$ belongs to the relative boundary of C, then by Theorem 7.3, there exists a supporting hyperplane to C that contains $\mathbf{x}$. This hyperplane will intersect $\Lambda[C]$ in a manifold H_{k-1} that is a hyperplane of dimension $k-1$ in $\Lambda[C]$ and that supports C at $\mathbf{x}$. The set $H_{k-1} \cap C$ is compact and convex and has dimension $\leqslant k - 1$. By the inductive hypotheses $\mathbf{x}$ is a convex combination of the extreme points of $H_{k-1} \cap C$. By Lemma 4.1 the extreme points of $H_{k-1} \cap C$ are extreme points of C. Thus $\mathbf{x} \in \text{co}(C_e)$ whenever $\mathbf{x}$ is a relative boundary point of C.

If $\mathbf{x} \in \text{ri}(C)$, let L be a line in $\Lambda[C]$ through the point $\mathbf{x}$. Then $L \cap \text{ri}(C)$ is a line segment $(\mathbf{y}, \mathbf{z})$ whose end points $\mathbf{y}$ and $\mathbf{z}$ are in the relative boundary of C. Since C is compact, $\mathbf{y} \in C$ and $\mathbf{z} \in C$. Since $\mathbf{x}$ is a convex combination of $\mathbf{y}$ and $\mathbf{z}$ and since $\mathbf{y}$ and $\mathbf{z}$ are in $\text{co}(C_e)$, it follows that $\mathbf{x} \in \text{co}(C_e)$, and (4) is established.

III

CONVEX FUNCTIONS

1. DEFINITION AND ELEMENTARY PROPERTIES

The simplest class of functions are real-valued functions that are translates of linear functions, or affine functions, f defined by the formula $f(\mathbf{x}) = \langle \mathbf{a}, \mathbf{x} \rangle + b$. One could consider the next class of functions in order of complexity to be the translates of the functions $y = \|\mathbf{x}\|^p$, $p \geqslant 1$. In $\mathbb{R}^3 = (\mathbf{x}, y)$, where $\mathbf{x} \in \mathbb{R}^2$, the graph of $y = \|\mathbf{x}\|$ is the upper nappe of a right circular cone with vertex at the origin and the graph of $y = \|\mathbf{x}\|^2$ is a paraboloid of revolution. The graphs of such functions are axially symmetric about a line parallel to the y-axis. Intuitively, it appears that all of the above functions have the property that the set of points $(\mathbf{x}, y)$ in $\mathbb{R}^{n+1}$ that lie on or above the graph is a convex set. It is therefore reasonable to consider the next level of functions to be the functions with this property. Such functions are called convex functions. It turns out that convex functions are important in optimization problems as well as in other areas of mathematics. We shall define convex functions using a different property and then show in Lemma 1.1 below that our definition is equivalent to the one suggested.

Definition 1.1. A real-valued function f defined in a convex set C in $\mathbb{R}^n$ is said to be *convex* if for every $\mathbf{x}_1$ and $\mathbf{x}_2$ in C and every $\alpha \geqslant 0$, $\beta \geqslant 0$, $\alpha + \beta = 1$,

$$f(\alpha \mathbf{x}_1 + \beta \mathbf{x}_2) \leqslant \alpha f(\mathbf{x}_1) + \beta f(\mathbf{x}_2). \tag{1}$$

The function f is said to be *strictly convex* if strict inequality holds in (1).

The geometric interpretation of this definition is that a convex function is a function such that if $\mathbf{z}_1$ and $\mathbf{z}_2$ are any two points in $\mathbb{R}^{n+1}$ on the graph of f, then points of the line segment $[\mathbf{z}_1, \mathbf{z}_2]$ joining $\mathbf{z}_1$ and $\mathbf{z}_2$ lie on or above the graph of f.

Figure 3.1 illustrates the definition. Let $C = (x_3, w)$, where $x_3 = \alpha x_1 + \beta x_2$, $\alpha > 0$, $\beta > 0$, $\alpha + \beta = 1$, be a point on the line segment joining the points $A = (x_1, f(x_1))$ and $B = (x_2, f(x_2))$ on the graph of f. Let $D = (x_2, w)$ and let

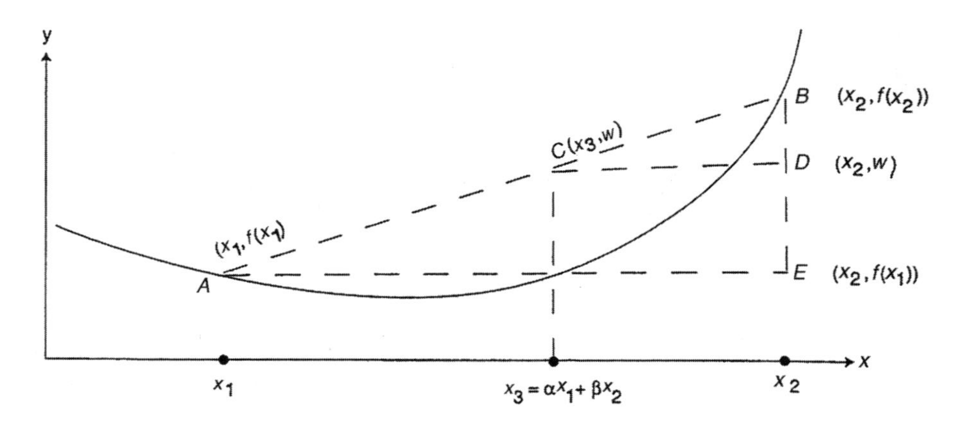

Figure 3.1.

$E = (x_2, f(x_1))$. Then

$$\frac{f(x_2) - w}{\alpha\|x_2 - x_1\|} = \frac{\|BD\|}{\|CD\|} = \frac{\|BE\|}{\|AE\|} = \frac{f(x_2) - f(x_1)}{\|x_2 - x_1\|}.$$

From the extreme terms of these equalities we get that

$$w = \alpha f(x_1) + (1 - \alpha) f(x_2) = \alpha f(x_1) + \beta f(x_2).$$

Thus (x_3, w) lies above or on the graph if and only if (1) holds.

Note that if the domain of f is not convex, then the left-hand side of (1) need not be defined. Therefore, when defining convex functions, it only makes sense to consider functions defined on convex sets. Henceforth we shall not repeat the term *real-valued* when considering convex functions; it shall be tacitly assumed that f is real valued.

Definition 1.2. A function f defined on a convex set C is said to be *concave* if $-f$ is convex.

Definition 1.3. Let f be a real-valued function defined on a set A. The *epigraph* of f, denoted by $\text{epi}(f)$, is the set

$$\text{epi}(f) = \{(\mathbf{x}, y) : \mathbf{x} \in A,\ y \in \mathbb{R},\ y \geqslant f(\mathbf{x})\}.$$

In Figure 3.2 the hatched areas represent the epigraphs of $y = x^2$, $-1 \leqslant x \leqslant 1$, and $y = x^3$, $-1 \leqslant x \leqslant 1$.

The epigraph of a function is the set of points $(\mathbf{x}, y)$ in $\mathbb{R}^{n+1}$ that lie on or above the graph of f.

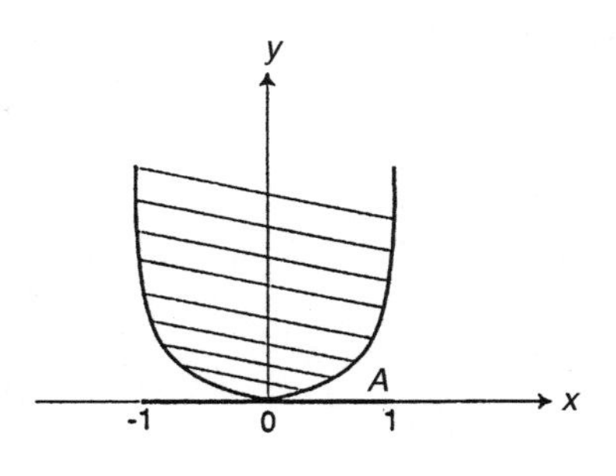

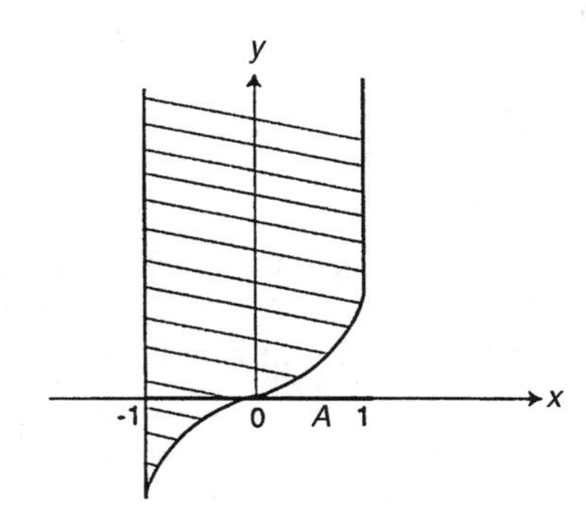

Figure 3.2.

LEMMA 1.1. *A function f defined on a convex set C is convex if and only if* epi(f) *is convex.*

Proof. Suppose that epi(f) is convex. Let $\mathbf{x}_1$ and $\mathbf{x}_2$ belong to C. Then $(\mathbf{x}_1, f(\mathbf{x}_2))$ and $(\mathbf{x}_2, f(\mathbf{x}_2))$ belong to epi(f). Since epi(f) is convex, for any $(\alpha, \beta) \in P_2$

$$\alpha(\mathbf{x}_1, f(\mathbf{x}_1)) + \beta(\mathbf{x}_2, f(\mathbf{x}_2)) = (\alpha\mathbf{x}_1 + \beta\mathbf{x}_2, \alpha f(\mathbf{x}_1) + \beta f(\mathbf{x}_2))$$

belongs to epi(f). By the definition of epi(f),

$$\alpha f(\mathbf{x}_1) + \beta f(\mathbf{x}_2) \geqslant f(\alpha\mathbf{x}_1 + \beta\mathbf{x}_2),$$

and so f is convex.

Now suppose f is convex. Let $(\mathbf{x}_1, y_1)$ and $(\mathbf{x}_2, y_2)$ be any two points in epi(f). Then $y_1 \geqslant f(\mathbf{x}_1)$ and $y_2 \geqslant f(\mathbf{x}_2)$. For any (α, β) in P_2 let

$$(\mathbf{x}_3, y) = \alpha(\mathbf{x}_1, y_1) + \beta(\mathbf{x}_2, y_2) = (\alpha\mathbf{x}_1 + \beta\mathbf{x}_2, \alpha y_1 + \beta y_2).$$

Since $(\mathbf{x}_1, y_1)$ and $(\mathbf{x}_2, y_2)$ are in epi(f), and since f is convex, we have

$$y = \alpha y_1 + \beta y_2 \geqslant \alpha f(\mathbf{x}_1) + \beta f(\mathbf{x}_2) \geqslant f(\alpha\mathbf{x}_1 + \beta\mathbf{x}_2) = f(\mathbf{x}_3).$$

Thus, $(\mathbf{x}_3, y) \in$ epi(f), and so epi(f) is convex.

LEMMA 1.2. *If f is a convex function defined on a convex set C in* $\mathbb{R}^n$, *then for any real γ the set $\{\mathbf{x} : f(\mathbf{x}) \leqslant \gamma\}$ is either empty or convex.*

We leave the proof of this lemma as an exercise for the reader.

The converse to this lemma is false, as can be seen from the function $f(x) = x^3$.

LEMMA 1.3. *Let* $\{f_\alpha\}$ *be a family of convex functions defined on a convex set* C *in* $\mathbb{R}^n$ *and let there exist a real-valued function* M *with domain* C *such that, for each* α, $f_\alpha(\mathbf{x}) \leqslant M(\mathbf{x})$ *for all* $\mathbf{x}$ *in* C. *Then the function* f *defined by* $f(\mathbf{x}) = \sup\{f_\alpha(\mathbf{x}) : \alpha\}$ *is convex.*

Proof. Since for each $\mathbf{x}$ in C $f_\alpha(\mathbf{x}) \leqslant M(\mathbf{x})$ for all α, it follows that $f(\mathbf{x}) \leqslant M(\mathbf{x})$, and so f is real valued. Also $\operatorname{epi}(f) = \bigcap_\alpha \operatorname{epi}(f_\alpha)$, so $\operatorname{epi}(f)$, being the intersection of convex sets, is convex. Therefore f is convex.

THEOREM 1.1 (JENSEN'S INEQUALITY). *Let* f *be a real-valued function defined on a convex set* C *in* $\mathbb{R}^n$. *Then* f *is convex if and only if for every finite set of points* $\mathbf{x}_1, \ldots, \mathbf{x}_k$ *in* C *and every* $\boldsymbol{\lambda} = (\lambda_1, \ldots, \lambda_k)$ *in* P_k

$$f(\lambda_1\mathbf{x}_1 + \cdots + \lambda_k\mathbf{x}_k) \leqslant \lambda_1 f(\mathbf{x}_1) + \cdots + \lambda_k f(\mathbf{x}_k). \tag{2}$$

Proof. If (2) holds for all positive integers k, then it holds for $k = 2$. But for $k = 2$, the relation (2) is the definition of convex function.

Now suppose that f is convex. We shall prove (2) by induction on k. For $k = 1$ the relation is trivial. For $k = 2$ the relation follows from the definition of a convex function. Suppose that $k > 2$ and that (2) has been established for $k - 1$. We shall show that (2) holds for k. If $\lambda_k = 1$, then there is nothing to prove. If $\lambda_k \neq 1$, set $\Lambda = \Sigma_{i=1}^{k-1} \lambda_i$. Then

$$\Lambda > 0, \qquad \Lambda + \lambda_k = 1, \qquad \sum_{i=1}^{k-1} \frac{\lambda_i}{\Lambda} = 1,$$

and we may write (compare the proof of Lemma II.2.6)

$$\begin{aligned} f\left(\sum_{i=1}^{k} \lambda_i \mathbf{x}_i\right) &= f\left(\Lambda\left[\sum_{i=1}^{k-1} \left(\frac{\lambda_i}{\Lambda}\right)\mathbf{x}_i\right] + \lambda_k \mathbf{x}_k\right) \\ &\leqslant \Lambda f\left(\sum_{i=1}^{k-1} \left(\frac{\lambda_i}{\Lambda}\right)\mathbf{x}_i\right) + \lambda_k f(\mathbf{x}_k) \\ &\leqslant \lambda_1 f(\mathbf{x}_1) + \cdots + \lambda_{k-1} f(\mathbf{x}_{k-1}) + \lambda_k f(\mathbf{x}_k), \end{aligned}$$

where the first inequality follows from the convexity of f and the second inequality follows from the inductive hypothesis.

The next lemma in conjunction with other criteria is sometimes useful in determining whether a given function is convex.

LEMMA 1.4. *Let* f *be a convex function defined on a convex set* C *in* $\mathbb{R}^n$. *Let* g *be a nondecreasing convex function defined on an interval* I *in* $\mathbb{R}$. *Let* $f(C) \subseteq I$. *Then the composite function* $g \circ f$ *defined by* $(g \circ f)(\mathbf{x}) = g(f(\mathbf{x}))$ *is convex on* C.

We leave the proof as an exercise.

Let C be a convex set. Choose an $\mathbf{x}$ in C and a $\mathbf{v}$ in $\mathbb{R}^n$. Let

$$\Lambda = \{\lambda : \lambda \in \mathbb{R}, \mathbf{x} + \lambda \mathbf{v} \in C\}.$$

Let $l_2 = \sup\{\lambda : \lambda \in \Lambda\}$ and let $l_1 = \inf\{\lambda : \lambda \in \Lambda\}$. Then $l_1 \leqslant 0 \leqslant l_2$. If $l_1 \neq l_2$, then since C is convex, $\mathbf{x} + \lambda\mathbf{v} \in C$ for all λ in the open interval (l_1, l_2). Note that, depending on the structure of the set C, $\mathbf{x} + \lambda\mathbf{v}$ can belong to C for all λ in $[l_1, l_2)$ or $[l_1, l_2]$ or $(l_1, l_2]$.

Let f be a convex function defined on a convex set C. From the discussion in the preceding paragraph it follows that for each $\mathbf{x}$ in C and each $\mathbf{v}$ in $\mathbb{R}^n$ there is an interval $I \subseteq \mathbb{R}$ that depends on $\mathbf{x}$ and $\mathbf{v}$ such that the function $\varphi(\ ; \mathbf{x}, \mathbf{v})$ defined by

$$\varphi(\lambda; \mathbf{x}, \mathbf{v}) = f(\mathbf{x} + \lambda\mathbf{v}) \tag{3}$$

has I as its domain of definition. The interval I always includes the origin and can degenerate to the single point $\{0\}$; it can be open, half open, or closed. The function $\varphi(\ ; \mathbf{x}, \mathbf{v})$ gives the one-dimensional section of the function f through the point $\mathbf{x}$ in the direction of $\mathbf{v}$. The next lemma states that f is convex if and only if all the nondegenerate one-dimensional functions $\varphi(\ ; \mathbf{x}, \mathbf{v})$ are convex.

LEMMA 1.5. *A function f defined on a convex set C is convex if and only if for each $\mathbf{x}$ in C and $\mathbf{v}$ in $\mathbb{R}^n$ such that $\varphi(\ ; \mathbf{x}, \mathbf{v})$ is defined on a nondegenerate interval I the function $\varphi(\ ; \mathbf{x}, \mathbf{v})$ is convex on I.*

Proof. Suppose that f is convex. Then for any λ_1, λ_2 in I and (α, β) in P_2,

$$\begin{aligned}
\varphi(\alpha\lambda_1 + \beta\lambda_2; \mathbf{x}, \mathbf{v}) &= f(\mathbf{x} + (\alpha\lambda_1 + \beta\lambda_2)\mathbf{v}) \\
&\underset{\alpha+\beta=1}{\underset{\uparrow}{=}} f(\alpha(\mathbf{x} + \lambda_1\mathbf{v}) + \beta(\mathbf{x} + \lambda_2\mathbf{v})) \\
&\underset{f\text{ convex}}{\underset{\uparrow}{\leqslant}} \alpha f(\mathbf{x} + \lambda_1\mathbf{v}) + \beta f(\mathbf{x} + \lambda_2\mathbf{v}) \\
&\underset{\text{def }\varphi}{\underset{\uparrow}{=}} \alpha\varphi(\lambda_1; \mathbf{x}, \mathbf{v}) + \beta\varphi(\lambda_2; \mathbf{x}, \mathbf{v}).
\end{aligned}$$

This proves the convexity of $\varphi(\ ; \mathbf{x}, \mathbf{v})$.

Now suppose that for each $\mathbf{x}$ in C and $\mathbf{v}$ in $\mathbb{R}^n$ such that $I \neq \{0\}$, $\varphi(\ ; \mathbf{x}, \mathbf{v})$ is convex. Let $\mathbf{x}_1$ and $\mathbf{x}_2$ be in C and let (α, β) be in P_2. Then

$$\begin{aligned}
f(\alpha\mathbf{x}_1 + \beta\mathbf{x}_2) &= f((1 - \beta)\mathbf{x}_1 + \beta\mathbf{x}_2) \\
&= f(\mathbf{x}_1 + \beta(\mathbf{x}_2 - \mathbf{x}_1)) = \varphi(\beta; \mathbf{x}_1, \mathbf{x}_2 - \mathbf{x}_1) \\
&= \varphi(\alpha\cdot 0 + \beta\cdot 1; \mathbf{x}_1, \mathbf{x}_2 - \mathbf{x}_1).
\end{aligned} \tag{4}$$

Since

$$\varphi(\lambda; \mathbf{x}_1, \mathbf{x}_2 - \mathbf{x}_1) = f(\mathbf{x}_1 + \lambda(\mathbf{x}_2 - \mathbf{x}_1))$$

and since C is convex, $\varphi(\lambda; \mathbf{x}_1, \mathbf{x}_2 - \mathbf{x}_1)$ is defined for $0 \leqslant \lambda \leqslant 1$ and

$$\varphi(0; \mathbf{x}_1, \mathbf{x}_2 - \mathbf{x}_1) = f(\mathbf{x}_1), \qquad \varphi(1; \mathbf{x}_1, \mathbf{x}_2 - \mathbf{x}_1) = f(\mathbf{x}_2). \tag{5}$$

From the convexity of $\varphi(\ ; \mathbf{x}_1, \mathbf{x}_2 - \mathbf{x}_1)$ and (5) we get that the rightmost side of (4) is less than or equal to

$$\alpha\varphi(0; \mathbf{x}_1, \mathbf{x}_2 - \mathbf{x}_1) + \beta\varphi(1; \mathbf{x}_1, \mathbf{x}_2 - \mathbf{x}_1) = \alpha f(\mathbf{x}_1) + \beta f(\mathbf{x}_2).$$

Hence f is convex.

Remark 1.1. An examination of the proof shows that f is strictly convex if and only if, for each $\mathbf{x}$ and $\mathbf{v}$, $\varphi(\ ; \mathbf{x}, \mathbf{v})$ is strictly convex.

We now present some results for convex functions defined on real intervals. The first result is the so-called three-chord property.

THEOREM 1.2. *Let g be a convex function defined on an interval $I \subset \mathbb{R}$. Then if $x < y < z$ are three points in I,*

$$\frac{g(y) - g(x)}{y - x} \leqslant \frac{g(z) - g(x)}{z - x} \leqslant \frac{g(z) - g(y)}{z - y}. \tag{6}$$

If g is strictly convex, then strict inequality holds.

Figure 3.3 illustrates the theorem and shows why the conclusion of the theorem is called the three-chord property.

Proof. Since $x < y < z$, there exists a $0 < t < 1$ such that $y = x + t(z - x)$. Since g is convex,

$$\begin{aligned} g(y) - g(x) &= g((1 - t)x + tz) - g(x) \\ &\leqslant (1 - t)g(x) + tg(z) - g(x) \\ &= t(g(z) - g(x)). \end{aligned} \tag{7}$$

If in (7) we now divide through by $t(z - x) > 0$ and note that $y - x = t(z - x)$, we get that

$$\frac{g(y) - g(x)}{y - x} \leqslant \frac{g(z) - g(x)}{z - x}.$$

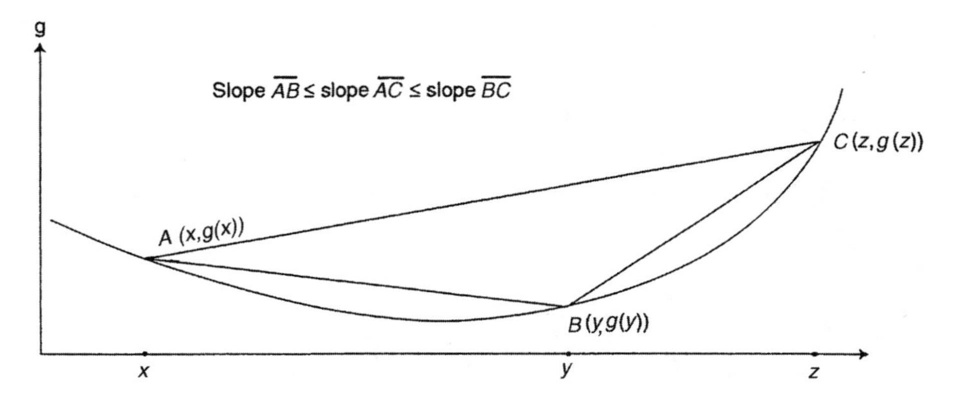

Figure 3.3.

To obtain the second inequality, we write

$$
\begin{aligned}
g(z) - g(y) &= g(z) - g(x + t(z - x)) \\
&\geqslant g(z) - (1 - t)g(x) - tg(z) \\
&= (1 - t)[g(z) - g(x)].
\end{aligned}
$$

If we now divide through by $z - y > 0$ and note that $z - y = (1 - t)(z - x)$, we obtain the second inequality.

COROLLARY 1. *For each c in I the slope function $s(\;; c)$ defined by*

$$s(w; c) = \frac{g(w) - g(c)}{w - c}, \qquad w \neq c, \qquad w \in I,$$

is an increasing function. If g is strictly convex, then *$s(\;; c)$ is strictly increasing.*

Proof. We must show that if $w_1 < w_2$, then

$$s(w_1; c) \leqslant s(w_2; c). \tag{8}$$

For two points w_1 and w_2 such that $c < w_1 < w_2$, inequality (8) follows from (6) by taking $x = c$, $y = w_1$, and $z = w_2$. For two points w_1 and w_2 such that $w_1 < w_2 < c$, inequality (8) follows from (6) by taking $z = c$, $x = w_1$, and $y = w_2$. Lastly, if $w_1 < c < w_2$, the inequality (8) follows from (6) by taking $y = c$, $x = w_1$, and $z = w_2$.

Let f be a real-valued function defined on an interval I. If at a point c in I

$$\lim_{h \to 0+} \frac{f(c+h) - f(c)}{h}$$

exists, then f is said to have a *right-hand derivative* at c whose value $f'_+(c)$ is the value of the above limit. Similarly, the *left-hand derivative* of f at c, denoted by $f'_-(c)$, is

$$\lim_{h \to 0-} \frac{f(c+h) - f(c)}{h}$$

provided the limit exists.

COROLLARY 2. *At each point c in* int(I) *the right- and left-hand derivatives $g'_+(c)$ and $g'_-(c)$ exist and $g'_-(c) \leqslant g'_+(c)$. Moreover, g'_+ and g'_- are increasing functions.*

Proof. In (6) take $y = c$. Then for $x < c < z$ we have

$$\frac{g(x) - g(c)}{x - c} \leqslant \frac{g(z) - g(c)}{z - c}.$$

It follows from this relation, from Corollary 1, and from Exercise I.6.6. that $g'_-(c)$ and $g'_+(c)$ exist and $g'_-(c) \leqslant g'_+(c)$.

Now let $c < d$ be two points in int(I) and let $h > 0$ be such that $c + h < d$ and $d - h > c$. Then by Theorem 1.2

$$\frac{g(c+h) - g(c)}{h} \leqslant \frac{g(d) - g(c)}{d - c} \leqslant \frac{g(d-h) - g(d)}{-h}.$$

If we now let $h \to 0^+$, we get $g'_+(c) \leqslant g'_-(d)$. Combining this with $g'_-(x) \leqslant g'_+(x)$ for any x in int(I), we get

$$g'_-(c) \leqslant g'_+(c) \leqslant g'_-(d) \leqslant g'_+(d),$$

which gives the desired monotonicity properties.

Remark 1.2. If c is a left-hand end point of I and g is defined at c, then from the monotonicity of $s(\,;c)$ it follows that $g'_+(c)$ always exists if we permit the value $-\infty$.

The next corollary is an immediate consequence of Corollary 2 and is a result that the reader encountered in elementary calculus.

COROLLARY 3. *If g is convex and twice differentiable on I, then g' is an increasing function on I and $g''(x) \geqslant 0$ for all x in I.*

We can now return to functions defined on $\mathbb{R}^n$.

Definition 1.4. Let f be a real-valued function defined on a set S in $\mathbb{R}^n$. Let $\mathbf{x}$ be a point in S and let $\mathbf{v}$ be a vector in $\mathbb{R}^n$ such that $\mathbf{x} + \lambda\mathbf{v}$ belongs to S for λ in some interval $[0, l)$. Then f is said to have a *directional derivative at* $\mathbf{x}$ *in the direction* $\mathbf{v}$ if

$$\lim_{\lambda \to 0+} \frac{f(\mathbf{x} + \lambda\mathbf{v}) - f(\mathbf{x})}{\lambda}$$

exists. We denote the directional derivative by $f'(\mathbf{x}; \mathbf{v})$.

LEMMA 1.6. *Let f be a convex function defined on a convex set C and let $\mathbf{x}$ be a point of C. If $\mathbf{x}$ is an interior point, then $f'(\mathbf{x}; \mathbf{v})$ exists for every $\mathbf{v}$ in $\mathbb{R}^n$. If $\mathbf{x}$ is a boundary point and $\mathbf{v}$ is a vector in $\mathbb{R}^n$ such that $\mathbf{x} + \lambda\mathbf{v}$ belongs to C for sufficiently small $\lambda > 0$, then $f'(\mathbf{x}; \mathbf{v})$ exists if we allow the value $-\infty$.*

Proof. If $\mathbf{x}$ is an interior point of C, then for each $\mathbf{v}$ in $\mathbb{R}^n$ there exists a positive number $\lambda_0 = \lambda_0(\mathbf{v})$ such that for $|\lambda| < \lambda_0$ the point $\mathbf{x} + \lambda\mathbf{v}$ is in C. Hence the function $\varphi(\ ; \mathbf{x}, \mathbf{v})$ of (3) is defined for $|\lambda| < \lambda_0$. By Lemma 1.5, $\varphi(\ ; \mathbf{x}, \mathbf{v})$ is convex, and so by Corollary 2 of Theorem 1.2, $\varphi'_+(0; \mathbf{x}, \mathbf{v})$ exists. But

$$\frac{\varphi(\lambda; \mathbf{x}, \mathbf{v}) - \varphi(0; \mathbf{x}, \mathbf{v})}{\lambda} = \frac{f(\mathbf{x} + \lambda\mathbf{v}) - f(\mathbf{x})}{\lambda},$$

and so $f'(\mathbf{x}, \mathbf{v})$ exists and equals $\varphi'_+(0; \mathbf{x}, \mathbf{v})$. If $\mathbf{x}$ is a boundary point of C, the assertion follows from similar arguments and Remark 1.1.

In the next lemma we record a useful inequality.

LEMMA 1.7. *Let f be a convex function defined on a convex set C. Let $\mathbf{x}$ and $\mathbf{y}$ be points in C and let $\mathbf{z} = \mathbf{x} + t(\mathbf{y} - \mathbf{x})$, $0 < t < 1$. Then*

$$f(\mathbf{y}) - f(\mathbf{x}) \geqslant \frac{f(\mathbf{z}) - f(\mathbf{x})}{t}. \tag{9}$$

Proof. The function $\varphi(\ ; \mathbf{x}, \mathbf{y} - \mathbf{x})$ in (3) is defined for $0 \leqslant t \leqslant 1$, with $\varphi(1; \mathbf{x}, \mathbf{y} - \mathbf{x}) = f(\mathbf{y})$, $\varphi(0; \mathbf{x}, \mathbf{y} - \mathbf{x}) = f(\mathbf{x})$ and $\varphi(t; \mathbf{x}, \mathbf{y} - \mathbf{x}) = f(\mathbf{z})$ for $0 <$

$t < 1$. By Corollary 1 to Theorem 1.2,

$$\varphi(1; \mathbf{x}, \mathbf{y} - \mathbf{x}) - \varphi(0; \mathbf{x}, \mathbf{y} - \mathbf{x}) \geqslant \frac{\varphi(t; \mathbf{x}, \mathbf{y} - \mathbf{x}) - \varphi(0; \mathbf{x}, \mathbf{y} - \mathbf{x})}{t},$$

which is the same as (9).

We note that (9) can also be established directly from the definition of a convex function.

We next show that a convex function is continuous on the interior of its domain of definition.

THEOREM 1.3. *Let f be a convex function defined on a convex set C in $\mathbb{R}^n$. If $\mathbf{x}$ is an interior point of C, then f is continuous at $\mathbf{x}$.*

If $\mathbf{x}$ is not an interior point of C, then f need not be continuous at $\mathbf{x}$, as the following example shows. Let $C = [-1, 1] \subseteq \mathbb{R}$ and let $f(x) = x^2$ for $-1 < x < 1$ and $f(-1) = f(1) = 2$. The function f is convex on $[-1, 1]$ since $\operatorname{epi}(f)$ is convex, but f is discontinuous at ± 1.

We now prove the theorem.

If $\mathbf{x}$ is an interior point of C, then there exists a $\rho > 0$ such that $\overline{B(\mathbf{x}, \rho)} \subset C$. Let $\delta = \rho/\sqrt{n}$. Let C_δ denote the cube with center at $\mathbf{x}$ and length of side equal to 2δ; that is,

$$C_\delta = \{\mathbf{y} : |y_i - x_i| \leqslant \delta,\ i = 1, \ldots, n\}.$$

Since

$$\|\mathbf{y} - \mathbf{x}\| = \left[\sum_{i=1}^{n} |x_i - y_i|^2\right]^{1/2},$$

it follows that if $|x_i - y_i| \leqslant \rho/\sqrt{n}$ for each i, then $\|\mathbf{y} - \mathbf{x}\| \leqslant \rho$. Thus $C_\delta \subseteq \overline{B(\mathbf{x}, \rho)}$. On the other hand, if $\|\mathbf{y} - \mathbf{x}\| \leqslant \delta$, then we cannot have $|y_j - x_j| > \delta$ for any index j in set $1, \ldots, n$. Hence $B(\mathbf{x}, \delta) \subseteq C_\delta$.

The first step in the proof is to show that f is bounded above on $B(\mathbf{x}, \delta)$. It is easier to show that f is bounded above on the larger set C_δ than to work directly with $B(\mathbf{x}, \delta)$. Let $C_{\delta e}$ denote the set of 2^n vertices of C_δ. By Exercise II.4.7(a), the set $C_{\delta e}$ is the set of extreme points of C_δ. By Theorem II.4.3, $C_\delta = \operatorname{co}(C_{\delta e})$. But $C_{\delta e}$ is compact, so by Lemma II.2.7, $\operatorname{co}(C_{\delta e})$ is compact. Hence

$$C_\delta = \overline{\operatorname{co}(C_{\delta e})} = \operatorname{co}(C_{\delta e}).$$

[We could also conclude this directly from Exercise II.4.7 (b).] Thus, for each

$\mathbf{y}$ in C_δ, there exists a $\mathbf{p}$ in $\mathbb{P}_{2^n}$ such that

$$\mathbf{y} = \sum_{i=1}^{2^n} p_i \mathbf{v}_i,$$

where the $\mathbf{v}_i$ are the vertices of C_δ.

Let

$$M = \max\{f(\mathbf{v}) : \mathbf{v} \in C_{\delta e}\}.$$

Then by the convexity of f and Theorem 1.1 (Jensen's inequality), for each $\mathbf{y}$ in C_δ

$$f(\mathbf{y}) \leqslant \sum_{i=1}^{2^n} p_i f(\mathbf{v}_i) \leqslant \sum_{i=1}^{2^n} p_i M = M.$$

Thus, since $\overline{B(\mathbf{x}, \delta)} \subseteq C_\delta$,

$$f(\mathbf{y}) \leqslant M \quad \text{for } \mathbf{y} \text{ in } \overline{B(\mathbf{x}, \delta)}. \tag{10}$$

Let $\mathbf{z} \neq \mathbf{x}$ be an arbitrary point in $B(\mathbf{x}, \delta)$, and consider the line determined by $\mathbf{x}$ and $\mathbf{z}$. This line will intersect the surface of $\overline{B(\mathbf{x}, \delta)}$ at two points, $\mathbf{x} + \mathbf{u}$ and $\mathbf{x} - \mathbf{u}$, where $\mathbf{u} = t(\mathbf{z} - \mathbf{x})$ for some $t > 1$ and $\|\mathbf{u}\| = \delta$. See Figure 3.4.

We shall use the bound on f and the convexity of f to estimate

$$|f(\mathbf{z}) - f(\mathbf{x})|$$

in terms of $\|\mathbf{z} - \mathbf{x}\|$ and thus prove the continuity of f at $\mathbf{x}$. To this end, we

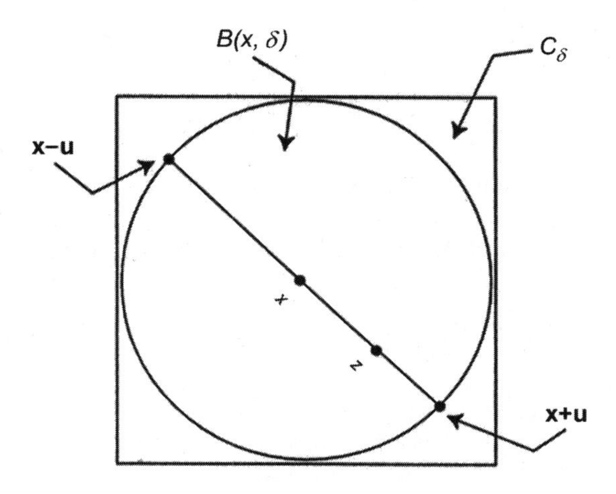

Figure 3.4.

first express $\mathbf{z}$ as a convex combination of $\mathbf{x}$ and $\mathbf{x} + \mathbf{u}$. Thus

$$\mathbf{z} = (1 - \lambda)\mathbf{x} + \lambda(\mathbf{x} + \mathbf{u}) = \mathbf{x} + \lambda\mathbf{u}, \qquad 0 < \lambda < 1. \tag{11}$$

Therefore, $\lambda = \|\mathbf{z} - \mathbf{x}\|\delta^{-1}$. We next express $\mathbf{x}$ as a convex combination of $\mathbf{z}$ and $\mathbf{x} - \mathbf{u}$. Thus

$$\mathbf{x} = (1 - t)(\mathbf{x} - \mathbf{u}) + t\mathbf{z}, \qquad 0 < t < 1.$$

Recalling relation (1) in Section 1 of Chapter II and using the relation $\lambda = \|\mathbf{z} - \mathbf{x}\|\delta^{-1}$, we see that

$$t(1 - t)^{-1} = \delta\|\mathbf{z} - \mathbf{x}\|^{-1} = \lambda^{-1},$$

and so $t = (1 + \lambda)^{-1}$. Therefore,

$$\mathbf{x} = \frac{1}{1 + \lambda}\mathbf{z} + \frac{\lambda}{1 + \lambda}(\mathbf{x} - \mathbf{u}). \tag{12}$$

Since f is convex, we have, using (10), (11), and (12),

$$f(\mathbf{z}) \leqslant (1 - \lambda)f(\mathbf{x}) + \lambda f(\mathbf{x} + \mathbf{u}) \leqslant (1 - \lambda)f(\mathbf{x}) + \lambda M,$$

$$f(\mathbf{x}) \leqslant \frac{1}{1 + \lambda}f(\mathbf{z}) + \frac{\lambda}{1 + \lambda}f(\mathbf{x} - \mathbf{u}) \leqslant \frac{1}{1 + \lambda}f(\mathbf{z}) + \frac{\lambda}{1 + \lambda}M.$$

From these inequalities we get

$$-\lambda[M - f(\mathbf{x})] \leqslant f(\mathbf{z}) - f(\mathbf{x}) \leqslant \lambda[M - f(\mathbf{x})].$$

Hence, for $\mathbf{z} \in B(\mathbf{x}, \delta)$

$$|f(\mathbf{z}) - f(\mathbf{x})| \leqslant \delta^{-1}[M - f(\mathbf{x})]\|\mathbf{z} - \mathbf{x}\|, \tag{13}$$

from which the continuity of f at $\mathbf{x}$ follows.

Remark 1.3. A function $\mathbf{f}$ defined on a set D in $\mathbb{R}^m$ with range in $\mathbb{R}^m$ is said to be *Lipschitz continuous on D* if there exists a constant $K > 0$ such that

$$\|\mathbf{f}(\mathbf{x}) - f(\mathbf{y})\| \leqslant K\|\mathbf{y} - \mathbf{x}\|$$

for all $\mathbf{x}, \mathbf{y}$ in D. Since a real-valued continuous function defined on a compact set is bounded above and below, relation (13) implies that a convex function defined on an open convex set C is Lipschitz continuous on every compact subset contained in C.

Remark 1.4. The proof of Theorem 1.3 can easily be modified to show that if f is convex on a convex set C, then f is continuous on $\text{ri}(C)$.

In Section 11 of Chapter I we saw that the existence of partial derivatives at a point does not imply differentiability. For convex functions, however, the existence of partial derivatives at a point does imply differentiability.

THEOREM 1.4. *Let f be a convex function defined on a convex set C. If the partial derivatives of f with respect to each variable exist at an interior point $\mathbf{x}$ of C, then f is differentiable at $\mathbf{x}$.*

Proof. Define the function η on $\mathbb{R}^n$ by

$$\eta(\mathbf{h}) = f(\mathbf{x} + \mathbf{h}) - f(\mathbf{x}) - \sum_{j=1}^{n} \frac{\partial f}{\partial x_j}(\mathbf{x})h_j.$$

To prove that f is differentiable at $\mathbf{x}$, we must show that

$$\frac{\eta(\mathbf{h})}{\|\mathbf{h}\|} \to 0 \quad \text{as } \|\mathbf{h}\| \to 0. \tag{14}$$

Let

$$\varepsilon_i = \frac{f(\mathbf{x} + nh_i\mathbf{e}_i) - f(\mathbf{x})}{nh_i} - \frac{\partial f}{\partial x_i}(\mathbf{x}).$$

Since f is convex and a linear function is concave and convex, the function η is convex. Thus

$$\begin{aligned} \eta(\mathbf{h}) &= \eta\left(\sum_{i=1}^{n} \frac{nh_i\mathbf{e}_i}{n}\right) \leqslant \frac{1}{n}\sum_{i=1}^{n} \eta(nh_i\mathbf{e}_i) \\ &= \sum_{i=1}^{n} \varepsilon_i h_i \leqslant \sum_{i=1}^{n} |\varepsilon_i|\,|h_i| \leqslant \left(\sum_{i=1}^{n} |\varepsilon_i|\right)\|\mathbf{h}\| \end{aligned} \tag{15}$$

for $i = 1, \ldots, n$. The existence of the partial derivatives implies that each of the $\varepsilon_i \to 0$ as $\|\mathbf{h}\| \to 0$. Also,

$$0 = \eta(\mathbf{0}) = \eta\left(\frac{\mathbf{h} + (-\mathbf{h})}{2}\right) \leqslant \frac{\eta(\mathbf{h}) + \eta(-\mathbf{h})}{2}.$$

Hence

$$\eta(\mathbf{h}) \geqslant -\eta(-\mathbf{h}) \geqslant -\left(\sum_{i=1}^{n} |\varepsilon_i'|\right)\|\mathbf{h}\|, \tag{16}$$

where, for each $i = 1, \ldots, n$, $\varepsilon_i' \to 0$ as $\|\mathbf{h}\| \to 0$. We now obtain (14) from (15) and (16).

Exercise 1.1. Show that if f and g are convex functions defined on a convex set C, then for any $\lambda > 0$, $\mu > 0$, the function $\lambda f + \mu g$ is convex.

Exercise 1.2. Let $\{f_n\}$ be a sequence of convex functions defined on a convex set C. Show that if $f(\mathbf{x}) = \lim_{n\to\infty} f_n(\mathbf{x})$ exists for all $x \in C$, then f is convex.

Exercise 1.3. Prove Lemma 1.2.

Exercise 1.4. Prove Lemma 1.4.

Exercise 1.5. Let $I = [a, b]$ be a real interval and let C be a convex set in $\mathbb{R}^n$. Let $f:(t, \mathbf{x}) \to f(t, \mathbf{x})$ be a real-valued function defined on $I \times C$ that is continuous on I for each fixed $\mathbf{x}$ in C and is convex on C for each fixed t in I. Show that

$$F(\mathbf{x}) = \int_a^b f(t, \mathbf{x})\, dt$$

is defined for each $\mathbf{x}$ in C and that F is convex on C.

Exercise 1.6. (a) Show that any norm v on $\mathbb{R}^n$ is convex on $\mathbb{R}^n$.

(b) Let S be a nonempty convex set, and let $\|\cdot\|$ denote the euclidean norm in $\mathbb{R}^n$. Let f be the distance function to S defined by

$$f(\mathbf{y}) = \inf\{\|\mathbf{y} - \mathbf{x}\| : \mathbf{x} \in S\}.$$

Show that f is convex on $\mathbb{R}^n$.

(c) Show that f is Lipschitz continuous on compact subsets of $\mathbb{R}^n$.

Exercise 1.7. Let f be a convex function defined on $\mathbb{R}^n$. Let $\mathbf{x}_0$ be a fixed point in $\mathbb{R}^n$. Show that the function $d(\cdot): \mathbb{R}^n \to \mathbb{R}$ defined by $d(\mathbf{v}) = f'(x_0; \mathbf{v})$ is (a) positively homogeneous [i.e., for $\lambda > 0$, $d(\lambda\mathbf{v}) = \lambda d(\mathbf{v})$] and (b) convex.

Exercise 1.8. Let g be a concave function defined on $\mathbb{R}^n$. Let $S = \{\mathbf{x} : g(\mathbf{x}) > 0\}$ be nonempty.

(a) Show that S is convex.

(b) Show that if $f(\mathbf{x}) = 1/g(\mathbf{x})$ for $\mathbf{x}$ in S, then f is convex on S.

Exercise 1.9. Let C be a convex cone (see Exercise II.2.10 for the definition). A real-valued function g defined on C is said to be a *gauge function on* C if

(a) $g(\mu\mathbf{x}) = \mu g(\mathbf{x})$ for all $\mathbf{x}$ in C and $\mu \geqslant 0$ and
(b) $g(\mathbf{x} + \mathbf{y}) \leqslant g(\mathbf{x}) + g(\mathbf{y})$ for all $\mathbf{x}, \mathbf{y}$ in C.

Show that g is convex.

Exercise 1.10. Let C be a convex set. The *support function* s of C is the real-valued function defined by

$$s(\mathbf{u}) = \sup\{\langle \mathbf{u}, \mathbf{x} \rangle : \mathbf{x} \in C\}$$

for all $\mathbf{u}$ such that the supremum is finite.

(a) Show that if the domain D of s is nonempty, then D is in a convex cone.
(b) Show that s is a gauge function on D.

Exercise 1.11. Let C be a compact convex set with $\mathbf{0} \in \operatorname{int}(C)$. The *Minkowski distance function* p determined by C is defined by

$$p(\mathbf{x}) = \inf\{\lambda : \lambda \geqslant 0, \mathbf{x} \in \lambda C\}.$$

Show that p is a gauge function.

Exercise 1.12. Let C be a compact convex set and let s be its support function. Show that the following statements hold:

(a) The domain of s is $\mathbb{R}^n$.
(b) For each $\mathbf{u} \neq \mathbf{0}$ there exists a point $\mathbf{x}_\mathbf{u}$ in C such that $s(\mathbf{u}) = \langle \mathbf{u}, \mathbf{x}_\mathbf{u} \rangle$.
(c) The hyperplane $H_\mathbf{u} = \{\mathbf{x} : \langle \mathbf{u}, \mathbf{x} \rangle = s(\mathbf{u})\}$ supports C at $\mathbf{x}_\mathbf{u}$.
(d) The distance from $H_\mathbf{u}$ to $\mathbf{0}$ is equal to $s(\mathbf{u}/\|\mathbf{u}\|)$.

Exercise 1.13. Let C be a nonempty closed convex set $C \neq \mathbb{R}^n$. Let s be the support function of C with nonempty domain D. Show that

$$C = \{\mathbf{x} : \langle \mathbf{u}, \mathbf{x} \rangle \leqslant s(\mathbf{u}) \text{ for all } \mathbf{u} \text{ in } D\}.$$

Hint: An alternate definition of C is

$$C = \bigcap_{\mathbf{u} \in D} \{\mathbf{x} : \langle \mathbf{u}, \mathbf{x} \rangle \leqslant s(\mathbf{u}),\ \mathbf{u} \text{ fixed in } D\},$$

and see Exercise II.4.6.

2. SUBGRADIENTS

If f is a real-valued function defined on a set C in $\mathbb{R}^n$, then the graph of f is the set of points $(\mathbf{x}, z)$ in $\mathbb{R}^{n+1}$ of the form $(\mathbf{x}, f(\mathbf{x}))$, where $\mathbf{x} \in C$. If f is a differentiable function, then given a point $\mathbf{x}_0$ in the interior of C, the graph of f has a tangent plane at $(\mathbf{x}_0, f(\mathbf{x}_0))$ whose normal vector is $(\nabla f(\mathbf{x}_0), -1)$.

If C is convex and f is convex, f need not be differentiable at an interior point of C, as the function f defined by $f(x) = |x|$ shows. In this section we shall introduce a generalization of the gradient vector, called the subgradient, that exists at points $(\mathbf{x}, f(\mathbf{x}))$ of the graph of a convex function f, where $\mathbf{x}$ is an interior point of C. We shall show that if f is a convex function defined on C and $\mathbf{x}$ is an interior point of C, then epi(f) has a supporting hyperplane at the point $(\mathbf{x}, f(\mathbf{x}))$. The supporting hyperplane need not be unique. In the next section we shall show that f is differentiable at $\mathbf{x}$ if and only if f has a unique subgradient at $\mathbf{x}$. This subgradient turns out to be $\nabla f(\mathbf{x})$. Thus, if f is differentiable at $\mathbf{x}$, there is a unique supporting hyperplane to epi(f) at $(\mathbf{x}, f(\mathbf{x}))$. This hyperplane is the tangent plane to the graph at $(\mathbf{x}, f(\mathbf{x}))$. If $(\boldsymbol{\xi}, -1)$ is a normal vector to a supporting hyperplane, then $\boldsymbol{\xi}$ has some of the properties that the gradient vector would have.

We now define the notion of subgradient.

Definition 2.1. A function f defined on a set C is said to have a *subgradient* at a point $\mathbf{x}_0$ in C if there exists a vector $\boldsymbol{\xi}$ in $\mathbb{R}^n$ such that

$$f(\mathbf{x}) \geqslant f(\mathbf{x}_0) + \langle \boldsymbol{\xi}, \mathbf{x} - \mathbf{x}_0 \rangle$$

for all $\mathbf{x} \in C$. The vector $\boldsymbol{\xi}$ is called a subgradient.

Geometrically, $\boldsymbol{\xi}$ is a subgradient of f at $\mathbf{x}_0$ if the graph of f in $\mathbb{R}^{n+1}$ lies on or above the graph of the hyperplane $y = f(\mathbf{x}_0) + \langle \boldsymbol{\xi}, \mathbf{x} - \mathbf{x}_0 \rangle$. Since $(\mathbf{x}_0, f(\mathbf{x}_0))$ is in this hyperplane, this hyperplane will be a supporting hyperplane to epi(f) at $(\mathbf{x}_0, f(\mathbf{x}_0))$. Thus, the existence of a subgradient is a statement about the existence of a nonvertical supporting hyperplane to epi(f). In Figure 3.5 we illustrate the definition with the function $f(x) = |x|$ at $x = 0$.

Any line through the origin with slope ξ satisfying $-1 \leqslant \xi \leqslant 1$ has the property that

$$|x| \geqslant \xi x = |0| + \xi(x - 0)$$

for all x. Thus any ξ in the interval $[-1, 1]$ is a subgradient of $|x|$ at $x = 0$. If $x_0 > 0$, then $\xi = 1$ is the only subgradient of $|x|$ at x_0; if $x_0 < 0$, then $\xi = -1$ is the only subgradient of $|x|$ at x_0.

Definition 2.2. The set of subgradients of a function f at a point $\mathbf{x}_0$ is called the *subdifferential* of f at $\mathbf{x}_0$ and is denoted by $\partial f(\mathbf{x}_0)$.

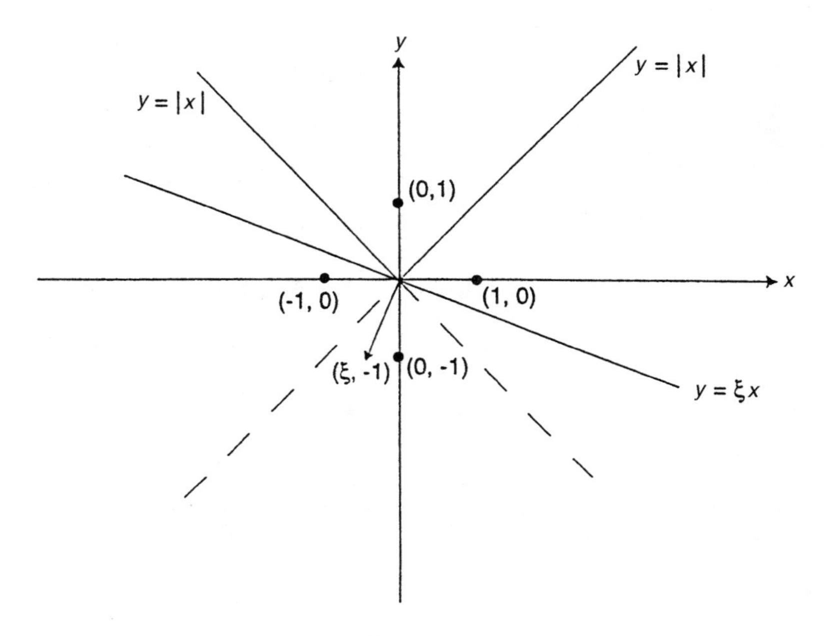

Figure 3.5.

The subdifferential of $|x|$ at $x = 0$ is the interval $[-1, 1]$. The subdifferential of $|x|$ at $x_0 > 0$ is the singleton $\{1\}$, and the subdifferential at $x_0 < 0$ is $\{-1\}$.

Example 2.1. The subdifferential will be the empty set if no subgradient exists. This may happen, as the following example shows. Let $C = [-1, 1]$ and let $f(x) = -\sqrt{1 - x^2}$. (Sketch the graph.) The graph of f has a nonvertical tangent line at any point x_0 in $(-1, 1)$. From the graph it appears that for such x_0 the graph lies above the tangent line, and so $\partial f(x_0) = \{f'(x_0)\}$. Although this can be verified in this example by relatively straightforward calculations, we shall not do so. The validity of the statement will follow from general theorems that we will prove. At the points $x = \pm 1$, the graph has a vertical tangent line, but the tangent line does not lie below the graph. Thus it appears that f does not have a subgradient at these points. We now show this analytically.

A number ξ is a subgradient of f at $x = -1$ if and only if

$$-\sqrt{1 - x^2} \geqslant \xi(x + 1) \quad \text{for all } x \text{ in } [-1, 1].$$

For x in $(-1, 1]$ this inequality is equivalent to

$$-\left(\frac{1 - x}{1 + x}\right)^{1/2} \geqslant \xi \quad x \text{ in } (-1, 1].$$

If we let x in $(-1, 1]$ tend to -1, the left-hand side of this inequality tends to $-\infty$. Hence there is no $\xi \in \mathbb{R}$ such that the subgradient inequality holds for all x in $[-1, 1]$.

A similar argument shows that $\partial f(1) = \varnothing$.

THEOREM 2.1. *Let f be a convex function defined on a convex set C. Then at each point $\mathbf{x}_0$ in $\mathrm{int}(C)$ there exists a vector $\boldsymbol{\xi}$ in $\mathbb{R}^n$ such that*

$$f(\mathbf{x}) \geqslant f(\mathbf{x}_0) + \langle \boldsymbol{\xi}, \mathbf{x} - \mathbf{x}_0 \rangle \tag{1}$$

for all $\mathbf{x}$ in C. In other words, $\partial f(\mathbf{x}_0) \neq \varnothing$ at every interior point $\mathbf{x}_0$ of C.

Example 2.1 shows that the requirement $\mathbf{x}_0 \in \mathrm{int}(C)$ is essential.

Proof. We shall show that the conclusion of the theorem is a consequence of the existence of a supporting hyperplane to $\mathrm{epi}(f)$ at $(\mathbf{x}_0, f(\mathbf{x}_0))$.

The point $(\mathbf{x}_0, f(\mathbf{x}_0))$ is a boundary point of $\mathrm{epi}(f)$. Since $\mathrm{epi}(f)$ is convex, we may apply Theorem II.4.1 with $C = \mathrm{epi}(f)$ and get that there exists a vector $(\mathbf{a}, \beta) \neq (\mathbf{0}, 0)$ in $\mathbb{R}^{n+1}$ such that

$$\langle (\mathbf{a}, \beta), (\mathbf{x}, y) \rangle \leqslant \langle (\mathbf{a}, \beta), (\mathbf{x}_0, f(\mathbf{x}_0)) \rangle$$

for all $(\mathbf{x}, y)$ in $\mathrm{epi}(f)$. This relation is equivalent to

$$\langle \mathbf{a}, \mathbf{x} \rangle + \beta y \leqslant \langle \mathbf{a}, \mathbf{x}_0 \rangle + \beta f(\mathbf{x}_0) \tag{2}$$

for all $(\mathbf{x}, y)$ in $\mathrm{epi}(f)$, that is, for all $\mathbf{x}$ in C and $y \geqslant f(\mathbf{x})$.

Since the right-hand side of (2) is a constant, if β were strictly positive, we could obtain a contradiction by letting y become arbitrarily large. Hence $\beta \leqslant 0$.

If $\beta = 0$, then $\mathbf{a} \neq \mathbf{0}$ and (2) becomes

$$\langle \mathbf{a}, \mathbf{x} \rangle \leqslant \langle \mathbf{a}, \mathbf{x}_0 \rangle \quad \text{all } \mathbf{x} \in C. \tag{3}$$

Since $\mathbf{x}_0 \in \mathrm{int}(C)$, there exists an $\varepsilon > 0$ such that $\mathbf{x}_1 = \mathbf{x}_0 + \varepsilon \mathbf{a}$ belongs to C. Setting $\mathbf{x} = \mathbf{x}_1$ in (3) gives $\varepsilon \langle \mathbf{a}, \mathbf{a} \rangle \leqslant 0$, which is a contradiction. Thus, $\beta < 0$.

In (2) we now divide through by $\beta < 0$ and set $y = f(\mathbf{x})$ to get

$$f(\mathbf{x}) \geqslant f(\mathbf{x}_0) + \left\langle \frac{\mathbf{a}}{\beta}, \mathbf{x}_0 - \mathbf{x} \right\rangle$$

for all $\mathbf{x} \in C$. Now set $\boldsymbol{\xi} = -\mathbf{a}/\beta$ to get (1).

COROLLARY 1. *If f is strictly convex, then strict inequality holds in* (1).

We leave the proof as an exercise.

The converse of Theorem 2.1 is false. To see this, let C be the square in $\mathbb{R}^2$, $\{\mathbf{x}: 0 \leqslant x_1 \leqslant 1,\ 0 \leqslant x_2 \leqslant 1\}$. Let $f(\mathbf{x}) = 0$ for $\mathbf{x}$ such that $0 \leqslant x_1 \leqslant 1$ and $0 < x_2 \leqslant 1$. Let $f(\mathbf{x}) = \frac{1}{2} - |x_1 - \frac{1}{2}|$ for $\mathbf{x}$ such that $0 \leqslant x_1 \leqslant 1$ and $x_2 = 0$. The function f has a unique subgradient $\boldsymbol{\xi} = \mathbf{0}$ at every point in int(C), yet f is not convex on C. The function f, however, is convex on int(C), and this phenomenon illustrates the general state of affairs.

THEOREM 2.2. *Let C be a convex set in $\mathbb{R}^n$. If at each point $\mathbf{x}_0$ in C f has a subgradient, then f is convex on C.*

Remark 2.1. If we merely assume that at each point in int(C) the subdifferential is not empty, then we can only conclude that f is convex on int(C).

Proof. Let $\mathbf{x}$ and $\mathbf{y}$ be two points in C and let $\mathbf{z} = \alpha\mathbf{x} + \beta\mathbf{y}$, $\alpha > 0$, $\beta > 0$, $\alpha + \beta = 1$. Since f has a subgradient at $\mathbf{z}$, say $\boldsymbol{\xi}$,

$$f(\mathbf{x}) \geqslant f(\mathbf{z}) + \langle \boldsymbol{\xi}, \mathbf{x} - \mathbf{z} \rangle = f(\mathbf{z}) + \beta \langle \boldsymbol{\xi}, \mathbf{x} - \mathbf{y} \rangle,$$

$$f(\mathbf{y}) \geqslant f(\mathbf{z}) + \langle \boldsymbol{\xi}, \mathbf{y} - \mathbf{z} \rangle = f(\mathbf{z}) - \alpha \langle \boldsymbol{\xi}, \mathbf{x} - \mathbf{y} \rangle.$$

If we now multiply the first inequality by $\alpha > 0$, the second inequality by $\beta > 0$, and add the results, we get

$$\alpha f(\mathbf{x}) + \beta f(\mathbf{y}) \geqslant f(\mathbf{z}) = f(\alpha\mathbf{x} + \beta\mathbf{y}).$$

Hence f is convex.

Remark 2.2. The proof of the theorem shows that if at each $\mathbf{x}_0$ in C there exists a subgradient $\boldsymbol{\xi}$ such that $f(\mathbf{x}) > f(\mathbf{x}_0) + \langle \boldsymbol{\xi}, \mathbf{x} - \mathbf{x}_0 \rangle$, for $\mathbf{x} \neq \mathbf{x}_0$, $\mathbf{x}$ in C, then f is strictly convex on C.

Exercise 2.1. Show that if $\partial f(\mathbf{x}_0)$ is not empty, then $\partial f(\mathbf{x}_0)$ is closed and convex.

Exercise 2.2. Prove Corollary 1 to Theorem 2.1.

Exercise 2.3. Let f be a convex function defined on a convex set C and let $\mathbf{x}_0 \in \operatorname{int}(C)$. Show that $\boldsymbol{\xi} \in \partial f(\mathbf{x}_0)$ if and only if $f'(\mathbf{x}_0, \mathbf{v}) \geqslant \langle \boldsymbol{\xi}, \mathbf{v} \rangle$ for all $\mathbf{v}$ in $\mathbb{R}^n$.

Exercise 2.4. Let f be convex and concave on $\mathbb{R}^n$. Show that f is affine, that is, $f(\mathbf{x}) = \langle \mathbf{a}, \mathbf{x} \rangle + b$.

Exercise 2.5. Let $f(\mathbf{x}) = \|\mathbf{x}\|$. Show that $\partial f(\mathbf{0}) = \{\boldsymbol{\xi}: \|\boldsymbol{\xi}\| \leqslant 1\}$. F (From Theorem 3.1 we will get that, for $\mathbf{x} \neq \mathbf{0}$, $\partial f(\mathbf{x}) = \{\mathbf{x}/\|\mathbf{x}\|\}$).

Exercise 2.6. Let $f(x_1, x_2) = |x_1 + x_2|$. Show that f is convex and calculate $\partial f((0, 0))$.

3. DIFFERENTIABLE CONVEX FUNCTIONS

We first show that if f is a convex function and differentiable [in the sense of relation (5) in Section 11 of Chapter I] at an interior point of its domain, then there is a unique subgradient at the point and the subgradient is the gradient vector, and conversely, if f has a unique subgradient at $\mathbf{x}$, then f is differentiable at $\mathbf{x}$.

THEOREM 3.1. *Let f be a convex function defined on a convex set C in $\mathbb{R}^n$, and let $\mathbf{x}$ be an interior point of C. Then f is differentiable at $\mathbf{x}$ if and only if f has a unique subgradient at $\mathbf{x}$. Moreover, the unique element in the subgradient is $\nabla f(\mathbf{x})$.*

Proof. Let f be differentiable at $\mathbf{x}$. Since f is convex and $\mathbf{x} \in \text{int}(C)$, by Theorem 2.1 the subgradient $\partial f(\mathbf{x})$ is not empty. Let $\xi \in \partial f(\mathbf{x})$. Then for any $\mathbf{h}$ in $\mathbb{R}^n$ and sufficiently small $t > 0$

$$f(\mathbf{x} + t\mathbf{h}) \geqslant f(\mathbf{x}) + \langle \xi, t\mathbf{h} \rangle.$$

Since f is differentiable at $\mathbf{x}$, we have that for $\mathbf{h} \neq \mathbf{0}$

$$f(\mathbf{x} + t\mathbf{h}) = f(\mathbf{x}) + \langle \nabla f(\mathbf{x}), t\mathbf{h} \rangle + \eta(t\mathbf{h}),$$

where $\eta(t\mathbf{h})/t\|\mathbf{h}\| \to 0$ as $t \to 0$. If we subtract the second relation from the first, we get

$$0 \geqslant t[\langle \xi, \mathbf{h} \rangle - \langle \nabla f(\mathbf{x}, h \rangle - \eta(t\mathbf{h})\|\mathbf{h}\|/t\|\mathbf{h}\|].$$

If we now divide by $t > 0$ and then let $t \to 0$, we get that $\langle \xi - \nabla f(\mathbf{x}), \mathbf{h} \rangle \leqslant 0$ for all $\mathbf{h} \neq \mathbf{0}$ in $\mathbb{R}^n$. In particular, if $\xi - \nabla f(\mathbf{x}) \neq \mathbf{0}$, then this inequality holds for $\mathbf{h} = \xi - \nabla f(\mathbf{x})$, whence $\|\xi - \nabla f(\mathbf{x})\|^2 \leqslant 0$. Thus, $\xi = \nabla f(\mathbf{x})$.

We now suppose that the subgradient $\partial f(\mathbf{x})$ consists of a unique element ξ. We will show that all the first-order partial derivatives of f exist at $\mathbf{x}$. The differentiability of f at $\mathbf{x}$ will then follow from Theorem 1.4.

To simplify the notation, we assume tha $\mathbf{x} = \mathbf{0}$ and that $f(\mathbf{x}) = 0$. There is no loss of generality in making this assumption, since this amounts to a translation of the origin in $\mathbb{R}^n \times \mathbb{R}$ to $(\mathbf{x}, f(\mathbf{x}))$. Let us now suppose that at least one partial derivative of f at $\mathbf{0}$ does not exist. For definiteness let us assume that it is the partial with respect to x_1. Let

$$g(t) = f(\mathbf{0} + t\mathbf{e}_1),$$

where $\mathbf{e}_1$ is the first standard basis vector $(1, 0, \ldots, 0)$. Since f is convex and $\mathbf{0} \in \text{int}(C)$, the function g is defined and convex on some maximal interval I containing $\mathbf{0}$ in its interior. Moreover, $\partial f/\partial x_1$ does not exist at $\mathbf{0}$ if and only if $g'(0)$ does not exist. Since g is convex, it follows from Corollary 2 to Theorem 1.2 that $g'_-(0)$ and $g'_+(0)$ exist and that $g'_-(0) \leqslant g'_+(0)$. Therefore, if $g'(0)$ does not exist, then $g'_-(0) < g'_+(0)$.

Let ξ_1 be any number satisfying

$$g'_-(0) < \xi_1 < g'_+(0). \tag{1}$$

We shall show that corresponding to each such ξ_1 we can find a $\boldsymbol{\xi}$ in $\partial f(\mathbf{0})$, with distinct ξ_1 determining distinct $\boldsymbol{\xi}$. This will contradict the hypothesis that $\partial f(\mathbf{0})$ consists of a unique element, and the proof will be accomplished.

It follows from (1) and Corollary 1.1 that for $h > 0$

$$\frac{g(-h) - g(0)}{-h} \leqslant g'_-(0) < \xi_1 < g'_+(0) \leqslant \frac{g(h) - g(0)}{h}.$$

Since $g(0) = f(\mathbf{0}) = 0$, we get from the preceding inequalities that for all h in I_1

$$f(h\mathbf{e}_1) = g(h) > \xi_1 h, \tag{2}$$

with equality holding only if $h = 0$. Let V_1 denote the one-dimensional subspace spanned by $\mathbf{e}_1$. Thus $V_1 = \{\mathbf{x} : \mathbf{x} = t\mathbf{e}_1, t \in \mathbb{R}\}$. Define a linear functional L_1 on V_1 by the formula

$$L_1(\mathbf{x}) = L_1(t\mathbf{e}_1) = \xi_1 t. \tag{3}$$

Since the representation $\mathbf{x} = t\mathbf{e}_1$ is unique for $\mathbf{x}$ in V_1, the linear functional L_1 is well defined. Note that L_1 depends on ξ_1. Combining (2) and (3), we get that for $\mathbf{x}$ in $V_1 \cap C$

$$f(\mathbf{x}) \geqslant L_1(\mathbf{x}) = \xi_1 t. \tag{4}$$

If $n = 1$, we are through. Otherwise, let V_2 denote the two-dimensional subspace spanned by $\mathbf{e}_1$ and $\mathbf{e}_2$. Thus,

$$V_2 = \{\mathbf{y} : \mathbf{y} = t\mathbf{e}_1 + u\mathbf{e}_2,\ t,\ u \in \mathbb{R}\} = \{\mathbf{y} : \mathbf{y} = \mathbf{x} + u\mathbf{e}_2,\ \mathbf{x} \in V_1,\ u \in \mathbb{R}\}.$$

We shall extend L_1 to a linear functional L_2 defined on V_2 and shall find a number ξ_2 such that for all $\mathbf{y}$ in $V_2 \cap C$

$$f(\mathbf{y}) \geqslant L_2(\mathbf{y}) = \xi_1 t + \xi_2 u. \tag{5}$$

Let $\mathbf{x}$ and $\mathbf{z}$ belong to $V_1 \cap C$ and let $\alpha > 0$, $\beta > 0$. It then follows from the linearity of L_1, from the convexity of $V_1 \cap C$, from (4), and from the convexity of f that

$$\begin{aligned}
\frac{\alpha}{\alpha+\beta} L_1(\mathbf{x}) + \frac{\beta}{\alpha+\beta} L_1(\mathbf{z}) &= L_1\left(\frac{\alpha}{\alpha+\beta}\mathbf{x} + \frac{\beta}{\alpha+\beta}\mathbf{z}\right) \\
&\underset{(4)}{\leqslant} f\left(\frac{\alpha}{\alpha+\beta}\mathbf{x} + \frac{\beta}{\alpha+\beta}\mathbf{z}\right) \\
&= f\left(\frac{\alpha}{\alpha+\beta}(\mathbf{x} - \beta\mathbf{e}_2) + \frac{\beta}{\alpha+\beta}(\mathbf{z} + \alpha\mathbf{e}_2)\right) \\
&\underset{f\text{ convex}}{\leqslant} \frac{\alpha}{(\alpha+\beta)} f(\mathbf{x} - \beta\mathbf{e}_2) + \frac{\beta}{\alpha+\beta} f(\mathbf{z} + \alpha\mathbf{e}_2).
\end{aligned}$$

Multiplying through by $\alpha + \beta$ gives

$$\alpha L_1(\mathbf{x}) + \beta L_1(\mathbf{z}) \leqslant \alpha f(\mathbf{x} - \beta\mathbf{e}_2) + \beta f(\mathbf{z} + \alpha\mathbf{e}_2),$$

and so

$$\frac{L_1(\mathbf{x}) - f(\mathbf{x} - \beta\mathbf{e}_2)}{\beta} \leqslant \frac{f(\mathbf{z} + \alpha\mathbf{e}_2) - L_1(\mathbf{z})}{\alpha}. \tag{6}$$

Denote the left-hand side of (6) by $\lambda(\mathbf{x}, \beta)$ and denote the right-hand side of (6) by $\rho(\mathbf{z}, \alpha)$. Hence

$$\sup\{\lambda(\mathbf{x}, \beta) : \mathbf{x} \in V_1 \cap C, \beta > 0\} \leqslant \inf\{\rho(\mathbf{z}, \alpha) : \mathbf{z} \in V_1 \cap C, \alpha > 0\}$$

and both the supremum and infimum are finite. Hence there exists a number ξ_2 such that

$$\frac{L_1(\mathbf{x}) - f(\mathbf{x} - \beta\mathbf{e}_2)}{\beta} \leqslant \xi_2 \leqslant \frac{f(\mathbf{x} + \alpha\mathbf{e}_2) - L_1(\mathbf{x})}{\alpha} \tag{7}$$

for all $\mathbf{x} \in V_1 \cap C$, $\alpha > 0$, $\beta > 0$.

Now let $\mathbf{y}$ be an element in $V_2 \cap C$ that is not in V_1. Then $\mathbf{y} = \mathbf{x} + u\mathbf{e}_2$ with $\mathbf{x} = t\mathbf{e}_1$ and $u \neq 0$. If $u > 0$, take $\alpha = u$ in (7); if $u < 0$, take $\beta = -u$ in (7). Then from (7) and (3) we get that for all $\mathbf{y}$ in $V_2 \cap C$

$$f(\mathbf{y}) = f(\mathbf{x} + u\mathbf{e}_2) \geqslant L_1(\mathbf{x}) + \xi_2 u = \xi_1 t + \xi_2 u,$$

which is (5).

Proceeding inductively, we get that for each ξ_1 satisfying (1) there exists a vector $\boldsymbol{\xi} = (\xi_1, \xi_2, \ldots, \xi_n)$ such that if $\mathbf{x} = (x_1, \ldots, x_n)$ is in C, then

$$f(\mathbf{x}) \geqslant \langle \boldsymbol{\xi}, \mathbf{x} \rangle$$

Since $f(\mathbf{0}) = 0$, this says that $\boldsymbol{\xi} \in \partial f(\mathbf{0})$.

The next theorem is an immediate consequence of Theorems 2.1, 2.2, and 3.1.

THEOREM 3.2. *Let C be an open convex set in $\mathbb{R}^n$ and let f be real valued and differentiable on C. Then f is convex if and only if for each $\mathbf{x}_0$ in C*

$$f(\mathbf{x}) \geqslant f(\mathbf{x}_0) + \langle \nabla f(\mathbf{x}_0), \mathbf{x} - \mathbf{x}_0 \rangle$$

for all $\mathbf{x}$ in C. Also, f is strictly convex if and only if for each $\mathbf{x}_0$ in C and all $\mathbf{x} \neq \mathbf{x}_0$ in C

$$f(\mathbf{x}) > f(\mathbf{x}_0) + \langle \nabla f(\mathbf{x}_0), \mathbf{x} - \mathbf{x}_0 \rangle.$$

Remark 3.1. If $n = 2$ and f is differentiable at $\mathbf{x}_0$, then the surface in $\mathbb{R}^3$ defined by $y = f(\mathbf{x})$ has a tangent plane at $\mathbf{x}_0$ whose equation can be written as

$$y - f(\mathbf{x}_0) = \langle \nabla f(\mathbf{x}_0), \mathbf{x} - \mathbf{x}_0 \rangle.$$

Thus, the geometric interpretation of Theorem 3.2 is that a function f that is differentiable on C is convex on C if and only if for each point $\mathbf{x}_0$ in C the surface defined by $y = f(\mathbf{x})$ lies above the tangent plane to the surface at $\mathbf{x}_0$. This is not surprising since we obtained the result in Theorem 3.2 by considering the supporting hyperplanes to epi(f), and for differentiable f, these are the tangent planes to the surface.

Let f be a real-valued function defined and of class $C^{(2)}$ on an open set D in $\mathbb{R}^n$. (See Section 11 in Chapter I for the definition of class $C^{(2)}$.) For such functions we have a computationally feasible criterion for convexity. In the ensuing discussion recall that, by Theorem I.11.3, f is differentiable at each point $\mathbf{x}_0$ in D.

We first develop our criterion if D is an interval $I \subseteq \mathbb{R}$.

LEMMA 3.1. *Let f be of class $C^{(2)}$ on an open interval $I \subseteq \mathbb{R}$. Then f is convex on I if and only if $f''(x) \geqslant 0$ for all x in I. If $f''(x) > 0$ or all x in I, then f is strictly convex.*

Note that if f is strictly convex, then we cannot conclude that $f''(x) > 0$ for all x, as the example $f(x) = x^4$ shows. The function f is strictly convex, yet $f''(0) = 0$.

Proof. If f is convex on I, then by Corollary 3 to Theorem 1.2, $f''(x) \geqslant 0$ on I. Conversely, suppose $f''(x) \geqslant 0$ on I. Let x_0 be a fixed point in I and let x be any other point in I. Then by Taylor's theorem

$$f(x) - f(x_0) - f'(x_0)(x - x_0) = \tfrac{1}{2}f''(x_*)(x - x_0)^2 \geqslant 0,$$

where x_* is a point between x_0 and x. It now follows from Theorem 3.2 that f is convex on I. If $f''(x) > 0$ on I it follows from Theorem 3.2 that f is strictly convex.

Recall that a symmetric $n \times n$ matrix A with entries (a_{ij}) is said to be *positive semidefinite* if the quadratic form

$$\mathbf{x}^t A\mathbf{x} = \langle \mathbf{x}, A\mathbf{x} \rangle = \sum_{i,j=1}^{n} a_{ij}x_{ij}$$

is nonnegative for all $\mathbf{x}$ in $\mathbb{R}^n$. The matrix A is said to be *positive definite* if $\langle \mathbf{x}, A\mathbf{x} \rangle$ is positive for all $\mathbf{x} \neq \mathbf{0}$ in $\mathbb{R}^n$.

THEOREM 3.3. *Let f be of class $C^{(2)}$ on an open convex set D. Then f is convex on D if and only if the Hessian matrix*

$$H(\mathbf{x}) = \left(\frac{\partial^2 f}{\partial x_j \, \partial x_i}(\mathbf{x})\right)$$

is positive semidefinite at each point $\mathbf{x}$ in D. If $H(\mathbf{x})$ is positive definite at each $\mathbf{x}$, then f is strictly convex.

Proof. Let $\mathbf{x}$ be a point in D. Since D is open, for each vector $\mathbf{v}$ in $\mathbb{R}^n$, $\mathbf{v} \neq \mathbf{0}$, there exists an $\varepsilon > 0$ such that the line segment $(\mathbf{x} - \varepsilon\mathbf{v}, \mathbf{x} + \varepsilon\mathbf{v})$ is in D. Hence the function $\varphi(\ ; \mathbf{x}, \mathbf{v})$ in relation (3) in Section 1 is defined on $(-\varepsilon, \varepsilon)$. Also, since f is of class $C^{(2)}$ on D, $\partial^2 f(\mathbf{x})/\partial x_j \, \partial x_i = \partial^2 f(\mathbf{x})/\partial x_i \, \partial x_j$ at each $\mathbf{x}$ in D, and so $H(\mathbf{x})$ is symmetric.

If f is convex, then by Lemma 1.5, the function $\varphi(\ ; \mathbf{x}, \mathbf{v})$ is convex on the interval $(-\varepsilon, \varepsilon)$. By Lemma 3.1, $\varphi''(0; \mathbf{x}, \mathbf{v}) \geqslant 0$. If $\mathbf{v} = (v_1, \ldots, v_n)$, then straightforward calculations as in relations (12) and (13) in Section 11 of Chapter I give

$$\varphi'(t; \mathbf{x}, \mathbf{v}) = \frac{d}{dt} f(\mathbf{x} + t\mathbf{v}) = \sum_{i=1}^{n} \frac{\partial f}{\partial x_i}(\mathbf{x} + t\mathbf{v})v_i,$$

$$\varphi''(t; \mathbf{x}, \mathbf{v}) = \sum_{i=1}^{n} \sum_{j=1}^{n} \frac{\partial^2 f}{\partial x_j \, \partial x_i} f(\mathbf{x} + t\mathbf{v})v_i v_j = \langle \mathbf{v}, H(\mathbf{x} + t\mathbf{v})\mathbf{v} \rangle. \tag{8}$$

Setting $t = 0$ gives

$$\langle \mathbf{v}, H(\mathbf{x})\mathbf{v} \rangle = \varphi''(0; \mathbf{x}, \mathbf{v}) \geqslant 0. \tag{9}$$

Since (9) holds for all $\mathbf{v}$, $H(\mathbf{x})$ is positive semidefinite.

Conversely, suppose now that $H(\mathbf{x})$ is positive semidefinite for each $\mathbf{x}$ in D. Then from (8), for each $\mathbf{x}$ in D and for each $\mathbf{v}$ in $\mathbb{R}^n$, $\varphi''(t; \mathbf{x}, \mathbf{v}) \geqslant 0$ for all t in $(-\varepsilon, \varepsilon)$. Hence by Lemma 3.1, $\varphi(\ ; \mathbf{x}, \mathbf{v})$ is convex. Since $\mathbf{x}$ and $\mathbf{v}$ are arbitrary, it follows from Lemma 1.5 that f is convex on D. The argument also shows that if $H(\mathbf{x})$ is positive definite for all $\mathbf{x}$ in D, then f is strictly convex.

We assume that the reader is familiar with the following facts. Given a symmetric matrix A, there exists an orthogonal matrix P such that $P^tAP = D$, where D is a diagonal matrix whose diagonal elements are the eigenvalues of A. Thus A is positive definite if and only if all the diagonal entries of D (eigenvalues of A) are positive and A is positive semidefinite if and only if they are nonnegative. For a survey of numerical methods for obtaining the eigenvalues of a symmetric matrix see Golub and Van Loan [1996].

Let A be an $n \times n$ matrix with entries a_{ij}. Let

$$\Delta_1 = a_{11}, \qquad \Delta_k = \det\begin{pmatrix} a_{11} & \cdots & a_{1k} \\ \vdots & \ddots & \vdots \\ a_{k1} & \cdots & a_{kk} \end{pmatrix}, \qquad k = 2, 3, 4, \ldots, n.$$

The determinants Δ_k are called the principal minors of A. Another criterion for positive definiteness of a symmetric matrix A with entries a_{ij} is the following.

LEMMA 3.2. *The symmetric matrix A is positive definite if and only if $\Delta_k > 0$ for $k = 1, \ldots, n$ and is negative definite if and only if $(-1)^k\Delta_k > 0$ for $k = 1, \ldots, n$.*

We shall prove this lemma in Exercise IV.2.10.

If we assume that $\Delta_k \geqslant 0$ for all $k = 1, 2, \ldots, n$ and not all the $\Delta_k = 0$, then there is difference between $n \geqslant 3$ and $n = 2$.

For $n \geqslant 3$, if $\Delta_k \geqslant 0$, $k = 1, \ldots, n$, and not all $\Delta_k = 0$, then A need not be positive semidefinite. To see this, consider the matrix

$$A = \begin{pmatrix} 1 & 0 & \cdots & 0 \\ 0 & 0 & \cdots & 0 \\ \vdots & \vdots & \ddots & \vdots \\ 0 & 0 & \cdots & -1 \end{pmatrix}.$$

Then $\langle \mathbf{x}, A\mathbf{x} \rangle = x_1^2 - x_n^2$, which is not nonnegative for all $\mathbf{x}$.

For $n = 2$, however, if $\Delta_1 \geqslant 0$, $\Delta_2 \geqslant 0$, and not both Δ_1 and Δ_2 are zero, then the symmetric matrix A is positive semidefinite. To see this, let

$$A = \begin{pmatrix} a_{11} & a \\ a & a_{22} \end{pmatrix}.$$

If $\Delta_1 = 0$, then $\Delta_2 = -a^2$. Then to have $\Delta_2 \geqslant 0$, we must have $a = 0$. But then $\Delta_2 = 0$. Hence we must have $a_{11} > 0$. Thus $\Delta_1 > 0$. Since $\Delta_2 \geqslant 0$,

$$a_{11}a_{22} \geqslant a^2. \tag{10}$$

Since $a_{11} > 0$, it follows that $a_{22} \geqslant 0$. For $\mathbf{x} = (x_1, x_2)$

$$\langle \mathbf{x}, A\mathbf{x} \rangle = a_{11}x_1^2 + 2ax_1x_2 + a_{22}x_2^2.$$

If $\mathbf{x} = (x_1, 0)$ with $x_1 \neq 0$, we get

$$\langle \mathbf{x}, A\mathbf{x} \rangle = a_{11}x_1^2 > 0.$$

If $\mathbf{x} = (x_1, x_2)$ with $x_2 \neq 0$, we have $x_1 = tx_2$ for some $-\infty < t < \infty$ and

$$\langle \mathbf{x}, A\mathbf{x} \rangle = [a_{11}t^2 + 2at + a_{22}]x_2^2.$$

Denote the term in square brackets by $q(t)$. From (10) we get that the discriminant of q, which equals $4a^2 - 4a_{11}a_{22}$, is nonpositive. Thus the quadratic q either has a double root or has no real roots. Since $a_{11} > 0$, it follows that $q(t) \geqslant 0$ for all t. Hence, $\langle \mathbf{x}, A\mathbf{x} \rangle \geqslant 0$ for all $\mathbf{x}$.

Exercise 3.1. Determine whether or not the following functions are convex on $\mathbb{R}^3$:

(i) $f(x, y, z) = x^2 + 2xy + 4xz + 3y^2 + yz + 7z^2$,
(ii) $f(x, y, z) = x^2 + 4xy + 4y^2 + 2xz + 4yz$.

Exercise 3.2. Determine the convex set in $\mathbb{R}^2$ on which the function

$$f(x, y) = x^2 - 2xy + \tfrac{1}{3}y^3 - 3y$$

is (i) convex and (ii) strictly convex.

Exercise 3.3. For what values of r is the function $f(x) = x^r - r \ln x$ convex on $(0, \infty)$.

Exercise 3.4. Show that $f(\mathbf{x}) = x_1^{r_1} + \cdots + x_n^{r_n}$, $r_i \geqslant 1$, is convex on $\{\mathbf{x} : \mathbf{x} \geqslant 0\}$.

Exercise 3.5. Show that the n-dimensional ellipsoid

$$\left\{\mathbf{x}: \sum_{i=1}^{n}\left(\frac{x_i}{a_i}\right)^2 \leqslant 1\right\}$$

is convex. *Hint:* There is a very easy solution.

Exercise 3.6. For what values of p is $f(\mathbf{x}) = \|\mathbf{x}\|^p$ convex on $\mathbb{R}^n$. *Hint:* Use Lemma 1.4.

Exercise 3.7. Show that if $c_i > 0$, $i = 1, \ldots, 4$, the function

$$f(\mathbf{x}) = \exp(c_1 x_1^2 + c_2 x_1^{-1} + c_3 x_2^2 + c_4 x_2^{-1})$$

is convex on $S = \{(x_1, x_2) : x_1 > 0,\ x_2 > 0\}$.

Exercise 3.8. Show that the function $f(\mathbf{x}) = (1 + \|\mathbf{x}\|^2)^{p/2}$, $p \geqslant 1$, is convex on $\mathbb{R}^n$.

Exercise 3.9. Show that the function $f(\mathbf{x}) = -(x_1 + x_2 + \cdots + x_n)^{1/n}$, $n \geqslant 1$, is convex on $S = \{\mathbf{x} : \mathbf{x} \geqslant \mathbf{0}\}$.

Exercise 3.10. A function f that is positive on a convex set C in $\mathbb{R}^n$ is said to be *logarithmically convex* if the function $g(\mathbf{x}) = \ln f(\mathbf{x})$ is convex on C:

(i) Show that if f is logarithmically convex, then f is convex.
(ii) Is it true that every positive convex function is logarithmically convex?

Exercise 3.11. Let f be differentiable on an open convex set X. Show that if f is strictly convex, then $\nabla f(\mathbf{x}) \neq \nabla f(\mathbf{y})$ for all distinct pairs of points $\mathbf{x}$, $\mathbf{y}$ in X.

Exercise 3.12. Let C be an open convex set and let f be real valued and differentiable on C. Prove that f is convex on C if and only if for each pair of points $\mathbf{x}_0$ and $\mathbf{x}_1$ in C

$$\langle \nabla f(\mathbf{x}_1) - \nabla f(\mathbf{x}_0), \mathbf{x}_1 - \mathbf{x}_0 \rangle \geqslant 0.$$

Show that f is strictly convex if and only if strict inequality holds.

4. ALTERNATIVE THEOREMS FOR CONVEX FUNCTIONS

The theorems of this section are alternative theorems for systems of inequalities involving convex and affine functions. They find application in optimization problems involving such functions.

Definition 4.1. A vector-valued function $\mathbf{f} = (f_1, \ldots, f_m)$ with domain a convex set C in $\mathbb{R}^n$ and range in $\mathbb{R}^m$ is said to be convex if each of the real-valued component functions f_i, $i = 1, \ldots, m$, is convex.

Affine functions or affine transformations were defined in Definition II.6.6. It was shown there that an affine function $\mathbf{h}$ with domain $\mathbb{R}^n$ and range $\mathbb{R}^m$ has the form

$$\mathbf{h}(\mathbf{x}) = A\mathbf{x} + \mathbf{b},$$

where A is an $m \times n$ matrix and $\mathbf{b}$ is a vector in $\mathbb{R}^m$.

Note that $\mathbf{h} = (h_1, \ldots, h_m)$, where $h_i = \langle \mathbf{a}_i, \mathbf{x} \rangle + b_i$, and $\mathbf{a}_i$ is the ith row of A.

THEOREM 4.1. *Let C be a convex set in $\mathbb{R}^n$ and let $\mathbf{f} = (f_1, \ldots, f_m)$ be defined and convex on C. Let $\mathbf{h} = (h_1, \ldots, h_k)$ be an affine mapping from $\mathbb{R}^n$ to $\mathbb{R}^k$. If*

$$\mathbf{f}(\mathbf{x}) < \mathbf{0}, \qquad \mathbf{h}(\mathbf{x}) = \mathbf{0} \tag{1}$$

has no solution $\mathbf{x}$ in C, then there exist a vector $\mathbf{p}$ in $\mathbb{R}^m$ and a vector $\mathbf{q}$ in $\mathbb{R}^k$ such that $\mathbf{p} \geqslant \mathbf{0}$, $(\mathbf{p}, \mathbf{q}) \neq \mathbf{0}$, and

$$\langle \mathbf{p}, \mathbf{f}(\mathbf{x}) \rangle + \langle \mathbf{q}, \mathbf{h}(\mathbf{x}) \rangle \geqslant 0 \tag{2}$$

for all $\mathbf{x}$ in C.

Proof. For each $\mathbf{x}$ in C define the set

$$S(\mathbf{x}) = \{(\mathbf{y}, \mathbf{z}) : \mathbf{y} \in \mathbb{R}^m,\ \mathbf{z} \in \mathbb{R}^k,\ \mathbf{y} > \mathbf{f}(\mathbf{x}),\ \mathbf{z} = \mathbf{h}(\mathbf{x})\}.$$

For each $\mathbf{x}$ in C, the set $S(\mathbf{x})$ is the set of solutions $(\mathbf{y}, \mathbf{z})$ in $\mathbb{R}^{m+k}$ of the system

$$\mathbf{y} > \mathbf{f}(\mathbf{x}), \qquad \mathbf{z} = \mathbf{h}(\mathbf{x}). \tag{3}$$

Thus for each $\mathbf{x}$, the origin in $\mathbb{R}^{m+k}$ is not in $S(\mathbf{x})$. Let

$$S = \bigcup_{\mathbf{x} \in C} S(\mathbf{x}).$$

Since the origin in $\mathbb{R}^{m+k}$ belongs to none of the sets $S(\mathbf{x})$, it does not belong to S. It is readily verified that S is convex. Hence by Theorem II.3.2 there is a hyperplane that separates S and $\mathbf{0}$ in $\mathbb{R}^{m+k}$. That is, there exist vectors $\mathbf{p}$ in $\mathbb{R}^m$ and $\mathbf{q}$ in $\mathbb{R}^k$ such that $(\mathbf{p}, \mathbf{q}) \neq \mathbf{0}$ and

$$\langle \mathbf{p}, \mathbf{y} \rangle + \langle \mathbf{q}, \mathbf{z} \rangle \geqslant 0 \tag{4}$$

for all $(\mathbf{y}, \mathbf{z})$ in S.

If $(\mathbf{y}, \mathbf{z})$ belongs to S, then $\mathbf{y} > \mathbf{f}(\mathbf{x})$ for some $\mathbf{x}$ in C. Hence if some component of $\mathbf{p}$, say p_i, were negative, we could take y_i to be arbitrarily large and obtain a contradiction to (4). Hence $\mathbf{p} \geqslant \mathbf{0}$.

Now let $\varepsilon > 0$ and let $\mathbf{e}$ denote the vector in $\mathbb{R}^m$ all of whose components are equal to 1. Then for $\mathbf{x} \in C$ the vector $(\mathbf{f}(\mathbf{x}) + \varepsilon\mathbf{e}, \mathbf{h}(\mathbf{x}))$ belongs to S, and so

$$\langle \mathbf{p}, \mathbf{f}(\mathbf{x})\rangle + \langle \mathbf{q}, \mathbf{h}(\mathbf{x})\rangle \geqslant -\varepsilon(\langle \mathbf{p}, \mathbf{e}\rangle).$$

We obtain (2) by letting $\varepsilon \to 0$.

Remark 4.1. From the proof it is clear that if the affine function $\mathbf{h}$ is absent, then the conclusion is that there exists a $\mathbf{p} \neq \mathbf{0}$, $\mathbf{p} \geqslant \mathbf{0}$ such that $\langle \mathbf{p}, \mathbf{f}(\mathbf{x})\rangle \geqslant 0$ for all $\mathbf{x}$ in C.

Remark 4.2. This theorem is not a true alternative theorem, as it does not assert that either the inequality (1) has a solution or the inequality (2) has a solution, but never both. If we strengthen the hypotheses of the theorem by assuming $C = \mathbb{R}^n$ and the rows of A to be linearly independent, then we do get a true alternative theorem. This is taken up in Exercise 4.2.

The next two theorems are corollaries of Theorem 4.1.

THEOREM 4.2. *Let* $\mathbf{f}$ *be a vector-valued convex function defined on a convex set* C *in* $\mathbb{R}^n$ *and with range in* $\mathbb{R}^m$. *Then either*

$$\text{I:} \quad \mathbf{f}(\mathbf{x}) < \mathbf{0}$$

has a solution $\mathbf{x}$ in C *or there exists a* $\mathbf{p}$ *in* $\mathbb{R}^m$, $\mathbf{p} \neq \mathbf{0}$, $\mathbf{p} \geqslant \mathbf{0}$, *such that*

$$\text{II:} \quad \langle \mathbf{p}, \mathbf{f}(\mathbf{x})\rangle \geqslant 0$$

for all $\mathbf{x}$ *in* C, *but never both.*

Proof. Suppose I has a solution $\mathbf{x}_0$. Then for all $\mathbf{p} \neq \mathbf{0}$, $\mathbf{p} \geqslant \mathbf{0}$, we have $\langle \mathbf{p}, \mathbf{f}(\mathbf{x}_0)\rangle < 0$, so II cannot hold.

Suppose now that I has no solution. Then by Theorem 4.1 and Remark 4.1 following the theorem, there exists a $\mathbf{p} \neq \mathbf{0}$, $\mathbf{p} \geqslant \mathbf{0}$ such that II holds for all $\mathbf{x}$.

Theorem 4.2 is a true generalization of Gordan's theorem, Theorem II.5.2. To see that this is so, let $C = \mathbb{R}^n$ and let $\mathbf{f}(\mathbf{x}) = A\mathbf{x}$. Then $A\mathbf{x} < \mathbf{0}$ has no solution $\mathbf{x}_0$ if and only if there exists a vector $\mathbf{p}$ in $\mathbb{R}^m$, $\mathbf{p} \neq \mathbf{0}$, $\mathbf{p} \geqslant \mathbf{0}$, such that $\langle \mathbf{p}, A\mathbf{x}\rangle \geqslant 0$ for all $\mathbf{x}$. If we take $\mathbf{x} = A^t\mathbf{p}$, we get $\langle A^t\mathbf{p}, A^t\mathbf{p}\rangle \geqslant 0$. If we take $\mathbf{x} = -A^t\mathbf{p}$, we get $\langle A^t\mathbf{p}, A^t\mathbf{p}\rangle \leqslant 0$. Hence $A^t\mathbf{p} = \mathbf{0}$. In other words we have shown that $A\mathbf{x} < \mathbf{0}$ has no solution if and only if there exists a $\mathbf{p} \neq \mathbf{0}$, $\mathbf{p} \geqslant \mathbf{0}$ such that $A^t\mathbf{p} = \mathbf{0}$, which is Gordan's theorem.

THEOREM 4.3. *Let C be a compact convex set in $\mathbb{R}^n$ and let $\{f_\alpha\}_{\alpha \in A}$ be a possibly infinite family of real-valued convex functions that are continuous on C. If the system*

$$f_\alpha(\mathbf{x}) \leqslant 0, \qquad \alpha \in A, \tag{5}$$

has no solution in C, then there exists a finite subfamily $f_{\alpha_1}, \ldots, f_{\alpha_m}$ and a vector $\mathbf{p} = (p_1, \ldots, p_m)$ such that $\mathbf{p} \neq \mathbf{0}$, $\mathbf{p} \geqslant \mathbf{0}$ and such that the real-valued function F defined by

$$F(\mathbf{x}) = \sum_{i=1}^{m} p_i f_{\alpha_i}(\mathbf{x}) \tag{6}$$

satisfies

$$\min_{\mathbf{x} \in C} F(\mathbf{x}) > 0. \tag{7}$$

The proof will be carried out by translating the statement of the nonexistence of solutions to the system (5) to a statement about the empty intersection of a collection of closed subsets of a compact set. The finite-intersection property will then be used to obtain a statement about the empty intersection of a finite collection of sets. The last statement will then be translated to a statement about the nonexistence of a solution to the system (1). An application of Theorem 4.1 will then yield (7).

Proof. Since for each $\alpha \in A$ the function f_α is continuous, it follows that for each $\alpha \in A$ and $\varepsilon > 0$ the set

$$C_{\alpha,\varepsilon} = \{\mathbf{x} : \mathbf{x} \in C, f_\alpha(\mathbf{x}) \leqslant \varepsilon\}$$

is closed and is contained in C. If the intersection of the closed sets $\{C_{\alpha,\varepsilon}\}$, where $\alpha \in A$ and $\varepsilon > 0$, were not empty, there would exist an $\mathbf{x}_0$ in C such that $f_\alpha(\mathbf{x}_0) \leqslant \varepsilon$ for all $\alpha \in A$ and all $\varepsilon > 0$. Hence $f_\alpha(\mathbf{x}_0) \leqslant 0$ for all $\alpha \in A$, contradicting the hypothesis that $f_\alpha(\mathbf{x}) \leqslant 0$ has no solution in C. Thus the intersection of the sets $C_{\alpha,\varepsilon}$ is empty. Since C is compact, by Remark I.7.1 and Theorem I.7.3 there exists a finite subcollection $C_{\alpha_1,\varepsilon_1}, C_{\alpha_2,\varepsilon_2}, \ldots, C_{\alpha_m,\varepsilon_m}$ of $\{C_{\alpha,\varepsilon}\}$ whose intersection is empty. That is, the system $f_{\alpha_i}(\mathbf{x}) - \varepsilon_i \leqslant 0$, $i = 1, \ldots, m$, has no solution $\mathbf{x}$ in C. Therefore the system

$$f_{\alpha_i}(\mathbf{x}) - \varepsilon_i < 0, \qquad i = 1, \ldots, m,$$

has no solution $\mathbf{x}$ in C. Therefore, by Theorem 4.1 and Remark 1, there exists a vector $\mathbf{p} = (p_1, \ldots, p_m)$ with $\mathbf{p} \neq \mathbf{0}$ and $p_i \geqslant 0$, $i = 1, \ldots, m$, such that

$$\sum_{i=1}^{m} p_i f_{\alpha_i}(\mathbf{x}) \geqslant \sum_{i=1}^{m} p_i \varepsilon_i > 0 \tag{8}$$

for all $\mathbf{x}$ in C. The function F defined in (6) is continuous on the compact set C and hence attains a minimum on C. From this and from (8) the conclusion follows.

Remark 4.3. The reader who is familiar with the notion of semicontinuity will see that the requirement that the functions f_α be continuous can be replaced by the less restrictive requirement that they be lower semicontinuous.

Exercise 4.1. Verify that the set S in Theorem 4.1 is convex.

Exercise 4.2. Prove the following corollary to Theorem 4.1. Let $\mathbf{f} = (f_1, \ldots, f_m)$ be defined and convex on $\mathbb{R}^n$. Let $\mathbf{h} = A\mathbf{x} - \mathbf{b}$ be an affine mapping from $\mathbb{R}^n$ to $\mathbb{R}^k$ and let the rows of A be linearly independent. Then either

$$\text{I:} \quad \mathbf{f}(\mathbf{x}) < \mathbf{0}, \qquad A\mathbf{x} - \mathbf{b} = \mathbf{0}$$

has a solution $\mathbf{x}$ in $\mathbb{R}^n$ or

$$\text{II:} \quad \langle \mathbf{p}, \mathbf{f}(\mathbf{x}) \rangle + \langle \mathbf{q}, A\mathbf{x} - \mathbf{b} \rangle \geqslant 0 \quad \text{all } \mathbf{x} \in \mathbb{R}^n$$

has a solution $(\mathbf{p}, \mathbf{q}) \neq \mathbf{0}$, $\mathbf{p} \in \mathbb{R}^m$, $\mathbf{p} \geqslant \mathbf{0}$, $\mathbf{q} \in \mathbb{R}^k$, but never both.

5. APPLICATION TO GAME THEORY

In this section we use Theorem 4.3 to prove the von Neumann minimax theorem and then apply the minimax theorem to prove the existence of value and saddle points in mixed strategies for matrix games. Our proof will require the concept of uniform continuity and a sufficient condition for uniform continuity.

Definition 5.1. A function $\mathbf{f}$ defined on a set S in $\mathbb{R}^n$ with range in $\mathbb{R}^m$ is said to be *uniformly continuous* on S if for each $\varepsilon > 0$ there exists a $\delta > 0$ such that if $\mathbf{x}$ and $\mathbf{x}'$ are points in S satisfying $\|\mathbf{x} - \mathbf{x}'\| < \delta$, then $\|\mathbf{f}(\mathbf{x}') - \mathbf{f}(\mathbf{x})\| < \varepsilon$.

THEOREM 5.1. *Let S be a compact set in $\mathbb{R}^n$ and let $\mathbf{f}$ be a mapping defined and continuous on S with range in $\mathbb{R}^m$. Then $\mathbf{f}$ is uniformly continuous on S.*

For a proof we refer the reader to Bartle and Sherbert [1999] or Rudin [1976].

THEOREM 5.2 (VON NEUMANN MINIMAX THEOREM). *Let $X \subset \mathbb{R}^m$ and $Y \subset \mathbb{R}^n$ be compact convex sets. Let $f : X \times Y \to \mathbb{R}$ be continuous on $X \times Y$. For each $\mathbf{x}$ in X let $f(\mathbf{x}, \ ) : Y \to \mathbb{R}$ be convex and for each $\mathbf{y} \in Y$ let $f(\ , \mathbf{y}) : X \to \mathbb{R}$ be concave.*

Then

$$\max_{\mathbf{x}\in X}\min_{\mathbf{y}\in Y} f(\mathbf{x}, \mathbf{y}) = \min_{\mathbf{y}\in Y}\max_{\mathbf{x}\in X} f(\mathbf{x}, \mathbf{y}), \tag{1}$$

and there exists a pair $(\mathbf{x}_*, \mathbf{y}_*)$ *with* $\mathbf{x}_*$ *in* X *and* $\mathbf{y}_*$ *in* Y *such that for all* $\mathbf{x}$ *in* X *and all* $\mathbf{y}$ *in* Y

$$f(\mathbf{x}, \mathbf{y}_*) \leqslant f(\mathbf{x}_*, \mathbf{y}_*) \leqslant f(\mathbf{x}_*, \mathbf{y}). \tag{2}$$

Moreover, if v *denotes the common value in* (1), *then* $f(\mathbf{x}_*, \mathbf{y}_*) = v$.

A pair $(\mathbf{x}_*, \mathbf{y}_*)$ as in (2) is called a *saddle point* of the function f.

The relation (1) is not true in general for continuous functions f defined in $X \times Y$, where X and Y are compact and convex. To see this, let $X = [0, 1]$, let $Y = [0, 1]$, and let $f(x, y) = (x - y)^2$. Then for fixed x, $\min\{f(x, y) : y \in Y\}$ is achieved at $y = x$ and equals zero. Hence in this example the left-hand side of (1) equals zero. On the other hand, for fixed y in the interval $[0, \frac{1}{2}]$

$$\max_x \{f(x, y) : x \in X\} = (1 - y)^2,$$

and for fixed y in the interval $[\frac{1}{2}, 1]$

$$\max_x \{f(x, y) : x \in X\} = y^2.$$

Hence in this example the right-hand side of (1) equals $\frac{1}{4}$.

The concavity in X and convexity in Y is a sufficient condition but not a necessary condition, as the example $X = [0, 1]$, $Y = [0, 1]$, and $f(x, y) = e^x e^y$ shows.

Before we take up the proof of Theorem 5.2, which will be carried out in several steps, we observe the following. If A and B are arbitrary sets and g is a real-valued bounded function defined on $A \times B$, then the quantities

$$\sup_{a\in A}\inf_{b\in B} g(a, b) \quad \text{and} \quad \inf_{b\in B}\sup_{a\in A} g(a, b)$$

are well defined, while the quantities

$$\max_{a\in A}\min_{b\in B} g(a, b) \quad \text{and} \quad \min_{b\in B}\max_{a\in A} g(a, b)$$

need not be. For a simple example, take $A = (0, 1)$, $B = (0, 1)$, and $g(a, b) = ab$. Thus, implicit in (1) is the statement, which must be proved, that the left- and right-hand sides of (1) exist. The proof of this statement will constitute the second step of our proof. The first step is the general observation given in Lemma 5.1.

LEMMA 5.1. *Let A and B be arbitrary sets and let $g : A \times B \to \mathbb{R}$ be bounded. Then*

$$\sup_{a \in A} \inf_{b \in B} g(a, b) \leqslant \inf_{b \in B} \sup_{a \in A} g(a, b).$$

Proof. Choose an element $b_1 \in B$. Then for each $a \in A$,

$$g(a, b_1) \leqslant \sup_{a \in A} g(a, b_1).$$

Since the choice of b_1 was arbitrary, this relationship holds for each a in A and each b in B. Hence, for each $a \in A$

$$\inf_{b \in B} g(a, b) \leqslant \inf_{b \in B} \sup_{a \in A} g(a, b).$$

The right-hand side is a fixed real number, so

$$\sup_{a \in A} \inf_{b \in B} g(a, b) \leqslant \inf_{b \in B} \sup_{a \in A} g(a, b).$$

We now proceed to the second step of the proof and show that the left- and right-hand sides of (1) exist.

For fixed $\mathbf{x}$ in X, the function $f(\mathbf{x}, \,) : Y \to \mathbb{R}$ defined by $f(\mathbf{x}, \mathbf{y})$ is continuous on the compact set Y. Hence

$$\psi(\mathbf{x}) = \min_{\mathbf{y} \in Y} f(\mathbf{x}, \mathbf{y})$$

is defined for each $\mathbf{x}$ in X. We assert that the function ψ is continuous on X. To prove the assertion, let $\varepsilon > 0$ be given. Since f is continuous on the compact set $X \times Y$, it is uniformly continuous on $X \times Y$. Hence there exists a $\delta > 0$ such that if $\|\mathbf{x}_1 - \mathbf{x}_2\| < \delta$, then

$$f(\mathbf{x}_2, \mathbf{y}) - \tfrac{1}{2}\varepsilon < f(\mathbf{x}_1, \mathbf{y}) < f(\mathbf{x}_2, \mathbf{y}) + \tfrac{1}{2}\varepsilon$$

for all $\mathbf{y}$ in Y. Upon taking the minimum over Y, we get that

$$\psi(\mathbf{x}_2) - \tfrac{1}{2}\varepsilon \leqslant \psi(\mathbf{x}_1) \leqslant \psi(\mathbf{x}_2) + \tfrac{1}{2}\varepsilon,$$

or $|\psi(\mathbf{x}_2) - \psi(\mathbf{x}_2)| < \varepsilon$ whenever $\|\mathbf{x}_2 - \mathbf{x}_1\| < \delta$. This proves the continuity of ψ.

Since ψ is continuous on the compact set X, there exists and $\mathbf{x}_*$ in X such that

$$\psi(\mathbf{x}_*) = \max_{\mathbf{x} \in X} \psi(\mathbf{x}) = \max_{\mathbf{x} \in X} \left[\min_{\mathbf{y} \in X} f(\mathbf{x}, \mathbf{y}) \right]. \tag{3}$$

It follows from (3) that the left-hand side of (1) is well defined.

From the definition of ψ we have

$$\psi(\mathbf{x}_*) = \min_{\mathbf{y} \in Y} f(\mathbf{x}_*, \mathbf{y}). \tag{4}$$

Similarly, for each $\mathbf{y}$ in Y we can set

$$\phi(\mathbf{y}) = \max_{\mathbf{x} \in X} f(\mathbf{x}, \mathbf{y})$$

and then obtain the existence of a $\mathbf{y}_*$ in Y such that

$$\phi(\mathbf{y}_*) = \min_{\mathbf{y} \in Y} \phi(\mathbf{y}) = \min_{\mathbf{y} \in Y} \left[\max_{\mathbf{x} \in X} f(\mathbf{x}, \mathbf{y}) \right] \tag{5}$$

and

$$\phi(\mathbf{y}_*) = \max_{\mathbf{x} \in X} f(\mathbf{x}, \mathbf{y}_*). \tag{6}$$

Thus the right-hand side of (1) is well defined.

The third step in the proof is to establish the equality in (1). Let

$$\alpha = \max_{\mathbf{x} \in X} \min_{\mathbf{y} \in Y} f(x, y), \qquad \beta = \min_{\mathbf{y} \in Y} \max_{\mathbf{x} \in X} f(x, y).$$

By Lemma 5.1 with $X = A$, $Y = B$, $f = g$ and by the second step, we have that $\alpha \leqslant \beta$. Hence to prove (1), we must show that $\alpha \geqslant \beta$.

Let $\varepsilon > 0$ be arbitrary. We assert that there *does not* exist a $\mathbf{y}_0$ in Y such that for all $\mathbf{x}$ in X

$$f(\mathbf{x}, \mathbf{y}_0) \leqslant \beta - \varepsilon.$$

For if such a $\mathbf{y}_0$ did exist, we would have

$$\beta = \min_{\mathbf{y} \in Y} \max_{\mathbf{x} \in X} f(\mathbf{x}, \mathbf{y}) \leqslant \max_{\mathbf{x} \in X} f(\mathbf{x}, \mathbf{y}_0) \leqslant \beta - \varepsilon,$$

which is not possible.

Let $\{g_{\mathbf{x}}\}_{\mathbf{x}\in X}$ be the family of functions defined on Y by the formula

$$g_{\mathbf{x}}(\mathbf{y}) = f(\mathbf{x}, \mathbf{y}), \qquad \mathbf{x}\in X.$$

Then for each $\mathbf{x}\in X$, the function $g_{\mathbf{x}}$ is continuous and convex on Y. Moreover, for arbitrary $\varepsilon > 0$ the system

$$g_{\mathbf{x}}(\mathbf{y}) - \beta + \varepsilon \leqslant 0, \qquad \mathbf{x}\in X, \tag{7}$$

has no solution $\mathbf{y}$ in Y.

Since (7) has no solution, it follows from Theorem 4.3 that there exist points $\mathbf{x}_1, \ldots, \mathbf{x}_k$ in X, a vector $\mathbf{p}\in\mathbb{R}^k$, $\mathbf{p} \geqslant \mathbf{0}$, $\mathbf{p} \neq \mathbf{0}$, and a $\delta > 0$ such that

$$G(\mathbf{y}) = \sum_{i=1}^{k} p_i[f(\mathbf{x}_i, \mathbf{y}) - \beta + \varepsilon] \geqslant \delta > 0 \tag{8}$$

for each $\mathbf{y}$ in Y. Since $\Sigma_{i=1}^{k} p_i > 0$, we may divide through by this quantity in (8) and assume that $\mathbf{p}\in P_k$. Therefore, since X is convex, $\Sigma_{i=1}^{k} p_i\mathbf{x}_i$ is in X. Using (8) and the concavity of $f(\ ,\mathbf{y})$ for each fixed $\mathbf{y}$, we may write

$$f\left(\sum_{i=1}^{k} p_i\mathbf{x}_i, \mathbf{y}\right) \geqslant \sum_{i=1}^{k} p_i f(\mathbf{x}_i, \mathbf{y}) > \beta - \varepsilon$$

for each $\mathbf{y}$ in Y. Therefore,

$$\begin{aligned}\alpha = \max_{\mathbf{x}\in X}\min_{\mathbf{y}\in Y} f(\mathbf{x}, \mathbf{y}) &\geqslant \min_{\mathbf{y}\in Y} f\left(\sum_{i=1}^{k} p_i\mathbf{x}_i, \mathbf{y}\right)\\ &\geqslant \min_{\mathbf{y}\in Y}\sum_{i=1}^{k} p_i f(\mathbf{x}_i, \mathbf{y}) > \beta - \varepsilon.\end{aligned}$$

Since $\varepsilon > 0$ is arbitrary, we get that $\alpha \geqslant \beta$, and (1) is established.

The final step in the proof is to show that $v = f(\mathbf{x}_*, \mathbf{y}_*)$ and that (2) holds. It follows from (1), (3), and (5) that

$$\psi(\mathbf{x}_*) = \phi(\mathbf{y}_*) = v.$$

It now follows from (4) and (6) that

$$f(\mathbf{x}, \mathbf{y}_*) \leqslant v \leqslant f(\mathbf{x}_*, \mathbf{y}) \tag{9}$$

for all $\mathbf{x}$ in X and $\mathbf{y}$ in Y. In particular, (9) holds for $\mathbf{x} = \mathbf{x}_*$ and $\mathbf{y} = \mathbf{y}_*$, so that $v = f(\mathbf{x}_*, \mathbf{y}_*)$. The relation (2) now follows from (9).

The simplest two-person zero-sum games are matrix games. These games are played by two players that we shall call Blue and Red. Blue selects a positive integer i from a finite set I of positive integers $\{1,\ldots,m\}$. Red, simultaneously and in ignorance of Blue's choice, selects a positive integer j from a finite set J of positive integers $\{1,\ldots,n\}$. The "payoff" or "reward" to Blue in the event that he chooses i and Red chooses j is a predetermined real number $a(i,j)$. The payoff to Red is $-a(i,j)$. Hence the name zero-sum game. A *strategy* for Blue is a rule which tells Blue which positive integer i in I to select. A *strategy* for Red is a rule which tells Red which positive integer j in J to select. Blue's objective is to choose i in I so as to maximize $a(i,j)$. Red's objective is to choose j in J so as to minimize $a(i,j)$, that is, to maximize $-a(i,j)$. If the strategy for Blue is to choose the number i_0 in I, then we call the strategy a *pure* strategy for Blue. Similarly, if the strategy for Red is to choose the number j_0 in J, then we call the strategy a *pure* strategy for Red. Later we shall consider strategies that are not pure.

Another way of describing a matrix game is as follows. An $m \times n$ matrix A with entries $a(i,j)$, $i = 1,\ldots,m$, $j = 1,\ldots,n$, is given. Blue selects a row i and Red selects a column j. The choices are made simultaneously and without knowledge of the opponent's choice. If Blue selects row i and Red selects column j, then the payoff to Blue is $a(i,j)$ and the payoff to Red is $-a(i,j)$. Blue's objective is to maximize the payoff, while Red's objective is to minimize the payoff.

At this point, the formulation of a matrix game may strike the reader as rather artificial and uninteresting. It turns out, however, that many decision problems in conflict situations can be modeled as matrix games. As an example, we shall formulate a tactical naval problem as a matrix game. The reader will find many additional interesting examples elsewhere [Dresher, 1961]. The naval game appears there as a land conflict game called Colonel Blotto.

Two Red cargo ships are protected by three escort frigates under the command of Captain Pluto. The entire convoy is subject to attack by a pack of four Blue submarines under the command of Captain Nemo. The ships carry a hazardous cargo and thus maintain a separation such that one submarine cannot attack both ships. Captain Nemo's problem is to allocate his submarines to attack the two ships. Thus, he may allocate all four of his submarines to attack one of the ships and none to attack the second ship. Or, he may allocate three submarines to attack one ship and one submarine to attack the second ship. Or, he may allocate two submarines to attack each of the ships. We may summarize the choices, or strategies, available to Captain Nemo by five ordered pairs, $(4,0)$, $(0,4)$, $(3,1)$, $(1,3)$, $(2,2)$, where (α,β) indicates that he has allocated α submarines to attack ship 1 and β submarines to attack ship 2. Captain Pluto's problem is to allocate his frigates to defend the ships. He has four choices, or strategies, each of which can be denoted by an ordered pair (γ,δ), $0 \leqslant \gamma \leqslant 3$, $0 \leqslant \delta \leqslant 3$, $\gamma + \delta = 3$, where γ frigates are allocated to defend ship 1 and δ frigates are allocated to defend ship 2. We assume that conditions are such that when Captain Nemo allocates his submarines, he cannot determine the disposition of the escort frigates, and that when Captain Pluto allocates his escort frigates, he cannot determine the disposition of the attacking submarines.

If at a given ship the number of attacking submarines exceeds the number of defending frigates, then the payoff to Blue at that ship equals the number of Red frigates at that ship plus 1. The "plus 1" is awarded for the cargo ship that is assumed to be sunk. If at a given ship the number of attacking submarines is less than the number of defending frigates, then the payoff to Blue at the ship is

$$-(\text{Number of attacking submarines} + 1).$$

The "plus 1" is deducted for allowing the cargo ship to proceed to its destination. If at a given ship the number of attacking submarines equals the number of defending ships, then the encounter is a draw, and the payoff to Blue at that ship is zero. The total payoff to Blue resulting from given choices by the two antagonists is the sum of the payoffs at each ship. The payoff to Red is the negative of the payoff to Blue.

The game just described can be cast as a matrix game with payoff matrix

$$\begin{array}{lc} & \text{Captain Pluto's strategies} \\ \begin{array}{ll} & \\ & (4,0) \\ \text{Captain Nemo's} & (0,4) \\ \text{strategies} & (3,1) \\ & (1,3) \\ & (2,2) \end{array} & \begin{pmatrix} (3,0) & (0,3) & (2,1) & (1,2) \\ 4 & 0 & 2 & 1 \\ 0 & 4 & 1 & 2 \\ 1 & -1 & 3 & 0 \\ -1 & 1 & 0 & 3 \\ -2 & -2 & 2 & 2 \end{pmatrix} \end{array}.$$

A strategy for Blue (Captain Nemo) is a choice of row; a strategy for Red (Captain Pluto) is a choice of column.

If Captain Nemo chooses row i, he is guaranteed a payoff of at least

$$\min\{a(i,j) : j = 1, \ldots, 4\}.$$

If he picks the row that will give the payoff

$$v^- \equiv \max_i \min_j a(i,j),$$

then he will be guaranteed to obtain v^-. Captain Nemo can thus pick either row 1 or row 2 and be guaranteed a payoff of at least zero. Thus $v^- = 0$. On the other hand, if Captain Pluto chooses column j, he will lose at most

$\max\{a(i, j) : i = 1, \ldots, 5\}$. If Captain Pluto picks the column that gives Blue the payoff

$$v^+ = \min_j \max_i a(i, j),$$

then he is guaranteed to lose no more than v^+. Captain Pluto can thus pick either column 3 or 4 and be guaranteed to lose no more than three. Thus $v^+ = 3$.

Note that if Captain Nemo knows beforehand that Captain Pluto will pick column 3, then he can pick row 3 and obtain a payoff of 3, which is greater than $v^- = 0$. On the other hand, if Captain Pluto knows beforehand that Captain Nemo will pick row 2, then he can pick column 1 and lose zero units, which is less of a loss than three units ($v^+ = 3$). An examination of the payoff matrix shows that if one of the commander's choice is known in advance by the second commander, than the second commander can usually improve his payoff if he bases his choice on this information. Thus, in the submarine–convoy game secrecy of choice is essential. The open question, however, in the submarine–convoy game is, "What is the best strategy for each player?"

In contrast to the submarine–convoy game, consider the game with payoff matrix

$$\begin{pmatrix} 3 & 7 & 4 & 3 \\ 13 & 10 & 7 & 8 \\ 10 & 4 & 1 & 9 \\ 3 & 5 & 6 & 7 \end{pmatrix}.$$

Here

$$\max_i \min_j a(i, j) = \min_j \max_i a(i, j) = 7.$$

The strategies $i = i_*$, where $i_* = 2$, and $j = j_*$, where $j_* = 3$, have the properties that $a(i_*, j_*) = 7$ and

$$a(i, j_*) \leqslant a(i_*, j_*) \leqslant a(i_*, j)$$

for all $i = 1, \ldots, 4$ and $j = 1, \ldots, 4$. We say that the value of the game is 7, that $i_* = 2$ is an optimal (pure) strategy for Blue, and that $j_* = 3$ is an optimal (pure) strategy for Red. If Blue chooses his optimal strategy $i_* = 2$, then the best that Red can do is to also choose his optimal strategy $j_* = 3$. Also, if Red chooses his optimal strategy $j_* = 3$, then the best that Blue can do is to choose his optimal strategy $i_* = 2$. Thus, if either player announces beforehand that he will choose an optimal strategy, the opponent cannot take advantage of this information, other than to choose an optimal strategy.

We now consider the general matrix game with $m \times n$ payoff matrix A with entries $a(i, j)$, $i = 1, \ldots, m$, $j = 1, \ldots, n$. We always have

$$v^- = \max_i \min_j a(i, j) \leqslant \min_j \max_i a(i, j) \equiv v^+.$$

If the reverse inequality $v^+ \geqslant v^-$ holds, then we set

$$v = \min_j \max_i a(i, j) = \max_i \min_j a(i, j)$$

and say that the matrix game has a *value* v in pure strategies. It is easy to show that if the matrix game has a value v in pure strategies, then there exists a pair (i_*, j_*) such that

$$v = a(i_*, j_*), \tag{10}$$

and for all $i = 1, \ldots, m$, $j = 1, \ldots, n$,

$$a(i, j_*) \leqslant a(i_*, j_*) \leqslant a(i_*, j). \tag{11}$$

The strategies i_* and j_* are called *optimal* (*pure*) *strategies*. We point out that optimal strategies need not be unique. A pair of strategies (i_*, j_*) such that (11) holds is called a *saddle point* in pure strategies. It is also easy to show that if a matrix game has a saddle point in pure strategies, then the game has value v and (10) holds.

If the matrix game has a value, then each player should use an optimal strategy. For then, Blue, the maximizer, is guaranteed a payoff of v and Red, the minimizer, is guaranteed to lose at most $-v$. If one player, say Blue, uses an optimal strategy and Red does not, then the payoff to Blue, who plays optimally, will, in general, be more favorable than it would be had Red played optimally. This is the content of (11).

The question that still thrusts itself upon us is, "How does one play if the game does not have a saddle point, as in the submarine–convoy game?" To answer this question, we introduce the notion of mixed strategy.

Definition 5.2. A *mixed strategy* for Blue in the game with $m \times n$ payoff matrix A is a probability distribution on the rows of A. Thus, a mixed strategy is a vector $\mathbf{x} = (x_1, \ldots, x_m)$ with $x_i \geqslant 0$, $i = 1, \ldots, m$, and $\Sigma_{i=1}^m x_i = 1$.

A mixed strategy for Red is a probability distribution on the columns of A. Thus, a mixed strategy for Red is a vector $\mathbf{y} = (y_1, \ldots, y_n)$ with $y_i \geqslant 0$, $i = 1, \ldots, n$, and $\Sigma_{i=1}^n y_i = 1$.

The set of pure strategies for a player is a subset of the set of mixed strategies. For example, the pure strategy for Blue in which row i is always chosen is also the mixed strategy in which row i is chosen with probability 1, that is, the vector $\mathbf{e}_i = (0, \dots, 0, 1, 0, \dots, 0)$, where the ith entry is 1 and all other entries are zero.

In actual play, Blue will select his row using a random device which produces the value i with probability x_i, $i = 1, \dots, m$. Blue chooses row i if the random device produces the value i. Red chooses column j in similar fashion. If mixed strategies are used, neither player tries to outguess his opponent and both trust to luck. Moreover, if Blue uses a mixed strategy, Red cannot determine Blue's choice beforehand and take advantage of this information. A similar statement holds if Red uses a mixed strategy.

If mixed strategies are allowed, then it is appropriate to define the payoff of the matrix game with mixed strategies as the expected value of the payoff. Thus, if we denote the payoff resulting from the choice of strategies $(\mathbf{x}, \mathbf{y})$ by $P(\mathbf{x}, \mathbf{y})$, we have

$$P(\mathbf{x}, \mathbf{y}) = \mathbf{x}^t A \mathbf{y} = \langle \mathbf{x}, A\mathbf{y} \rangle. \tag{12}$$

We define the matrix game with matrix A in which mixed strategies are used as follows. Blue selects a vector $\mathbf{x}$ in P_m, and Red selects a vector $\mathbf{y}$ in P_n. The payoff to Blue is given by (12), and the payoff to Red is $-P(\mathbf{x}, \mathbf{y})$. We say that the matrix game with mixed strategies has a *value* if

$$\max_{\mathbf{x} \in P_m} \min_{\mathbf{y} \in P_n} P(\mathbf{x}, \mathbf{y}) = \min_{\mathbf{y} \in P_n} \max_{\mathbf{x} \in P_m} P(\mathbf{x}, \mathbf{y}). \tag{13}$$

We denote the number in (13) by v and call it the *value* of the game. We say that the game has a *saddle point* $(\mathbf{x}_*, \mathbf{y}_*)$, that $\mathbf{x}_*$ is an *optimal mixed strategy* for Blue, and that $\mathbf{y}_*$ is an *optimal mixed strategy* for Red if

$$P(\mathbf{x}, \mathbf{y}_*) \leqslant P(\mathbf{x}_*, \mathbf{y}_*) \leqslant P(\mathbf{x}_*, \mathbf{y}) \tag{14}$$

for all $\mathbf{x}$ in P_m and $\mathbf{y}$ in P_n.

THEOREM 5.3. *A matrix game has a value and a saddle point in mixed strategies.*

This theorem is a corollary of Theorem 5.2. Take $X = P_m$, $Y = P_n$, and $f = P$. The function P is continuous on $P_m \times P_n$. The sets P_m and P_n are compact and convex. Since P is bilinear, it is concave in $\mathbf{x}$ and convex in $\mathbf{y}$. Theorem 5.3 follows by applying Theorem 5.2 to P.

Note that the saddle point condition (13) implies that

$$v = P(\mathbf{x}_*, \mathbf{y}_*).$$

Exercise 5.1. Verify that in the submarine–convoy game the value in mixed strategies is $\frac{14}{9}$ and that

$$\mathbf{x}_* = (\tfrac{4}{9}, \tfrac{4}{9}, 0, 0, \tfrac{1}{9}), \qquad \mathbf{y}_* = (\tfrac{1}{18}, \tfrac{1}{18}, \tfrac{4}{9}, \tfrac{4}{9})$$

is a saddle point.

Exercise 5.2. Show that if a matrix game has a value in pure strategies, then (10) and (11) hold. Conversely, show that if a matrix game has a saddle point in pure strategies, then the game has a value v and (10) holds.

Exercise 5.3. Let X_* denote the set of optimal mixed strategies for Blue and let Y_* denote the set of optimal mixed strategies for Red. Show that X_* and Y_* are compact convex sets.

IV

OPTIMIZATION PROBLEMS

1. INTRODUCTION

In Section 9 of Chapter I we stated the basic problem in optimization theory as follows: Given a set S (not necessarily in $\mathbb{R}^n$) and a real-valued function f defined on S, does there exist an element s_* in S such that $f(s_*) \leqslant f(s)$ for all s in S, and if so find it. We also showed that the problem of maximizing f over S is equivalent to the problem of minimizing $-f$ over S. Thus, there is no loss of generality in only considering minimization problems, as we shall do henceforth.

If S is a set in $\mathbb{R}^n$ and f is a real-valued function on S, we say that a point s_0 in S is a *local minimizer* or that f has a *local minimum at* s_0 if there exists a $\delta > 0$ such that $f(s_0) \leqslant f(s)$ for all s in $B(s_0, \delta) \cap S$. If the strict inequality $f(s_0) < f(s)$ holds, then s_0 is said to be a *strict local minimizer*. If S is open, then $B(s_0, \delta) \cap S$ is an open set contained in S. Hence there exists a $\delta' > 0$ such that, for all s in $B(s_0, \delta')$, $f(s_0) \leqslant f(s)$. Therefore, if S is open, for s_0 to be a local minimizer, we need to find a $\delta > 0$ such that $f(s_0) \leqslant f(s)$ for all s in $B(s_0, \delta)$ and can omit the intersection with S. A point s_* that is a minimizer is also a local minimizer. A point s_0 that is a local minimizer need not be a minimizer.

In this chapter we shall develop necessary conditions and sufficient conditions for a point s_* to be a minimizer or local minimizer for various classes of problems. The determination of the set E of points that satisfy the necessary conditions does not, of course, solve the optimization problem. All we learn is that the set of solutions is to be found in this set. The determination of those points in E that are solutions to the problem requires further analysis. Sufficient conditions, if applicable to the problem at hand, are useful in this connection.

If S is an open set in $\mathbb{R}^n$, the optimization problem is often called an *unconstrained* problem.

A *programming problem* is a problem of the following type. A set X_0 in $\mathbb{R}^n$ is given as are functions f and $\mathbf{g}$ with domain X_0 and ranges in $\mathbb{R}^1$ and $\mathbb{R}^m$

respectively. Let

$$X = \{\mathbf{x} : \mathbf{x} \in X_0, \mathbf{g}(\mathbf{x}) \leqslant \mathbf{0}\}.$$

Problem P

Minimize f over the set X. That is, find an $\mathbf{x}_*$ in X such that $f(\mathbf{x}_*) \leqslant f(\mathbf{x})$ for all $\mathbf{x}$ in X.

The set X is called the *feasible set*, and elements of X are called *feasible points* or *feasible vectors*.

The problem is sometimes stated as follows: Minimize $f(\mathbf{x})$ subject to $\mathbf{x} \in X_0$ and $\mathbf{g}(\mathbf{x}) \leqslant \mathbf{0}$.

If $X_0 = \mathbb{R}^n$, A is an $m \times n$ matrix, and

$$f(\mathbf{x}) = \langle -\mathbf{b}, \mathbf{x} \rangle, \qquad \mathbf{g}(\mathbf{x}) = \begin{pmatrix} \mathbf{A}\mathbf{x} - \mathbf{c} \\ -\mathbf{x} \end{pmatrix}, \tag{1}$$

the problem is said to be a *linear programming* problem. We have written $-\mathbf{b}$ rather than $\mathbf{b}$ because in the linear programming literature the problem is usually stated as follows: Maximize $\langle \mathbf{b}, \mathbf{x} \rangle$ subject to $A\mathbf{x} \leqslant \mathbf{c}$ and $\mathbf{x} \geqslant \mathbf{0}$.

If X_0 is convex and the functions f and $\mathbf{g}$ are convex, the problem is said to be a *convex* programming problem.

2. DIFFERENTIABLE UNCONSTRAINED PROBLEMS

We will now denote the set on which the function f is defined by X rather than S. We first assume that X is an open interval in $\mathbb{R}^1$ and that f is differentiable on X. We will derive necessary conditions and sufficient conditions for a point x_0 in X to be a local minimizer. Although the reader surely knows these conditions from elementary calculus, it is instructive to review them.

Let x_0 in X be a local minimizer of f and let f be differentiable at x_0. Then there exists a $\delta > 0$ such that $f(x) \geqslant f(x_0)$ for x in $(x_0 - \delta, x_0 + \delta)$. Therefore, for $x_0 < x < x_0 + \delta$

$$\frac{f(x) - f(x_0)}{x - x_0} \geqslant 0, \tag{1}$$

and for $x_0 - \delta < x < x_0$

$$\frac{f(x) - f(x_0)}{x - x_0} \leqslant 0. \tag{2}$$

Since f is differentiable at x_0,

$$\lim_{x \to x_0^+} \frac{f(x) - f(x_0)}{x - x_0} = \lim_{x \to x_0^+} \frac{f(x) - f(x_0)}{x - x_0} = f'(x_0).$$

From this and from (1) we get $f'(x_0) \geqslant 0$, while from (2) we get $f'(x_0) \leqslant 0$. Hence $f'(x_0) = 0$.

Let us now assume further that f is of class $C^{(2)}$ on $(x_0 - \delta, x_0 + \delta)$. By Taylor's theorem, for each $x \neq x_0$ in $(x_0 - \delta, x_0 + \delta)$ there exists a point $\bar{x}$ that lies between x_0 and x such that

$$f(x) = f(x_0) + f'(x_0)(x - x_0) + \tfrac{1}{2}f''(\bar{x})(x - x_0)^2. \tag{3}$$

Since x_0 is a local minimizer, $f'(x_0) = 0$ and therefore

$$\tfrac{1}{2}f''(\bar{x})(x - x_0)^2 = f(x) - f(x_0) \geqslant 0.$$

Thus

$$f''(\bar{x}) = 2\,\frac{f(x) - f(x_0)}{(x - x_0)^2} \geqslant 0.$$

If we now let $x \to x_0$, then $\bar{x} \to x_0$ and $f''(\bar{x}) \to f''(x_0)$. Hence $f''(x_0) \geqslant 0$.

We have proved the following theorem.

THEOREM 2.1. *Let X be an open interval in $\mathbb{R}$, let f be a real-valued function defined on $\mathbb{R}$, and let x_0 in X be a local minimizer for f. If f is differentiable at x_0, then $f'(x_0) = 0$. If f is of class $C^{(2)}$ on some interval about x_0, then $f''(x_0) \geqslant 0$.*

A point c in X such that $f'(c) = 0$ is called a *critical point* of f. We have shown that for a differentiable function a local minimizer is a critical point.

If we replace the necessary condition $f''(x_0) \geqslant 0$ by the stronger condition $f''(x_0) > 0$, we obtain a sufficient condition for a critical point to be a local minimizer.

THEOREM 2.2. *Let X be an interval in $\mathbb{R}$ and let f be real-valued and of class $C^{(2)}$ on some interval about a point x_0 in X. If $f'(x_0) = 0$ and $f''(x_0) > 0$, then x_0 is a strict local minimizer for f.*

Proof. Since f'' is continuous on some interval about x_0 and since $f''(x_0) > 0$, there exists a $\delta > 0$ such that $f''(x) > 0$ for all x in $(x_0 - \delta, x_0 + \delta)$ (See Exercise I.5.4.) By Taylor's Theorem, for each $x \neq x_0$ in $(x_0 - \delta, x_0 + \delta)$ there exists an $\bar{x}$ lying between x_0 and x such that (3) holds. Since $f'(x_0) = 0$, we get that for each $x \neq x_0$ in $(x_0 - \delta, x_0 + \delta)$

$$f(x) - f(x_0) = \tfrac{1}{2}f''(\bar{x})(x - x_0)^2 > 0. \tag{4}$$

Thus x_0 is a strict local minimizer.

In elementary calculus texts Theorem 2.2 is often called the *second derivative test*.

We must have $f''(x_0) > 0$ rather than $f''(x_0) \geqslant 0$ for the conclusion of Theorem 2.2 to hold. To see this, take $X = \mathbb{R}$ and take $f(x) = x^3$. Then $f'(0) = 0$, $f''(0) = 0$, but zero is not a local minimizer. If, however, we assume that $f''(x) \geqslant 0$ for *all* x in X, then all the statements in the proof of Theorem 2.2 hold for each x in X, except that (4) now reads

$$f(x) - f(x_0) = \tfrac{1}{2} f''(\bar{x})(x - x_0)^2 \geqslant 0.$$

Thus x_0 is a minimizer. If we assume that $f''(x) \geqslant 0$ on some interval about x_0, then x_0 is a local minimizer. Also, if we assume that $f(x) > 0$ on all of X, then x_0 is a strict minimizer.

We summarize this discussion in the following corollary to Theorem 2.2

COROLLARY 1. *Let $f'(x_0) = 0$. If $f''(x) \geqslant 0$ on X, then x_0 is a minimizer. If $f''(x) > 0$ on X, then x_0 is a strict minimizer. If $f''(x) \geqslant 0$ on some interval about x_0, then x_0 is a local minimizer.*

We now take X to be an open set in $\mathbb{R}^n$. The next two theorems are the n-dimensional generalizations of Theorems 2.1 and 2.2.

THEOREM 2.3. *Let X be an open set in $\mathbb{R}^n$, let f be a real-valued function defined on X, and let $\mathbf{x}_0$ be a local minimizer. If f has first-order partial derivatives at $\mathbf{x}_0$, then*

$$\frac{\partial f}{\partial x_i}(\mathbf{x}_0) = 0, \qquad i = 1, \ldots, n. \tag{5}$$

If f is of class $C^{(2)}$ on some open ball $B(\mathbf{x}_0, \delta)$ centered at $\mathbf{x}_0$, then

$$H(\mathbf{x}_0) = \left(\frac{\partial^2 f}{\partial x_j \partial x_i}(\mathbf{x}_0) \right)$$

is positive semidefinite.

Proof. Since X is open, there exists a $\delta > 0$ such that for each $i = 1, \ldots, n$ the points

$$\mathbf{x}_0 + t\mathbf{e}_i = (x_{0,1}, \ldots, x_{0,i-1}, x_{0,i} + t, x_{0,i+1}, \ldots, x_{0,n})$$

are in X whenever $|t| < \delta$. Therefore for each $i = 1, \ldots, n$ the function φ_i defined by

$$\varphi_i(t) = f(\mathbf{x}_0 + t\mathbf{e}_i), \qquad |t| < \delta,$$

is well defined and has a local minimum at $t = 0$. Since the partial derivatives of f exist at $\mathbf{x}_0$, the chain rule implies that for each $i = 1, \ldots, n$ the derivative $\varphi_i'(0)$ exists and is given by

$$\varphi_i'(0) = \frac{\partial f}{\partial x_i}(\mathbf{x}_0).$$

Since each φ_i has a local minimum at $t = 0$, $\varphi_i'(0) = 0$. Thus (5) holds.

Now suppose that f is of class $C^{(2)}$ on some open ball $B(\mathbf{x}_0, \delta)$. Then for each unit vector $\mathbf{u}$ in $\mathbb{R}^n$ the function $\varphi(\ ;\mathbf{u})$ defined by

$$\varphi(t; \mathbf{u}) = f(\mathbf{x}_0 + t\mathbf{u}), \qquad |t| < \delta,$$

is well defined, is of class $C^{(2)}$, and has a local minimum at $t = 0$. By Theorem 2.1, $\varphi'(0, \mathbf{u}) = 0$ and $\varphi''(0; \mathbf{u}) \geqslant 0$. By a straightforward calculation as in relation (13) in Section 11 of Chapter I, we get

$$\varphi''(t; \mathbf{u}) = \langle \mathbf{u}, H(\mathbf{x}_0 + t\mathbf{u})\mathbf{u} \rangle.$$

Setting $t = 0$, we get that $\langle \mathbf{u}, H(\mathbf{x}_0)\mathbf{u} \rangle \geqslant 0$. Since $\mathbf{u}$ is an arbitrary unit vector, we get that $H(\mathbf{x}_0)$ is positive semidefinite.

In $\mathbb{R}^n$ points $\mathbf{c}$ such that $\nabla f(\mathbf{c}) = \mathbf{0}$ are called *critical points* of f.

THEOREM 2.4. *Let X be an open set in $\mathbb{R}^n$ and let f be a real-valued function of class $C^{(2)}$ on some ball $B(\mathbf{x}_0, \delta_1)$. If $H(\mathbf{x}_0)$ is positive definite and $\nabla f(\mathbf{x}_0) = \mathbf{0}$, then $\mathbf{x}_0$ is a strict local minimizer. If X is convex, f is of class $C^{(2)}$ on X, and $H(\mathbf{x})$ is positive semidefinite for all $\mathbf{x}$ in X, then $\mathbf{x}_0$ is a minimizer for f. If $H(\mathbf{x})$ is positive definite for all $\mathbf{x}$ in X, then $\mathbf{x}_0$ is a strict minimizer.*

Proof. We shall only prove the local statement and leave the proof of the last two to the reader.

If $H(\mathbf{x}_0)$ is positive definite at $\mathbf{x}_0$, then it follows from the continuity of the second partial derivatives of f and Lemma III.3.2 that there exists a $\delta > 0$ such that $H(\mathbf{x})$ is positive definite for all $\mathbf{x}$ in $B(\mathbf{x}_0, \delta)$. Let $\mathbf{x}$ be any point in $B(\mathbf{x}_0, \delta)$ and let $\mathbf{v} = \mathbf{x} - \mathbf{x}_0$. Then the function $\varphi(\ ;\mathbf{v})$ defined by

$$\varphi(t; \mathbf{v}) = f(\mathbf{x}_0 + t\mathbf{v})$$

is well defined and is $C^{(2)}$ on some real interval $(-1 - \eta, 1 + \eta)$, where $\eta > 0$. Moreover

$$\varphi(0; \mathbf{v}) = f(\mathbf{x}_0), \qquad \varphi(1; \mathbf{v}) = f(\mathbf{x}). \tag{6}$$

By Taylor's theorem, there exists a $\bar{t}$ in the interval (0, 1) such that

$$\varphi(1; \mathbf{v}) = \varphi(0; \mathbf{v}) + \varphi'(0; \mathbf{v}) + \tfrac{1}{2}\varphi''(\bar{t}; \mathbf{v}). \tag{7}$$

Again, straightforward calculations as in relations (12) and (13) in Section 11 of Chapter I give

$$\varphi'(t; \mathbf{v}) = \langle \nabla f(\mathbf{x}_0 + t\mathbf{v}), \mathbf{v} \rangle,$$

$$\varphi''(t; \mathbf{v}) = \langle \mathbf{v}, H(\mathbf{x}_0 + t\mathbf{v})\mathbf{v} \rangle, \qquad -1 - \eta < t < 1 + \eta.$$

If we set $t = 0$ in the first of these relations, we get that $\varphi'(0; \mathbf{v}) = \langle \nabla f(\mathbf{x}_0), \mathbf{v} \rangle$. If we then set $t = \bar{t}$ in the second of these relations, use the hypothesis $\nabla f(\mathbf{x}_0) = 0$, and substitute into equation (7), we get

$$\varphi(1; \mathbf{v}) - \varphi(0; \mathbf{v}) = \tfrac{1}{2}\langle \mathbf{v}, H(\mathbf{x}_0 + \bar{t}\mathbf{v})\mathbf{v} \rangle.$$

From this relation, from (6), and from the fact that $H(\mathbf{x})$ is positive definite for all $\mathbf{x}$ in $B(\mathbf{x}_0, \delta)$, we get that

$$f(\mathbf{x}) - f(\mathbf{x}_0) = \tfrac{1}{2}\langle \mathbf{v}, H(\mathbf{x}_0 + \bar{t}\mathbf{v})\mathbf{v} \rangle > 0$$

for all $\mathbf{x}$ in $B(\mathbf{x}_0, \delta)$ Hence $\mathbf{x}_0$ is a strict local minimizer.

Remark 2.1. Theorems 2.1 and 2.3 suggest the well-known strategy of solving unconstrained optimization problems by first finding the critical points. Some of these critical points can then be eliminated by considering the second derivative or Hessian at these points. By then considering strengthened conditions on the second derivative or the Hessian as in Theorems 2.2 and 2.4, one can show in some cases that a critical point is indeed a local minimizer or minimizer.

Exercise 2.1. Let X be a closed interval $[a, b]$ in $\mathbb{R}^1$ and let f be a real-valued function defined on X. Show that if $x = a$ is a local minimizer for f and if the right-hand derivative $f'_+(a)$ exists, then $f'_+(a) \geqslant 0$. Show that if $x = b$ is a local minimizer and the left-hand derivative $f'_-(b)$ exists, then $f'_-(b) \leqslant 0$.

Exercise 2.2. Find local minimizers, local maximizers, minimizers, and maximizers, if they exist, for the following functions. Exercise I.9.3 may be useful in some cases:

(a) $f(x_1, x_2) = x_1^2 - 4x_1 + 2x_2^2 + 7$,
(b) $f(x_1, x_2, x_3) = 3x_1^2 + 2x_2^2 + 2x_3^2 + 2x_1x_2 + 2x_2x_3 + 2x_1x_3$,
(c) $f(x_1, x_2) = x_1x_2 + 1/x_1 + 1/x_2$, $(x_1, x_2) \neq (0, 0)$, and
(d) $f(x_1, x_2) = (x_1 + x_2)(x_1x_2 + 1)$.

Exercise 2.3. Let

$$f(x_1, x_2) = x_1^2 + x_2^3, \qquad g(x_1, x_2) = x_1^2 + x_2^4.$$

Show that $\nabla f(\mathbf{0}) = \nabla g(\mathbf{0}) = \mathbf{0}$ and that both f and g have positive semidefinite Hessians at $\mathbf{0}$, but $\mathbf{0}$ is a local minimizer for g but for for f. This shows that the condition of positive definiteness of $H(\mathbf{x}_0)$ in Theorem 2.4 cannot be replaced by positive semidefiniteness.

Exercise 2.4. Complete the proof of Theorem 2.4 by proving the last two statements.

Exercise 2.5. Let

$$f(x, y) = x^4 + y^4 - 32y^2.$$

Minimize f on $\mathbb{R}^2$. (*Hint*: See Exercise I.9.3.)

Exercise 2.6. (a) Show that for no value of a does the function

$$f(x, y) = x^3 - 3axy + y^3$$

have a minimizer.

(b) For each value of a find local minimizers and local maximizers, if any exist, of this function.

Exercise 2.7. Let f be defined on $\mathbb{R}^2$ by

$$f(x, y) = x^3 + e^{3y} - 3xe^y.$$

Show that there is a unique point (x_0, y_0) at which $\nabla f(x, y_0) = 0$ and that this point is a local minimizer but not a minimizer.

Exercise 2.8 (Linear Regression). In the linear regression problem n points $(x_1, y_1), (x_2, y_2), \ldots, (x_n, y_n)$ are given in the xy-plane and it is required to "fit" a straight line $y = ax + b$ to these points in such a way that the sum of the squares of the vertical distances of the given points from the line is minimized. That is, a and b are to be chosen so that

$$f(a, b) = \sum_{i=1}^{n} (ax_i + b - y_i)^2$$

is minimized. The resulting line is called the linear regression line for the given points.

Show that the coefficients a and b of the linear regression line are given by

$$b = \bar{y} - a\bar{x}, \qquad a = \frac{n\bar{x}\bar{y} - \Sigma_{i=1}^{n} x_i y_i}{n(\bar{x})^2 - \Sigma_{i=1}^{n} x_i^2},$$

where

$$\bar{x} = \frac{1}{n} \sum_{i=1}^{n} x_i, \qquad \bar{y} = \frac{1}{n} \sum_{i=1}^{n} y_i.$$

Exercise 2.9 (Best Mean-Square Approximation by Orthogonal Functions). Let $\varphi_1, \ldots, \varphi_n$ be continuous and not identically zero on an interval $[a, b]$ and such that

$$\int_a^b \varphi_j(x)\varphi_k(x)\, dx = 0 \quad \text{if } j \neq k.$$

Given a piecewise continuous function f on $[a, b]$ show that the values of $c_1, \ldots, c_n$ that minimize

$$F(\mathbf{c}) = \int_a^b \left| f - \sum_{i=1}^{n} c_i \varphi_i \right|^2 dx$$

are given by the formulas

$$c_i = \frac{\langle \varphi_i, f \rangle}{\langle \varphi_i, \varphi_i \rangle}, \qquad i = 1, \ldots, n,$$

where

$$\langle \varphi, \psi \rangle = \int_a^b \varphi(x)\psi(x)\, dx.$$

Exercise 2.10. In this exercise we shall establish a series of results, some of which are of interest in their own right, that lead to a proof of Lemma III.3.2.

(a) Let $Q(\mathbf{x}) = \langle \mathbf{x}, A\mathbf{x} \rangle$, where A is a symmetric matrix. Show that if Q is positive definite, then A is nonsingular. *Hint*: If A were singular, there would exist an $\mathbf{x}_0 \neq \mathbf{0}$ such that $A\mathbf{x}_0 = \mathbf{0}$.

(b) Show that if Q is positive semidefinite and A is nonsingular, then Q is positive definite. *Hint*: If A is positive semidefinite, then there exists an $\mathbf{x}_0 \neq \mathbf{0}$ that minimizes Q.

(c) If Q is positive definite, then $\det A > 0$. *Hint*: Let I denote the identity matrix and let $0 \leqslant \lambda \leqslant 1$. For each $0 \leqslant \lambda \leqslant 1$ consider the quadratic form

$$P(\mathbf{x}, \lambda) = (1 - \lambda)\|\mathbf{x}\|^2 + \lambda Q(\mathbf{x}) = \langle \mathbf{x}, [(1 - \lambda)I + \lambda A]\mathbf{x} \rangle.$$

Let $d(\lambda) = \det[(1 - \lambda)I + \lambda A]$. Show that d is a continuous function, that $d(\lambda) \neq 0$, and that $d(0) = 1$.

(d) Let the function f with domain $\mathbb{R}^n$ be defined by

$$f(\mathbf{x}) = \langle \mathbf{x}, A\mathbf{x} \rangle + 2\langle \mathbf{b}, \mathbf{x} \rangle + c,$$

where A is an $n \times n$ positive-definite matrix and $\mathbf{b}$ is an n-vector. Show that $\mathbf{x}_0 = -A^{-1}\mathbf{b}$ in the unique minimizer of f and that

$$f(\mathbf{x}_0) = \langle \mathbf{b}, \mathbf{x}_0 \rangle + c = \frac{\det\begin{pmatrix} A & \mathbf{b} \\ \mathbf{b}^t & c \end{pmatrix}}{\det A}$$

Hint:

$$\det\begin{pmatrix} A & \mathbf{b} \\ \mathbf{b}^t & c \end{pmatrix} = \det\begin{pmatrix} A & A\mathbf{x}_0 + \mathbf{b} \\ \mathbf{b}^t & \langle \mathbf{b}, \mathbf{x}_0 \rangle + c \end{pmatrix}.$$

Why?

(e) Let

$$g(\mathbf{x}, y) = \langle \mathbf{x}, A\mathbf{x} \rangle + 2y\langle \mathbf{b}, \mathbf{x} \rangle + cy^2 = (\mathbf{x}, y)^t B(\mathbf{x}, y),$$

where

$$B = \begin{pmatrix} A & \mathbf{b} \\ \mathbf{b}^t & c \end{pmatrix}.$$

Show that the quadratic form g is positive definite if and only if A is positive definite and $\det B > 0$. *Hints*:

(i) Since $g(\mathbf{x}, 0) = \langle \mathbf{x}, A\mathbf{x} \rangle$, if g is positive definite, then A is positive definite.

(ii) For $y \neq 0$,

$$g(\mathbf{x}, y) = y^2 g\left(\frac{\mathbf{x}}{y}, 1\right)$$

(iii) If A is positive definite, then

$$g(\mathbf{x}, 1) = f(\mathbf{x}) \geqslant f(\mathbf{x}_0).$$

(f) Prove Lemma III.3.2. *Hint*: If A is positive definite, use (e) and induction. If A is negative definite, then $-A$ is positive definite.

3. OPTIMIZATION OF CONVEX FUNCTIONS

The points at which a convex function attains a maximum or minimum have important properties not shared by all functions.

THEOREM 3.1. *Let f be a convex function defined on a convex set C.*

(i) *If f attains a local minimum at $\mathbf{x}_0$, then f attains a minimum at $\mathbf{x}_0$.*
(ii) *The set of points at which f attains a minimum is either empty or convex.*
(iii) *If f is strictly convex and f attains a minimum at $\mathbf{x}_*$, then $\mathbf{x}_*$ is unique.*
(iv) *If f is not a constant function and if f attains a maximum at some point $\mathbf{x}$ in C, then $\mathbf{x}$ must be a boundary point of C.*

Proof. (i) Since f attains a local minimum at $\mathbf{x}_0$, there exists a $\delta > 0$ such that for all $\mathbf{z}$ in $B(\mathbf{x}_0, \delta) \cap C$

$$f(\mathbf{z}) - f(\mathbf{x}_0) \geqslant 0. \tag{1}$$

Let $\mathbf{y}$ be an arbitrary point in C. Then $[\mathbf{x}_0, \mathbf{y}]$ belongs to C. Hence for $0 < t < \delta/\|\mathbf{y} - \mathbf{x}_0\|$ the point $\mathbf{z} = \mathbf{x}_0 + t(\mathbf{y} - \mathbf{x}_0)$ belongs to $B(\mathbf{x}_0, \delta) \cap C$. Hence, from Lemma III.1.7 and (1) we get

$$f(\mathbf{y}) - f(\mathbf{x}_0) \geqslant \frac{f(\mathbf{z}) - f(\mathbf{x}_0)}{t} \geqslant 0.$$

(ii) If f attains a minimum at a unique point, then there is nothing to prove. Suppose that f attains a minimum at two distinct points $\mathbf{x}_1$ and $\mathbf{x}_2$. Let $\mu = f(\mathbf{x}_1) = f(\mathbf{x}_2) = \min\{f(\mathbf{x}) : \mathbf{x} \in C\}$. Then for any (α, β) in P_2 with $\alpha > 0$, $\beta > 0$,

$$\mu \leqslant f(\alpha\mathbf{x}_1 + \beta\mathbf{x}_2) \leqslant \alpha f(\mathbf{x}_1) + \beta f(\mathbf{x}_2) = (\alpha + \beta)\mu = \mu.$$

Hence f attains a minimum at each point of $[\mathbf{x}_1, \mathbf{x}_2]$.

(iii) Suppose that f attains a minimum at two distinct points $\mathbf{x}_1$ and $\mathbf{x}_2$. Let μ be as in (ii). Then for any (α, β) in P_2,

$$\mu \leqslant f(\alpha\mathbf{x}_1 + \beta\mathbf{x}_2) < \alpha f(\mathbf{x}_1) + \beta f(\mathbf{x}_2) = \mu.$$

Thus, we get a contradiction, $\mu < \mu$, and so $\mathbf{x}_1$ and $\mathbf{x}_2$ cannot be distinct.

(iv) Suppose f attains a maximum at $\mathbf{x}$. If $\mathbf{x}$ were not a boundary point, then $\mathbf{x}$ would be an interior point. Since f is not a constant function, there exists a point $\mathbf{y}$ in C such that $f(\mathbf{y}) < f(\mathbf{x})$. Since C is convex, the line segment $[\mathbf{x}, \mathbf{y}]$ belongs to C. Since $\mathbf{x}$ is an interior point, there is a line segment $[\mathbf{z}, \mathbf{y}]$ such that $\mathbf{x}$ is an interior point of $[\mathbf{z}, \mathrm{y}]$. Then $\mathbf{x} = \alpha\mathbf{z} + \beta\mathbf{y}$ for some (α, β) in

P_2 with $\alpha > 0$, $\beta > 0$, and

$$f(\mathbf{x}) \leqslant \alpha f(\mathbf{z}) + \beta f(\mathbf{y}) < \alpha f(\mathbf{x}) + \beta f(\mathbf{x}) = f(\mathbf{x}).$$

Thus we have the contradiction $f(\mathbf{x}) < f(\mathbf{x})$.

The next theorem and its corollary characterize points $\mathbf{x}_*$ at which a function attains a minimum.

THEOREM 3.2. *Let f be a real-valued function defined on a set C. Then f attains a minimum at $\mathbf{x}_*$ in C if and only if $\mathbf{0} \in \partial f(\mathbf{x}_*)$.*

Proof. If f attains a minimum at $\mathbf{x}_*$, then for all $\mathbf{x}$ in C,

$$f(\mathbf{x}) \geqslant f(\mathbf{x}_*) = f(\mathbf{x}_*) + \langle \mathbf{0}, \mathbf{x} - \mathbf{x}_* \rangle. \tag{2}$$

Hence $\mathbf{0} \in \partial f(\mathbf{x}_*)$. Conversely, if $\mathbf{0} \in \partial f(\mathbf{x}_*)$, then (2) holds and f attains a minimum at $\mathbf{x}_*$.

Note that the theorem does not require f or C to be convex. If, however, f is not convex, it can fail to have a subgradient at all points of its domain. Such a function therefore cannot attain a minimum. The logarithm function $\ln x$, or any other logarithm function, is an example. The next corollary does require f and C to be convex.

COROLLARY 1. *Let f be convex and differentiable on an open convex set C. Then f attains a minimum at $\mathbf{x}_*$ in C if and only if $\nabla f(\mathbf{x}_*) = \mathbf{0}$.*

Proof. Let f attain a minimum at $\mathbf{x}_*$. That $\nabla f(\mathbf{x}_*) = \mathbf{0}$ follows from Theorem 2.3, whether or not f and C are convex. If f is convex and differentiable on an open convex set C and $\nabla f(\mathbf{x}_*) = \mathbf{0}$, then by Theorem III.3.1

$$\{\mathbf{0}\} = \{\nabla f(\mathbf{x}_*)\} = \partial f(\mathbf{x}_*).$$

Thus, by the theorem, f attains a minimum at $\mathbf{x}_*$.

Remark 3.1. In geometric terms, Theorem 3.2 states that for a convex function f defined on a convex set C, a point $\mathbf{x}_*$ in C furnishes a minimum if and only if the epigraph of f has a horizontal supporting hyperplane at $(\mathbf{x}_*, f(\mathbf{x}_*))$. The corollary states that for a differentiable convex function defined on an open convex set C, a point $\mathbf{x}_*$ furnishes a minimum if and only if the tangent plane to the graph of f at $\mathbf{x}_*$ is horizontal.

The corollary also points out the true content of Theorem 2.4. By Theorem III.3.3, the positive definiteness or semidefiniteness of $H(\mathbf{x})$ on X implies that

f is convex. It is the convexity of f that is the important element in Theorem 2.4. The alert reader may have observed that much of the proof of Theorem 2.4 was a repetition of the proof of Theorem III.3.3.

Exercise 3.1. Show that $f(\mathbf{x}) = x_1^2 + x_2^2 - 3x_1 - 7x_2$ attains a minimum on $\mathbb{R}^2$ at a unique point and find the minimum.

Exercise 3.2. Maximize $f(\mathbf{x}) = x_1^2 + x_1x_2 + x_2^2$ subject to

$$-3x_1 - 2x_2 + 6 \leqslant 0, \qquad -x_1 + x_2 - 3 \leqslant 0, \qquad x_1 - 2 \leqslant 0.$$

(a) Sketch the feasible set.
(b) Show that a solution exists.
(c) Find the solution.

4. LINEAR PROGRAMMING PROBLEMS

Linear programming is used extensively in resource allocation problems. To illustrate, we formulate a simple job-scheduling problem.

An office furniture manufacturer makes n different types of desks. His profit per desk of type j is b_j. Thus, if each week he produces x_j desks of type j, then his profit for the week is

$$b_1x_1 + \cdots + b_nx_n.$$

The manufacturer's objective is to maximize his profit. The number of desks that he can produce per week is not unlimited but is constrained by the capacity of his plant. Suppose that the manufacture of a type 1 desk requires a_{11} hours per week on machine 1, that the manufacture of a type 2 desk requires a_{12} hours of time on machine 1, and the manufacture of a type j desk requires a_{1j} hours per week of time on machine 1. If x_j, $j = 1, \ldots, n$, denotes the number of desks of type j manufactured per week, then the number of hours that machine 1 is used per week is

$$a_{11}x_1 + a_{12}x_2 + \cdots + a_{1n}x_n.$$

Because of maintenance requirements, readjustments, and the number of shifts employed, the total time that machine 1 is in operation cannot exceed some fixed preassigned number c_1. Thus we have a constraint

$$a_{11}x_1 + \cdots + a_{1n}x_n \leqslant c_1.$$

Similarly, if the manufacture of one desk of type j requires a_{ij} hours per week on machine i and the total number of hours per week that machine i can operate is c_i, then

$$a_{i1}x_1 + \cdots + a_{in}x_n \leqslant c_i, \qquad i = 2, 3, \ldots, n.$$

Also the number of disks of type j made is nonnegative, so $x_j > 0$.

The problem for the desk manufacturer is to manufacture x_j desks of type j so as to maximize

$$\sum_{j=1}^{n} b_j x_j$$

subject to

$$\sum_{j=1}^{n} a_{ij}x_j \leqslant c_i, \qquad i = 1, \ldots, m, \; x_j \geqslant 0, \qquad j = 1, \ldots, n.$$

This is a linear programming problem in standard form.

For a brief account of the origin and early days of linear programming by a pioneer in the field, see Dantzig [1963]. Some early applications are also discussed there.

In this section we shall derive a necessary and sufficient condition that a point $\mathbf{x}_*$ be a solution to a linear programming problem. In practice the use of this condition to solve the problem is precluded by the dimension of the space $\mathbb{R}^n$. Instead, a direct method, the *simplex method*, is used. The simplex method will be discussed in Chapter VI. Nevertheless, we derive the necessary conditions for linear programming problems to illustrate the simplest case of the use of Farkas's theorem (and hence separation theorems) to derive necessary conditions in constrained optimization problems. The basic idea in this proof is a theme that runs throughout much of optimization theory.

The linear programming problem in *standard form* (SLP) is given as

$$\text{Minimize } \langle -\mathbf{b}, \mathbf{x} \rangle$$
$$\text{subject to } A\mathbf{x} \leqslant \mathbf{c}, \qquad \mathbf{x} \geqslant \mathbf{0},$$

where $\mathbf{x} \in \mathbb{R}^n$, A is an $m \times n$ matrix, $\mathbf{b} \neq \mathbf{0}$ is a vector in $\mathbb{R}^n$, and $\mathbf{c}$ is a vector in $\mathbb{R}^m$.

The linear programming problem SLP can also be written as follows: Minimize $\langle -\mathbf{b}, \mathbf{x} \rangle$ subject to

$$\begin{pmatrix} A \\ -I \end{pmatrix} \mathbf{x} \leqslant \begin{pmatrix} \mathbf{c} \\ \mathbf{0} \end{pmatrix}, \tag{1}$$

where I is the $n \times n$ identity matrix and $\mathbf{0}$ is the zero vector in $\mathbb{R}^n$. If we denote the matrix on the left in (2) by $\tilde{A}$ and the vector on the right by $\tilde{\mathbf{c}}$ and then drop the tildes, we can state the linear programming problem (LPI) as

$$\text{minimize } \langle -\mathbf{b}, \mathbf{x} \rangle$$
$$\text{subject to } A\mathbf{x} \leqslant \mathbf{c},$$

where A is an $m \times n$ matrix, $\mathbf{b} \in \mathbb{R}^n$, and $\mathbf{c} \in \mathbb{R}^m$. Here, *we have incorporated the constraint* $\mathbf{x} \geqslant \mathbf{0}$ *in the matrix* A *and the vector* $\mathbf{c}$.

THEOREM 4.1. *Let* $\mathbf{x}_*$ *be a solution of LPI. Then there exists a* $\boldsymbol{\lambda} \neq \mathbf{0}$ *in* $\mathbb{R}^m$ *such that* $(\mathbf{x}_*, \boldsymbol{\lambda})$ *satisfies the following:*

$$\begin{aligned} &(i) \quad A\mathbf{x}_* \leqslant \mathbf{c}, && (ii) \quad A^t\boldsymbol{\lambda} = \mathbf{b}, \\ &(iii) \quad \boldsymbol{\lambda} \geqslant \mathbf{0}, && (iv) \quad \langle A\mathbf{x}_* - \mathbf{c}, \boldsymbol{\lambda} \rangle = 0, \\ &(v) \quad \langle \mathbf{b}, \mathbf{x}_* \rangle = \langle \mathbf{c}, \boldsymbol{\lambda} \rangle. \end{aligned} \tag{2}$$

If there exist an $\mathbf{x}_* \in \mathbb{R}^n$ *and a* $\boldsymbol{\lambda} \neq \mathbf{0}$ *in* $\mathbb{R}^m$ *that satisfy* (i)–(iv), *then* $\mathbf{x}_*$ *is a solution of LPI.*

Proof. The feasible set for LPI is $X = \{\mathbf{x} : A\mathbf{x} \leqslant \mathbf{c}\}$. Since $\mathbf{x}_*$ is a solution to LPI, $\mathbf{x}_* \in X$, and so (i) holds.

We now prove (ii) and (iii). If $\mathbf{x}_*$ in X is a solution to LPI, then $\langle -\mathbf{b}, \mathbf{x}_* \rangle \leqslant \langle -\mathbf{b}, \mathbf{x} \rangle$ for all $\mathbf{x}$ in X. Hence $\langle \mathbf{b}, \mathbf{x} \rangle \leqslant \langle \mathbf{b}, \mathbf{x}_* \rangle$ for all $\mathbf{x}$ in X. If we set $\alpha = \langle \mathbf{b}, \mathbf{x}_* \rangle$, then we have that X is contained in the closed negative half space determined by the hyperplane $H_{\mathbf{b}}^{\alpha}$. By Exercise II.2.16, for $\mathbf{x}$ in $\operatorname{int}(X)$, we have $\langle \mathbf{b}, \mathbf{x} \rangle < \alpha$, and so $\mathbf{x}_*$ cannot be an interior point of X.

Let $\mathbf{a}_i$ denote the ith row of the matrix A. Let

$$E = \{i : \langle \mathbf{a}_i, \mathbf{x}_* \rangle = c_i\}, \qquad I = \{i : \langle \mathbf{a}_i, \mathbf{x}_* \rangle < c_i\}.$$

Since $\mathbf{x}_*$ is not an interior point of X, the set $E \neq \emptyset$. Let A_E denote the submatrix of A consisting of those rows $\mathbf{a}_i$ with i in E, and let A_I denote the submatrix of A consisting of those rows $\mathbf{a}_i$ with i in I.

We claim that the system

$$A_E\mathbf{x} \leqslant \mathbf{0}, \qquad \langle \mathbf{b}, \mathbf{x} \rangle > 0, \tag{3}$$

has no solution $\mathbf{x}$ in $\mathbb{R}^n$.

Intuitively this is clear, for if $\mathbf{x}_0$ would be a solution, then so would $\alpha\mathbf{x}_0$ for $\alpha > 0$, and we would have $A_E(\mathbf{x}_* + \alpha\mathbf{x}_0) \leqslant \mathbf{c}_E$. Also for α sufficiently small, by continuity, we would have $A_I(\mathbf{x}_* + \alpha\mathbf{x}_0) < \mathbf{c}_I$. Thus, if (3) had a solution, we would be able to move into the feasible set and at the same time increase $\langle \mathbf{b}, \mathbf{x}_* \rangle$ or decrease $\langle -\mathbf{b}, \mathbf{x}_* \rangle$. This would contradict the optimality of $\mathbf{x}_*$.

We now make this argument precise. Again, if $\mathbf{x}_0$ were a solution of (3), then $\alpha\mathbf{x}_0$ would be a solution of (3) for all $\alpha > 0$, and

$$\langle -\mathbf{b}, \mathbf{x}_* + \alpha\mathbf{x}_0\rangle = \langle -\mathbf{b}, \mathbf{x}_*\rangle - \alpha\langle \mathbf{b}, \mathbf{x}_0\rangle < \langle -\mathbf{b}, \mathbf{x}_*\rangle \tag{4}$$

for all $\alpha > 0$. Also,

$$A_E(\mathbf{x}_* + \alpha\mathbf{x}_0) = \mathbf{c}_E + \alpha A_E\mathbf{x}_0 \leqslant \mathbf{c}_E, \qquad \alpha > 0, \tag{5}$$

and

$$A_I(\mathbf{x}_* + \alpha\mathbf{x}_0) = A_I\mathbf{x}_* + \alpha A_I\mathbf{x}_0 = (\mathbf{c}_I - \mathbf{p}) + \alpha A_I\mathbf{x}_0,$$

where $\mathbf{p} > \mathbf{0}$ (Recall that $\mathbf{p} > \mathbf{0}$ means that all components of $\mathbf{p}$ are strictly positive.) By taking $\alpha > 0$ to be sufficiently small, we can make $\alpha A_I\mathbf{x}_0 - \mathbf{p} < \mathbf{0}$. Hence for $\alpha > 0$ and sufficiently small

$$A_I(\mathbf{x}_* + \alpha\mathbf{x}_0) = \mathbf{c}_I + (\alpha A_I\mathbf{x}_0 - \mathbf{p}) < \mathbf{c}_I. \tag{6}$$

In summary, from (5) and (6), we get that for sufficiently small $\alpha > 0$ the vectors $\mathbf{x}_* + \alpha\mathbf{x}_0$ are in X. From this and from (4), we get that $\mathbf{x}_*$ is not a solution to the linear programming problem. This contradiction shows that $\mathbf{x}_0$ cannot be a solution of (3).

Since the system (3) has no solution, it follows, from Theorem II.5.1 (Farkas's lemma) that there exists a vector $\mathbf{y}$ in $\mathbb{R}^{m_E}$, where m_E is the row dimension of A_E, such that

$$A_E^t\mathbf{y} = \mathbf{b}, \qquad \mathbf{y} \geqslant \mathbf{0}.$$

Therefore, if m_I is the row dimension of A_I, then

$$A_E^t\mathbf{y} + A_I^t\cdot\mathbf{0} = \mathbf{b},$$

where $\mathbf{0}$ is the zero vector in $\mathbb{R}^{m_I}$. If we now take $\boldsymbol{\lambda} = (\mathbf{y}, \mathbf{0})$, then $\boldsymbol{\lambda}$ is in $\mathbb{R}^m$ and satisfies

$$A^t\boldsymbol{\lambda} = (A_E^t, A_I^t)\begin{pmatrix}\mathbf{y}\\ \mathbf{0}\end{pmatrix} = \mathbf{b}, \qquad \boldsymbol{\lambda} \geqslant 0. \tag{7}$$

Thus, (ii) and (iii) are established.

From (7) we see that those components of $\boldsymbol{\lambda}$ corresponding to indices in I are zero. By the definition of E, $A_E\mathbf{x}_* - \mathbf{c}_E = \mathbf{0}$. Hence

$$\langle A\mathbf{x}_* - \mathbf{c}, \boldsymbol{\lambda}\rangle = \langle (A_E\mathbf{x}_* - \mathbf{c}_E, A_I\mathbf{x}_* - \mathbf{c}_I), (\mathbf{y}, \mathbf{0})\rangle = 0, \tag{8}$$

which is (iv).

From (iv) and (ii) we get

$$\langle \mathbf{c}, \boldsymbol{\lambda} \rangle = \langle A\mathbf{x}_*, \boldsymbol{\lambda} \rangle = \langle \mathbf{x}_*, A^t\boldsymbol{\lambda} \rangle = \langle \mathbf{x}_*, \mathbf{b} \rangle,$$

which is (v).

The proof of sufficiency is elementary and does not involve separation theorems. Let $\mathbf{x}$ be any feasible point. Then

$$\begin{aligned} \langle -\mathbf{b}, \mathbf{x} \rangle - \langle -\mathbf{b}, \mathbf{x}_* \rangle &= \langle -A^t\boldsymbol{\lambda}, \mathbf{x} \rangle + \langle \mathbf{c}, \boldsymbol{\lambda} \rangle \\ &= \langle \boldsymbol{\lambda}, -A\mathbf{x} + \mathbf{c} \rangle = -\langle A\mathbf{x} - \mathbf{c}, \boldsymbol{\lambda} \rangle \geqslant 0. \end{aligned}$$

Hence $\mathbf{x}_*$ is a solution to LPI.

Remark 4.1. An alternate proof of (iv) and (v) is to establish (v) first and then use (v) to establish (iv). Thus

$$\begin{aligned} \langle \mathbf{b}, \mathbf{x}_* \rangle &= \langle A^t\boldsymbol{\lambda}, \mathbf{x}_* \rangle = \langle \boldsymbol{\lambda}, A\mathbf{x}_* \rangle = \left\langle (\mathbf{y}, \mathbf{0}), \begin{pmatrix} A_E \\ A_I \end{pmatrix} \mathbf{x}_* \right\rangle \\ &= \langle \mathbf{y}, A_E\mathbf{x}_* \rangle = \langle \mathbf{y}, \mathbf{c}_E \rangle = \langle (\mathbf{y}, \mathbf{0}), (\mathbf{c}_E, \mathbf{c}_I) \rangle \\ &= \langle \boldsymbol{\lambda}, \mathbf{c} \rangle, \end{aligned}$$

which is (v). To establish (iv), we write

$$\begin{aligned} 0 &= \langle \mathbf{b}, \mathbf{x}_* \rangle - \langle \mathbf{c}, \boldsymbol{\lambda} \rangle = \langle A^t\boldsymbol{\lambda}, \mathbf{x}_* \rangle - \langle \mathbf{c}, \boldsymbol{\lambda} \rangle \\ &= \langle \boldsymbol{\lambda}, A\mathbf{x}_* \rangle - \langle \mathbf{c}, \boldsymbol{\lambda} \rangle = \langle A\mathbf{x}_* - \mathbf{c}, \boldsymbol{\lambda} \rangle. \end{aligned}$$

Remark 4.2. It follows from (i) and (iii) that (iv) holds if and only if for each $i = 1, \ldots, m$

$$(\langle \mathbf{a}_i, \mathbf{x}_* \rangle - c_i)\lambda_i = 0. \qquad \text{(iv)}'$$

That is, (iv)$'$ is equivalent to (iv). The condition (iv)$'$ is sometimes called the *complementary slackness condition.* It states that if there is "slack" in the ith constraint at $\mathbf{x}_*$, that is, if $\langle \mathbf{a}_i, \mathbf{x}_* \rangle < c_i$, then $\lambda_i = 0$.

Remark 4.3. Let the linear programming problem be reformulated with X_0 taken to be an arbitrary open set instead of $\mathbb{R}^n$. It is clear from the proof that if $\mathbf{x}_*$ is a solution to this problem, then the necessary conditions hold. Conversely, if there is a point $\mathbf{x}_*$ in X_0 at which the necessary conditions hold, then $\mathbf{x}_*$ is feasible and the minimum is achieved at $\mathbf{x}_*$.

The inequality constraints $A\mathbf{x} \leqslant \mathbf{c}$ can be converted to equality constraints by introducing a nonnegative vector $\boldsymbol{\xi}$ whose components are called *slack variables* and writing

$$A\mathbf{x} + \boldsymbol{\xi} = \mathbf{c}, \qquad \boldsymbol{\xi} \geqslant \mathbf{0}.$$

Thus the linear programming problem formulated in (1) can be written as

$$\begin{aligned} &\text{Minimize } \langle(-\mathbf{b}, \mathbf{0}), (\mathbf{x}, \boldsymbol{\xi})\rangle \\ &\text{subject to } (A, I)\begin{pmatrix}\mathbf{x}\\ \boldsymbol{\xi}\end{pmatrix} = \mathbf{c}, \qquad \mathbf{x} \geqslant \mathbf{0}, \boldsymbol{\xi} \geqslant \mathbf{0}. \end{aligned}$$

That is, the problem LPI can be written as a *canonical linear programming (CLP) problem*:

$$\begin{aligned} &\text{Minimize } \langle -\mathbf{b}, \mathbf{x}\rangle \\ &\text{subject to } A\mathbf{x} = \mathbf{c}, \qquad \mathbf{x} \geqslant \mathbf{0}. \end{aligned}$$

The simplex method is applied to problems in canonical form.

Conversely, the problem CLP can be written as a problem LPI by writing the equality constraint $A\mathbf{x} = \mathbf{c}$ as two inequality constraints, $A\mathbf{x} \leqslant \mathbf{c}$ and $-A\mathbf{x} \leqslant -\mathbf{c}$, and then incorporating the constraint $\mathbf{x} \geqslant \mathbf{0}$ in the matrix. That is, we write CLP as

$$\begin{aligned} &\text{Minimize } \langle -\mathbf{b}, \mathbf{x}\rangle \\ &\text{subject to } \begin{pmatrix} A \\ -A \\ -I \end{pmatrix}\mathbf{x} \leqslant \begin{pmatrix} \mathbf{c} \\ -\mathbf{c} \\ \mathbf{0} \end{pmatrix}. \end{aligned}$$

Remark 4.4. We warn the reader that the definitions "standard form" and "canonical form" are not uniform throughout the linear programming literature. Problems that we have designated as being in canonical form are designated as being in standard form by some authors who use the designation "canonical form" differently from ours. In consulting the literature the reader should always check the definitions used.

Exercise 4.1. Prove the following corollaries to Theorem 4.1.

COROLLARY 1. *Let $\mathbf{x}_*$ be a solution of the standard linear programming problem SLP. Then there exists a vector $\boldsymbol{\lambda} \neq \mathbf{0}$ in $\mathbb{R}^m$ such that $(\mathbf{x}_*, \boldsymbol{\lambda})$ satisfies*

(i) $A\mathbf{x}_* \leqslant \mathbf{c}$, $\mathbf{x}_* \geqslant 0$;
(ii) $\Sigma_{i=1}^{m} \lambda_i a_{ij} \geqslant b_j$, $j = 1, \ldots, n$, *with equality holding if* $\mathbf{x}_{*j} > 0$;
(iii) $\boldsymbol{\lambda} \geqslant \mathbf{0}$;
(iv) $\langle A\mathbf{x}_* - \mathbf{c}, \boldsymbol{\lambda} \rangle = 0$; *and*
(v) $\langle \mathbf{b}, \mathbf{x}_* \rangle = \langle \mathbf{c}, \boldsymbol{\lambda} \rangle$.

If there exist an $\mathbf{x}_*$ *in* $\mathbb{R}^n$ *and a* $\boldsymbol{\lambda} \neq \mathbf{0}$ *in* $\mathbb{R}^m$ *such that (i)–(v) hold, then* $\mathbf{x}_*$ *is a solution of SLP.*

COROLLARY 2. *Let* $\mathbf{x}_*$ *be a solution of the canonical linear programming problem CLP. Then there exists a vector* $\boldsymbol{\lambda} \neq \mathbf{0}$ *in* $\mathbb{R}^m$ *such that* $(\mathbf{x}_*, \boldsymbol{\lambda})$ *satisfies*

(i) $A\mathbf{x}_* = \mathbf{c}$, $\mathbf{x}_* \geqslant \mathbf{0}$;
(ii) $\Sigma_{i=1}^{m} \lambda_i a_{ij} \geqslant b_j$, $j = 1, \ldots, n$, *with equality holding if* $\mathbf{x}_{*j} > 0$;
(iii) $\langle A\mathbf{x}_* - \mathbf{c}, \boldsymbol{\lambda} \rangle = 0$; *and*
(iv) $\langle \mathbf{b}, \mathbf{x}_* \rangle = \langle \mathbf{c}, \boldsymbol{\lambda} \rangle$.

If there exist an $\mathbf{x}_*$ *in* $\mathbb{R}^n$ *and a* $\boldsymbol{\lambda} \neq \mathbf{0}$ *in* $\mathbb{R}^m$ *such that (i)–(iv) hold, then* $\mathbf{x}_*$ *is a solution of CLP.*

Note that in Corollary 2 the condition $\boldsymbol{\lambda} \geqslant \mathbf{0}$ is absent.

Exercise 4.2. Show that the problem LPI can be written as a problem in standard form.

5. FIRST-ORDER CONDITIONS FOR DIFFERENTIABLE NONLINEAR PROGRAMMING PROBLEMS

In this section we shall derive first-order necessary conditions for differentiable programming problems. As noted in the introduction of this chapter, the set of points that satisfy the necessary conditions need not furnish a solution to the programming problem. If a solution exists, then it belongs to this set. Solution of the necessary conditions corresponds to "bringing in the usual suspects" in a criminal investigation. Further investigation is required to determine which, if any, of these points, or "suspects," furnish a solution. In this vein, we shall show that if the data of the problem are convex, then a point that satisfies the necessary conditions is a solution.

Problem PI

Let X_0 be an open convex set in $\mathbb{R}^n$. Let f, $\mathbf{g}$, and $\mathbf{h}$ be C^1 functions with domain X_0 and ranges in $\mathbb{R}^1$, $\mathbb{R}^m$, and $\mathbb{R}^k$, respectively. Let

$$X = \{\mathbf{x} : \mathbf{x} \in X_0, \mathbf{g}(\mathbf{x}) \leqslant \mathbf{0}, \mathbf{h}(\mathbf{x}) = \mathbf{0}\}.$$

Minimize f over X.

The problem is often stated as

$$\begin{aligned}&\text{Minimize } f(\mathbf{x})\\&\text{subject to } \mathbf{g}(\mathbf{x}) \leqslant \mathbf{0}, \qquad \mathbf{h}(\mathbf{x}) = \mathbf{0}.\end{aligned}$$

If the constraints $\mathbf{g}(\mathbf{x}) \leqslant \mathbf{0}$ and $\mathbf{h}(\mathbf{x}) = \mathbf{0}$ are absent, the problem becomes an unconstrained problem, which was treated in Section 2. We saw there that a necessary condition for a point $\mathbf{x}_*$ to be a solution is that all first-order partial derivatives of f vanish at $\mathbf{x}_*$. For $n = 2, 3$ and $1 \leqslant k < n$, the problem with equality constraints and no inequality constraints is also usually treated in elementary calculus courses. It is stated there, with or without proof, that if $\mathbf{x}_*$ is a solution of the problem and if the k gradient vectors $\nabla h_j(\mathbf{x}_*)$ are linearly independent, then there exist scalars μ_j, $j = 1, \ldots, k$, such that for each $i = 1, \ldots, n$

$$\frac{\partial f}{\partial x_i}(\mathbf{x}_*) + \sum_{j=1}^{k} \mu_j \frac{\partial h_j(\mathbf{x}_*)}{\partial x_i} = 0$$

and

$$h_j(\mathbf{x}_*) = 0, \qquad j = 1, \ldots, k.$$

This is the Lagrange multiplier rule.

We now give a first-order necessary condition that is satisfied by a solution of problem PI. The condition includes a more general version of the Lagrange multiplier rule in the case that inequality constraints are absent.

We recall the notation established in Section 11 of Chapter I. If f is a real-valued function and $\mathbf{g} = (g_1, \ldots, g_m)$ is a vector-valued function that is differentiable at $\mathbf{x}_0$, then

$$\nabla f(\mathbf{x}_0) = \left(\frac{\partial f}{\partial x_1}(\mathbf{x}_0), \frac{\partial f}{\partial x_2}(\mathbf{x}_0), \ldots, \frac{\partial f}{\partial x_n}(\mathbf{x}_0)\right)$$

and

$$\nabla \mathbf{g}(\mathbf{x}_0) = \begin{pmatrix} \nabla g_1(\mathbf{x}_0) \\ \nabla g_2(\mathbf{x}_0) \\ \vdots \\ \nabla g_m(\mathbf{x}_0) \end{pmatrix} = \left(\frac{\partial g_i(\mathbf{x}_0)}{\partial x_j}\right).$$

THEOREM 5.1. *Let $\mathbf{x}_*$ be a solution of problem PI. Then there exists a real number $\lambda_0 \geqslant 0$, a vector $\boldsymbol{\lambda} \geqslant \mathbf{0}$ in $\mathbb{R}^m$, and a vector $\boldsymbol{\mu}$ in $\mathbb{R}^k$ such that*

(i) $(\lambda_0, \boldsymbol{\lambda}, \boldsymbol{\mu}) \neq \mathbf{0}$,
(ii) $\langle \boldsymbol{\lambda}, \mathbf{g}(\mathbf{x}_*) \rangle = 0$, *and*
(iii) $\lambda_0 \nabla f(\mathbf{x}_*) + \boldsymbol{\lambda}^t \nabla \mathbf{g}(\mathbf{x}_*) + \boldsymbol{\mu}^t \nabla \mathbf{h}(\mathbf{x}_*) = \mathbf{0}$.

Vectors $(\lambda_0, \boldsymbol{\lambda}, \boldsymbol{\mu})$ having the properties stated in the theorem are called *multipliers*, as are the components of these vectors.

Definition 5.1. A point $\mathbf{x}_*$ at which the conclusion of Theorem 5.1 holds will be called a *critical point* for problem PI.

Thus, Theorem 5.1 says that a solution $\mathbf{x}_*$ of problem PI is a critical point.

Remark 5.1. The necessary condition asserts that $\lambda_0 \geqslant 0$ and not the stronger statement $\lambda_0 > 0$. If $\lambda_0 > 0$, then we may divide through by λ_0 in (ii) and (iii) and relabel $\boldsymbol{\lambda}/\lambda_0$ and $\boldsymbol{\mu}/\lambda_0$ as $\boldsymbol{\lambda}$ and $\boldsymbol{\mu}$, respectively, and thus obtain statements (i)–(iii) with $\lambda_0 = 1$. In the absence of further conditions that would guarantee that $\lambda_0 > 0$, we cannot assume that $\lambda_0 = 1$. We shall return to this point later.

Remark 5.2. If $\mathbf{g} \equiv \mathbf{0}$, that is, the inequality constraints are absent, Theorem 5.1 becomes the Lagrange multiplier rule.

Remark 5.3. Since $\boldsymbol{\lambda} \geqslant 0$ and $\mathbf{g}(\mathbf{x}_*) \leqslant \mathbf{0}$, condition (ii) is equivalent to

$$\lambda_i g_i(\mathbf{x}_*) = 0, \qquad i = 1, \ldots, m. \tag{ii)$'$}$$

Condition (iii) is a system of n equations

$$\lambda_0 \frac{\partial f}{\partial x_j}(\mathbf{x}_*) + \sum_{i=1}^{m} \lambda_i \frac{\partial g_i(\mathbf{x}_*)}{\partial x_j} + \sum_{i=1}^{k} \mu_i \frac{\partial h_i(\mathbf{x}_*)}{\partial x_j} = 0, \qquad j = 1, \ldots, n. \tag{iii)$'$}$$

Remark 5.4. The necessary conditions are also necessary conditions for a local minimum. For if $\mathbf{x}_*$ is a local minimum, then there exists a $\delta > 0$ such that $f(\mathbf{x}_*) \leqslant f(\mathbf{x})$ for all $\mathbf{x}$ that are in $B(\mathbf{x}_*, \delta) \cap X_0$ and that satisfy the constraints $\mathbf{g}(\mathbf{x}) \leqslant \mathbf{0}$ and $\mathbf{h}(\mathbf{x}) = \mathbf{0}$. Thus, $\mathbf{x}_*$ is a global solution to the problem in which X_0 is replaced by $X_0 \cap B(\mathbf{x}_*, \delta)$.

Remark 5.5. Theorem 5.1, under additional hypotheses guaranteeing that $\lambda_0 > 0$, was proved in Kuhn and Tucker [1951]. This paper had great impact since interest in nonlinear programming was developing at that time in economics and other areas of application. The theorem with $\lambda_0 \geqslant 0$ had been proved in John [1948], and prior to that, again under conditions ensuring that $\lambda_0 > 0$, in Karush, [1939]. The results of John and of Karush were unnoticed when they appeared, probably because interest in programming problems had not yet come about. These papers were looked upon only as mathematical generalizations of the Lagrange multiplier rule. In the literature, Theorem 5.1, with additional hypotheses guaranteeing that $\lambda_0 > 0$, is often referred to as the Karush–Kuhn–Tucker Theorem. Theorem 5.1, as stated here, is referred to as the Fritz John theorem. We shall adopt this terminology and refer to Theorem 5.1 as the *FJ theorem* and call the multipliers $(\lambda_0, \boldsymbol{\lambda}, \boldsymbol{\lambda})$ FJ multipliers.

The next example, due to Fiacco and McCormick [1968], illustrates the use of Theorem 5.1 and will be used to illustrate some subsequent theorems.

Example 5.1. Let $X_0 = R^2$, let

$$f(\mathbf{x}) = (x_1 - 1)^2 + x_2^2,$$

and let

$$g(\mathbf{x}) = 2kx_1 - x_2^2 \leqslant 0, \qquad k > 0.$$

For each value of $k > 0$ we obtain a programming problem π_k. The feasible set X_k for problem π_k is the set of points on and exterior to the graph of the parabola $x_2^2 = 2kx_1$, and the function to be minimized is the distance from $\mathbf{x}$ to the point $(1, 0)$. In Figure 4.1 we have sketched some level curves of f and the parabolas for two values of k. Although the solution can be read from Figure 4.1, we will apply the appropriate theorems in this section and in the next section to solve this problem.

By straightforward calculation

$$\nabla f = (2(x_1 - 1),\ 2x_2), \qquad \nabla g = (2k,\ -2x_2).$$

From Theorem 5.1 we get that if a point $\mathbf{x}$ is a solution to the programming problem π_k, then there exist a $\lambda_0 \geqslant 0$ and a $\lambda \geqslant 0$ with $(\lambda_0, \lambda) \neq (0, 0)$ such that (λ_0, λ) and $\mathbf{x}$ satisfy

$$\lambda_0(2(x_1 - 1),\ 2x_2) + \lambda(2k,\ -2x_2) = (0,\ 0), \tag{1}$$

$$\lambda(2kx_1 - x_2^2) = 0. \tag{2}$$

If $\lambda_0 = 0$, then since $\lambda \neq 0$, equation (1) would imply that $k = 0$, contradicting the assumption that $k > 0$. Hence we may take $\lambda_0 = 1$. If $\lambda = 0$, then (1) would imply that $x_1 = 1$ and $x_2 = 0$. The point $(1, 0)$ is not feasible for any value of $k > 0$, so we may assume that $\lambda \neq 0$ in (1). It then follows from (2) that $2kx_1 - x_2^2 = 0$; that is, the point $\mathbf{x}$ must lie in the parabola. Thus we may replace (1) and (2) by

$$((x_1 - 1),\ x_2) + \lambda(k,\ -x_2) = (0,\ 0), \qquad x_2^2 = 2kx_1. \tag{3}$$

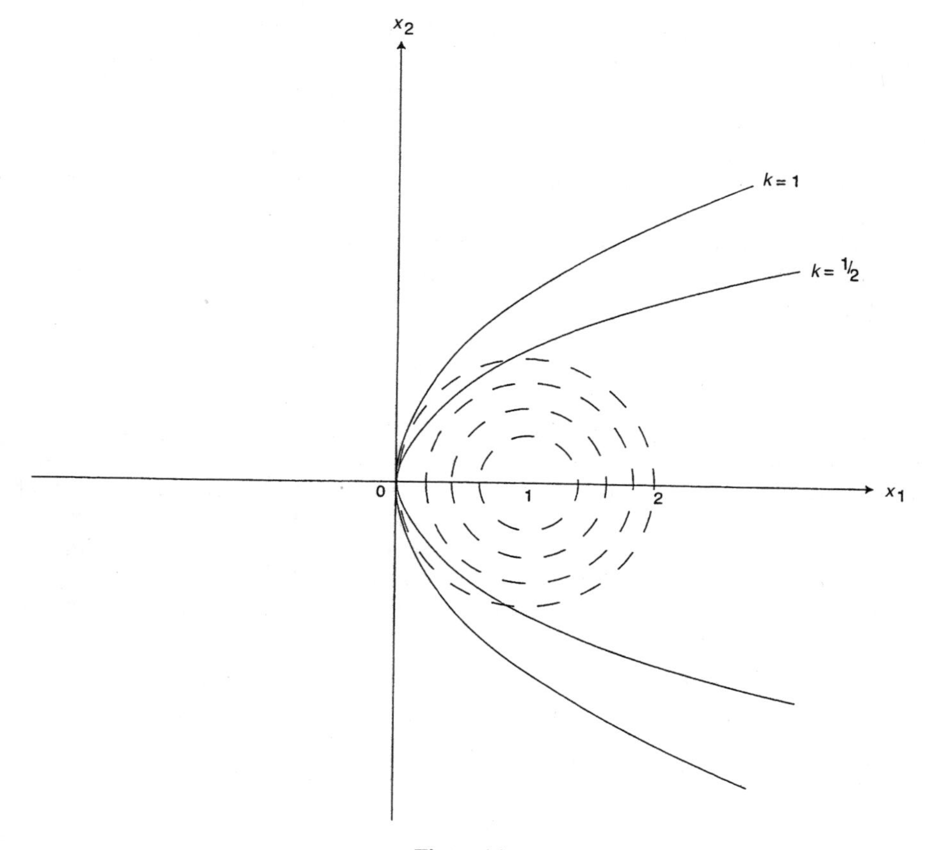

Figure 4.1.

If $x_2 \neq 0$, then we must have $\lambda = 1$, and so

$$x_1 = 1 - k, \qquad x_2 = \pm\sqrt{2k(1-k)}, \qquad 0 < k < 1.$$

If $x_2 = 0$, then from the second equation in (3) we get that $x_1 = 0$, and consequently $\lambda = 1/k$.

In summary, we have shown that if $0 < k < 1$, then the only points $\mathbf{x} = (x_1, x_2)$ and multipliers $(1, \lambda)$ that satisfy the necessary conditions for problem π_k are

$$\lambda = 1, \qquad x_1 = 1 - k \qquad x_2 = \pm\sqrt{2k(1-k)}, \tag{4}$$

$$\lambda = \frac{1}{k} \qquad x_1 = 0 \qquad x_2 = 0. \tag{5}$$

If $k \geqslant 1$, the necessary conditions are only satisfied by (5). Note that if $k = 1$, then (4) and (5) both give $\lambda = 1$ and $x_1 = x_2 = 0$.

We emphasize that Theorem 5.1 is a necessary condition, so that without further arguments we cannot determine which, if any, of the points (4) and (5) furnish a minimum In the problem π_k we are minimizing the distance from the point $(1, 0)$ to the closed set X_k. By Lemma II.3.3 a solution to this problem exists. Since for $k \geqslant 1$ the only point that satisfies the necessary conditions is the origin, this point is the solution to problem π_k for $k \geqslant 1$. If we denote the two points in (4) by $\mathbf{x}_1$ and $\mathbf{x}_2$, then

$$f(\mathbf{x}_1) = f(\mathbf{x}_2) = -k^2 + 2k.$$

The point in (5) is the origin, and $f(\mathbf{0}) = 1$. Since $-k^2 + 2k < 1$ for $k \neq 1$, a solution to problem π_k for $0 < k < 1$ is given by each of the points $\mathbf{x}_1$ and $\mathbf{x}_2$.

Before we present our proof of Theorem 5.1, we shall present the principal idea of a proof which, with variations, is popular in the literature. We suppose that the equality constraints have been converted into inequality constraints, $\mathbf{h}(\mathbf{x}) \leqslant \mathbf{0}$ and $-\mathbf{h}(\mathbf{x}) \leqslant \mathbf{0}$, so that the only constraints are $\mathbf{g}(\mathbf{x}) \leqslant \mathbf{0}$. By a translation of coordinates, we can assume that $\mathbf{x}_* = \mathbf{0}$ and that $f(\mathbf{x}_*) = f(\mathbf{0}) = 0$. We also assume that $\nabla f(\mathbf{0}) \neq \mathbf{0}$.

As in the linear programming problem, let

$$E = \{i : g_i(\mathbf{0}) = 0\}, \qquad I = \{i : g_i(\mathbf{0}) < 0\}. \tag{6}$$

The constraints $g_i(\mathbf{c}) = 0$ with i in E are said to be *active* or *binding* at $\mathbf{0}$; those with $i \in I$ are said to be *inactive* at $\mathbf{0}$.

Since f and $\mathbf{g}$ are C^1, they are differentiable by Theorem I.11.2. Therefore, since $f(\mathbf{0}) = 0$ and $\mathbf{g}_E(\mathbf{0}) = \mathbf{0}$,

$$f(\mathbf{x}) = \langle \nabla f(\mathbf{0}), \mathbf{x} \rangle + o(\|\mathbf{x}\|),$$

$$\mathbf{g}_E(\mathbf{x}) = \nabla \mathbf{g}_E(\mathbf{0})\mathbf{x} + \mathbf{o}(\|\mathbf{x}\|),$$

where $o(\|\mathbf{x}\|)$ and $\mathbf{o}(\|\mathbf{x}\|)$ represent terms such that $o(\|\mathbf{x}\|)/\|\mathbf{x}\| \to 0$ and $\mathbf{o}(\|\mathbf{x}\|)/\|\mathbf{x}\| \to \mathbf{0}$ as $\|\mathbf{x}\| \to \mathbf{0}$. Thus, $o(\|\mathbf{x}\|)$ and $\mathbf{o}(\|\mathbf{x}\|)$ represent *higher order terms.*

Since $\mathbf{g}$ is continuous and $\mathbf{g}_I(\mathbf{0}) < \mathbf{0}$, there exists a $\delta > 0$ such that $\mathbf{g}_I(\mathbf{x}) < \mathbf{0}$ whenever $\|\mathbf{x}\| < \delta$. The origin is therefore a solution of the local problem: Minimize $f(\mathbf{x})$ subject to $\mathbf{g}_E(\mathbf{x}) \leqslant \mathbf{0}$ and $\|\mathbf{x}\| < \delta$. If we drop the terms $o(\|\mathbf{x}\|)$ and $\mathbf{o}(\|\mathbf{x}\|)$, that is, if we linearize the problem about the solution $\mathbf{0}$ and assume that the problem is the linearized problem, then $\mathbf{x}_* = \mathbf{0}$ is the solution of the *linear* programming problem:

$$\text{Minimize } \langle \nabla f(\mathbf{0}), \mathbf{x} \rangle = \langle -(-\nabla f(\mathbf{0})), \mathbf{x} \rangle,$$

$$\text{subject to } \nabla \mathbf{g}_E(\mathbf{0})\mathbf{x} \leqslant \mathbf{0}, \qquad \|\mathbf{x}\| < \delta. \tag{7}$$

By Remark 4.3, we can now apply Theorem 4.1 with $A = \nabla \mathbf{g}_E(\mathbf{0})$ and $\mathbf{b} = -\nabla f(\mathbf{0})$ to the linearized problem and obtain the existence of a vector $\mathbf{v}_E \neq \mathbf{0}$ in $\mathbb{R}^{m_E}$ such that $\mathbf{v}_E \geqslant \mathbf{0}$ and

$$\mathbf{v}_E^t \nabla \mathbf{g}_E(\mathbf{0}) = -\nabla f(\mathbf{0}).$$

If we now set $\boldsymbol{\lambda} = (\mathbf{v}_E, \mathbf{0}_I)$, then $\boldsymbol{\lambda} \geqslant 0$ and we can rewrite this equation as

$$\nabla f(\mathbf{0}) + \mathbf{v}_E^t \nabla \mathbf{g}_E(\mathbf{0}) + \mathbf{0}_I^t \nabla \mathbf{g}_I(\mathbf{0}) = \nabla f(0) + \boldsymbol{\lambda}^t \nabla \mathbf{g}(\mathbf{0}) = \mathbf{0}.$$

The rightmost equation is statement (iii) of Theorem 5.1 with $\lambda_0 = 1$. (Recall that we are taking all the constraints to be inequality constraints.) Since $\lambda_0 = 1$, $(\lambda_0, \boldsymbol{\lambda}) \neq \mathbf{0}$. Also

$$\langle \boldsymbol{\lambda}, \mathbf{g}(\mathbf{0}) \rangle = \langle (\mathbf{v}_E, \mathbf{0}_I), (\mathbf{g}_E(\mathbf{0}), \mathbf{g}_I(\mathbf{0})) \rangle = 0,$$

which is (ii). Note that if the linearization is valid, then we can get a set of multipliers $(\lambda_0, \boldsymbol{\lambda})$ with $\lambda_0 = 1$.

To complete the proof, one must show that the terms $o(\|\mathbf{x}\|)$ and $\mathbf{o}(\|\mathbf{x}\|)$ can indeed be neglected. That is, it must be shown that if the terms represented by $o(\|\mathbf{x}\|)$ and $\mathbf{o}(\|\mathbf{x}\|)$ are absent, then $\mathbf{0}$ is a solution of (7). This part of the proof becomes quite technical and is carried out in different ways by different authors depending on the additional assumptions that are made. In Exercise 5.10 below the reader will be asked to carry out one such proof, the original proof by Kuhn and Tucker.

Our proof of Theorem 5.1, which we now present, will be a "penalty function" proof and is due to McShane [1973].

Proof. As noted in the preceding discussion, we may assume that $\mathbf{x}_* = \mathbf{0}$ and that $f(\mathbf{x}_*) = 0$.

Let E and I be as in (6). To simplify the notation, we assume that $E = \{1, 2, \ldots, r\}$ and that $I = \{r+1, \ldots, m\}$. Since E or I can be empty, we have $0 \leqslant r \leqslant m$.

Let ω be a function from $(-\infty, \infty)$ to $\mathbb{R}^1$ such that (i) ω is strictly increasing on $(0, \infty)$, (ii) $\omega(u) = 0$ for $u \leqslant 0$, (iii) ω is C^1, and (iv) $\omega(u) > 0$ for $u > 0$. Note that $\omega'(u) > 0$ for $u > 0$.

Since $\mathbf{g}$ is continuous and X_0 is open, there exists an $\varepsilon_0 > 0$ such that $B(\mathbf{0}, \varepsilon_0) \subset X_0$ and $\mathbf{g}_I(\mathbf{x}) < \mathbf{0}$ for $\mathbf{x} \in B(\mathbf{0}, \varepsilon_0)$.

Define a *penalty function* F as follows:

$$F(\mathbf{x}, p) = f(\mathbf{x}) + \|\mathbf{x}\|^2 + p \left\{ \sum_{i=1}^{r} \omega(g_i(\mathbf{x})) + \sum_{i=1}^{k} [h_i(\mathbf{x})]^2 \right\}, \tag{8}$$

where $\mathbf{x} \in X_0$ and p is a positive integer. We assert that for each ε satisfying $0 < \varepsilon < \varepsilon_0$, there exists a positive integer $p(\varepsilon)$ such that for $\mathbf{x}$ with $\|\mathbf{x}\| = \varepsilon$

$$F(\mathbf{x}, p(\varepsilon)) > 0. \tag{9}$$

We prove the assertion by assuming it to be false and arriving at a contradiction. If the assertion were false, then there would exist an ε' with $0 < \varepsilon' < \varepsilon_0$ such that for each positive integer p there exists a point $\mathbf{x}_p$ with $\|\mathbf{x}_p\| = \varepsilon'$ and $F(\mathbf{x}_p, p) \leqslant 0$. Hence, from (8)

$$f(\mathbf{x}_p) + \|\mathbf{x}_p\|^2 \leqslant -p\left\{\sum_{i=1}^{r} \omega(g_i(\mathbf{x}_p)) + \sum_{i=1}^{k} [h_i(\mathbf{x}_p)]^2\right\}. \tag{10}$$

Since $\|\mathbf{x}_p\| = \varepsilon'$ and since $S(\mathbf{0}, \varepsilon') = \{\mathbf{y}: \|\mathbf{y}\| = \varepsilon'\}$ is compact, there exist subsequences, which we relabel as $\mathbf{x}_p$ and p, and a point $\mathbf{x}_0$ with $\|\mathbf{x}_0\| = \varepsilon'$ such that $\mathbf{x}_p \to \mathbf{x}_0$. Since f, $\mathbf{g}$, and $\mathbf{h}$ are continuous,

$$f(\mathbf{x}_p) \to f(\mathbf{x}_0), \qquad \mathbf{g}_E(\mathbf{x}_p) \to \mathbf{g}_E(\mathbf{x}_0), \qquad \mathbf{h}(\mathbf{x}_p) \to \mathbf{h}(\mathbf{x}_0).$$

Therefore, if in (10) we divide through by $-p$ and then let $p \to \infty$, we get

$$0 \geqslant \sum_{i=1}^{r} \omega(g_i(\mathbf{x}_0)) + \sum_{i=1}^{k} [h_i(\mathbf{x}_0)]^2 \geqslant 0.$$

Hence for each $i = 1, \ldots, r$ we have $g_i(\mathbf{x}_0) \leqslant 0$, and for each $i = 1, \ldots, k$ we have $h_i(\mathbf{x}_0) = 0$. Since $\|\mathbf{x}_0\| = \varepsilon' < \varepsilon_0$, we have $g_I(\mathbf{x}_0) < \mathbf{0}$. Thus $\mathbf{x}_0$ is a feasible point and so

$$f(\mathbf{x}_0) \geqslant f(\mathbf{0}) = 0. \tag{11}$$

On the other hand, from (10) and from $\|\mathbf{x}_p\| = \varepsilon'$ we get that $f(\mathbf{x}_p) \leqslant -(\varepsilon')^2$, and so

$$f(\mathbf{x}_0) \leqslant -(\varepsilon')^2 < 0,$$

which contradicts (11). This proves the assertion.

For each ε in $(0, \varepsilon_0)$ the function $F(\ , p(\varepsilon))$ is continuous on the closed ball $\overline{B(\mathbf{0}, \varepsilon)}$. Since $\overline{B(\mathbf{0}, \varepsilon)}$ is compact, $F(\ , p(\varepsilon))$ attains its minimum on $\overline{B(\mathbf{0}, \varepsilon)}$ at some point $\mathbf{x}_\varepsilon$ with $\|\mathbf{x}_\varepsilon\| \leqslant \varepsilon$.

Since $F(\mathbf{x}, p(\varepsilon)) > 0$ for $\mathbf{x}$ with $\|\mathbf{x}\| = \varepsilon$, and since $F(\mathbf{0}, p(\varepsilon)) = f(\mathbf{0}) = 0$, it follows that $F(\ , p(\varepsilon))$ attains its minimum on $\overline{B(\mathbf{0}, \varepsilon)}$ at an *interior* point $\mathbf{x}_\varepsilon$ of $\overline{B(\mathbf{0}, \varepsilon)}$. Hence,

$$\frac{\partial F}{\partial x_j}(\mathbf{x}_\varepsilon, p(\varepsilon)) = 0, \qquad j = 1, \ldots, n.$$

Calculating $\partial F/\partial x_j$ from (8) gives

$$\frac{\partial f}{\partial x_j}(\mathbf{x}_\varepsilon) + 2x_{\varepsilon_j} + \sum_{i=1}^{r} p(\varepsilon)\omega'(g_i(\mathbf{x}_\varepsilon))\frac{\partial g_i}{\partial x_j}(\mathbf{x}_\varepsilon) + \sum_{i=1}^{k} 2p(\varepsilon)h_i(\mathbf{x}_\varepsilon)\frac{\partial h_i}{\partial x_j}(\mathbf{x}_\varepsilon) = 0 \quad \text{for } j = 1, \ldots, n. \tag{12}$$

Define

$$\begin{aligned}
L(\varepsilon) &= 1 + \sum_{i=1}^{r} [p(\varepsilon)\omega'(g_i(\mathbf{x}_\varepsilon))]^2 + \sum_{i=1}^{k} [2p(\varepsilon)h_i(\mathbf{x}_\varepsilon)]^2, \\
\lambda_0(\varepsilon) &= \frac{1}{\sqrt{L(\varepsilon)}}, \\
\lambda_i(\varepsilon) &= \frac{p(\varepsilon)\omega'(g_i(\mathbf{x}_\varepsilon))}{\sqrt{L(\varepsilon)}}, \qquad i = 1, \ldots, r, \\
\lambda_i(\varepsilon) &= 0, \qquad i = r+1, \ldots, m, \\
\mu_i(\varepsilon) &= \frac{2p(\varepsilon)h_i(\mathbf{x}_\varepsilon)}{\sqrt{L(\varepsilon)}}, \qquad i = 1, \ldots, k.
\end{aligned}$$

Note that

$$\begin{aligned}
&\text{(i)} \quad \lambda_0(\varepsilon) > 0, \\
&\text{(ii)} \quad \lambda_i(\varepsilon) \geqslant 0, \qquad i = 1, \ldots, r, \\
&\text{(iii)} \quad \lambda_i(\varepsilon) = 0, \qquad i = r+1, \ldots, m, \\
&\text{(iv)} \quad \|(\lambda_0(\varepsilon), \boldsymbol{\lambda}(\varepsilon), \boldsymbol{\mu}(\varepsilon))\| = 1,
\end{aligned} \tag{13}$$

where $\boldsymbol{\lambda}(\varepsilon) = (\lambda_1(\varepsilon), \ldots, \lambda_m(\varepsilon))$ and $\boldsymbol{\mu}(\varepsilon) = (\mu_1(\varepsilon), \ldots, \mu_k(\varepsilon))$.

If we divide through by $\sqrt{L(\varepsilon)}$ in (12), we get

$$\lambda_0(\varepsilon)\frac{\partial f}{\partial x_j}(\mathbf{x}_\varepsilon) + \frac{2x_{\varepsilon_j}}{\sqrt{L(\varepsilon)}} + \sum_{i=1}^{r} \lambda_i(\varepsilon)\frac{\partial g_i(\mathbf{x}_\varepsilon)}{\partial x_j} + \sum_{i=1}^{k} \mu_i(\varepsilon)\frac{\partial h_i(\mathbf{x}_\varepsilon)}{\partial x_j} = 0. \tag{14}$$

Now let $\varepsilon \to 0$ through a sequence of values ε_k. Then since $\|\mathbf{x}_\varepsilon\| < \varepsilon$, we have that

$$\mathbf{x}_{\varepsilon_k} \to \mathbf{0}. \tag{15}$$

Since the vectors $(\lambda_0(\varepsilon), \boldsymbol{\lambda}(\varepsilon), \boldsymbol{\mu}(\varepsilon))$ are all unit vectors [see (13)], there exists a subsequence, that we again denote as ε_k, and a unit vector $(\lambda_0, \boldsymbol{\lambda}, \boldsymbol{\mu})$ such that

$$(\lambda_0(\varepsilon_k), \boldsymbol{\lambda}(\varepsilon_k), \boldsymbol{\mu}(\varepsilon_k)) \to (\lambda_0, \boldsymbol{\lambda}, \boldsymbol{\mu}). \tag{16}$$

Since $(\lambda_0, \boldsymbol{\lambda}, \boldsymbol{\mu})$ is a unit vector, it is different from zero.

From (14), (15), (16), and the continuity of the partials of f, $\mathbf{g}$, and $\mathbf{h}$, we get

$$\lambda_0 \frac{\partial f}{\partial x_j}(\mathbf{0}) + \sum_{i=1}^{r} \lambda_i \frac{\partial g_i(\mathbf{0})}{\partial x_j} + \sum_{i=1}^{k} \mu_i \frac{\partial h_i(\mathbf{0})}{\partial x_j} = 0. \tag{17}$$

From (13) and (16) we see that $\lambda_i \geqslant 0$ for $i = 0, 1, \ldots, r$ and $\lambda_i = 0$ for $i = r+1, \ldots, m$. Thus, $\lambda_0 \geqslant 0$ and $\boldsymbol{\lambda} \geqslant \mathbf{0}$. Since $g_i(\mathbf{0}) = 0$ for $i = 1, \ldots, r$ and $\lambda_i = 0$ for $i = r+1, \ldots, m$, we have that $\lambda_i g_i(\mathbf{0}) = 0$ for $i = 1, \ldots, m$. Also, we can take the upper limit in the second term in (17) to be m and write

$$\lambda_0 \frac{\partial f}{\partial x_j}(\mathbf{0}) + \sum_{i=1}^{m} \lambda_i \frac{\partial g_i(\mathbf{0})}{\partial x_j} + \sum_{i=1}^{k} \mu_i \frac{\partial h_i(\mathbf{0})}{\partial x_j} = 0.$$

This completes the proof of the theorem.

We now present conditions that ensure that $\lambda_0 = 0$ cannot occur in Theorem 5.1. Such conditions are called *constraint qualifications* and are statements about the geometry of the feasible set in a neighborhood of $\mathbf{x}_*$. There are many constraint qualifications in the literature. We have selected two that are relatively simple and easy to apply. The second is easier to apply, but the first, given in Definition 5.2 below, will capture more points at which $\lambda_0 = 0$ cannot occur. All of the constraint qualifications that are appropriate to problem PI have the defect that, in general, they require knowledge of the point $\mathbf{x}_*$ rather than requiring knowledge of global properties of the functions involved. The reader who is interested in other constraint qualifications and the relationships among them is referred to Bazzara, Sherali, and Shetty [1993] and Mangasarian [1969].

Definition 5.2. The functions $\mathbf{g}$ and $\mathbf{h}$ satisfy the *constraint qualification CQ* at a feasible point $\mathbf{x}_0$ if

(i) the vectors $\nabla h_1(\mathbf{x}_0), \ldots, \nabla h_k(\mathbf{x}_0)$ are linearly independent and
(ii) the system

$$\nabla \mathbf{g}_E(\mathbf{x}_0)\mathbf{z} < \mathbf{0}, \qquad \nabla \mathbf{h}(\mathbf{x}_0)\mathbf{z} = 0, \tag{18}$$

has a solution $\mathbf{z}$ in $\mathbb{R}^n$. Here, $E = \{i : g_i(\mathbf{x}_0) = 0\}$.

If the equality constraints are absent, then the statements involving $\nabla \mathbf{h}(\mathbf{x}_0)$ are to be deleted.

Condition (ii) of the constraint qualification says that "to first order" we can move from $\mathbf{x}_0$ into the feasible set by moving along the surface defined by the equality constraints $\mathbf{h}(\mathbf{x}) = \mathbf{0}$ and into the interior of the set defined by $\mathbf{g}(\mathbf{x}) \leqslant \mathbf{0}$. It turns that this implies that we can actually move from $\mathbf{x}_0$ into the feasible set in this fashion. We shall prove this in the corollary to Lemma 6.1 below.

Note that CQ does not impose any condition on the point $\mathbf{x}_0$ other than that it be feasible. Also note that $\mathbf{z} = \mathbf{0}$ cannot be a solution of (18).

The constraint qualification CQ of Definition 5.2 was introduced by Mangasarian and Fromovitz [1967].

THEOREM 5.2. *Let $\mathbf{x}_*$ be a solution of problem PI and let CQ hold at $\mathbf{x}_*$. Then $\lambda_0 > 0$ and there exists a vector $\boldsymbol{\lambda} \geqslant 0$ in $\mathbb{R}^m$ and a vector $\boldsymbol{\mu}$ in $\mathbb{R}^k$ such that*

(*i*) $\langle \boldsymbol{\lambda}, \mathbf{g}(\mathbf{x}_*) \rangle = 0$ and
(*ii*) $\nabla f(\mathbf{x}_*) + \boldsymbol{\lambda}^t \nabla \mathbf{g}(\mathbf{x}_*) + \boldsymbol{\mu}^t \nabla \mathbf{h}(\mathbf{x}_*) = \mathbf{0}$.

We shall refer to the necessary conditions of Theorem 5.2 as the *Karush–Kuhn–Tucker* (KKT) conditions and to the multipliers of Theorem 5.2 as the *KKT multipliers.*

We now prove the theorem. First we will show that if CQ holds at $\mathbf{x}_*$, then λ_0 in Theorem 5.1 cannot be zero. Hence if $(\lambda_0, \boldsymbol{\lambda}, \boldsymbol{\mu})$ is as in Theorem 5.1, so is $(1, \boldsymbol{\lambda}/\lambda_0, \boldsymbol{\mu}/\lambda_0)$ and Theorem 5.2 follows. Note that once we have $\lambda_0 > 0$, then $(\lambda_0, \boldsymbol{\lambda}, \boldsymbol{\mu}) \neq (0, \mathbf{0}_m, \mathbf{0}_k)$.

We now show that $\lambda_0 = 0$ cannot occur. If $\lambda_0 = 0$ did occur, then by (iii) of Theorem 5.1

$$\boldsymbol{\lambda}^t \nabla \mathbf{g}(\mathbf{x}_*) + \boldsymbol{\mu}^t \nabla h(\mathbf{x}_*) = \mathbf{0}. \tag{19}$$

If $\boldsymbol{\lambda} = \mathbf{0}$, then $\boldsymbol{\mu} \neq \mathbf{0}$, since $(\lambda_0, \boldsymbol{\lambda}, \boldsymbol{\mu}) = (0, \mathbf{0}, \boldsymbol{\mu}) \neq \mathbf{0}$. But then $\boldsymbol{\mu}^t \nabla h(\mathbf{x}_*) = \mathbf{0}$, contradicting the linear independence of $\nabla h_1(\mathbf{x}_*), \ldots, \nabla h_k(\mathbf{x}_*)$. Thus, $\boldsymbol{\lambda} \neq \mathbf{0}$.

Since $\boldsymbol{\lambda} = (\boldsymbol{\lambda}_E, \boldsymbol{\lambda}_I)$ and $\boldsymbol{\lambda}_I = \mathbf{0}_I$, we have that $\boldsymbol{\lambda}_E \neq \mathbf{0}$. We may therefore rewrite (19) as

$$\boldsymbol{\lambda}_E^t \nabla \mathbf{g}_E(\mathbf{x}_*) + \boldsymbol{\mu}^t \nabla h(\mathbf{x}_*) = \mathbf{0}. \tag{20}$$

Multiplying on the right by a vector $\mathbf{z}$ that is a solution of (18) gives

$$\boldsymbol{\lambda}_E^t \nabla g_E(\mathbf{x}_*)\mathbf{z} = 0.$$

Since $\boldsymbol{\lambda}_E \neq \mathbf{0}$, $\boldsymbol{\lambda}_E \geqslant \mathbf{0}$, and $\nabla \mathbf{g}_E(\mathbf{x}_*)\mathbf{z} < \mathbf{0}$, we have a contradiction. Thus $\lambda_0 \neq 0$.

COROLLARY 1. *Let $\mathbf{x}_*$ be a solution of problem PI such that the set E defined in (6) is $\{1, \ldots, r\}$ and such that the vectors*

$$\nabla g_1(\mathbf{x}_*), \ldots, \nabla g_r(\mathbf{x}_*), \nabla h_1(\mathbf{x}_*), \ldots, \nabla h_k(\mathbf{x}_*) \tag{21}$$

are linearly independent. Then the conclusion of Theorem 5.2 holds.

Proof. If $\lambda_0 = 0$ in Theorem 5.1, then (20) holds. Since the vectors in (21) are linearly independent, this implies that $(\boldsymbol{\lambda}_E, \boldsymbol{\mu}) = \mathbf{0}$. Hence $(\lambda_0, \boldsymbol{\lambda}, \boldsymbol{\mu}) = \mathbf{0}$, which cannot occur, and so $\lambda_0 > 0$.

We leave to the reader the modifications of the arguments in the proofs of Theorem 5.2 and Corollary 1 if the equality constraints are absent.

Remark 5.6. We noted in Remark 5.3 that the necessary condition (i) of Theorem 5.2 is equivalent to the m equations

$$\lambda_i g_i(\mathbf{x}_*) = 0, \qquad i = 1, \ldots, m. \tag{i)$'$}$$

Condition (ii) in component form gives the system of n equations

$$\frac{\partial f}{\partial x_j}(\mathbf{x}_*) + \sum_{i=1}^{m} \lambda_i \frac{\partial g_i(\mathbf{x}_*)}{\partial x_j} + \sum_{i=1}^{r} \mu_i \frac{\partial h_i}{\partial x_j} = 0, \qquad i = 1, \ldots, n. \tag{ii)$'$}$$

Since $\mathbf{x}_*$ is a solution, it is feasible, and so we obtain an additional k equations

$$h_i(\mathbf{x}_*) = 0, \qquad i = 1, \ldots, k.$$

Thus we have a system of $m + n + k$ equations in the $m + n + k$ unknowns

$$\boldsymbol{\lambda} = (\lambda_1, \ldots, \lambda_m), \qquad \mathbf{x}_* = (x_{*1}, \ldots, x_{*n}), \qquad \boldsymbol{\mu} = (\mu_1, \ldots, \mu_k).$$

Remark 5.7. If the inequality constraints are absent and we only have the equality constraints $\mathbf{h}(\mathbf{x}) = \mathbf{0}$, then CQ becomes "$\nabla h_1(\mathbf{x}_0), \ldots, \nabla h_k(\mathbf{x}_0)$ are linearly independent." The constraint in Corollary 1 reduces to the same statement. Under these circumstances, the multiplier $(1, \boldsymbol{\mu})$ is unique. This follows from the observation that if $(1, \boldsymbol{\mu}')$ with $\boldsymbol{\mu} \neq \boldsymbol{\mu}'$ is another multiplier, then $(0, \boldsymbol{\mu} - \boldsymbol{\mu}')$ will be a multiplier. This, however, is impossible since $\lambda_0 = 0$ cannot be. A similar argument shows that if all the inequality constraints $g_i(\mathbf{x}) \leqslant 0$ are inactive at $\mathbf{x}_*$, then the multiplier $(1, 0, \boldsymbol{\mu})$ is unique.

In Remark 5.7 we have established the form of the Lagrange multiplier rule that is usually given in multivariate calculus courses. We state it as a corollary to Theorems 5.1 and 5.2.

COROLLARY 2. *Let $\mathbf{x}_*$ be a solution to the problem of minimizing $f(\mathbf{x})$ subject to the equality constraints $\mathbf{h}(\mathbf{x}) = \mathbf{0}$ and such that the vectors $\nabla h_1(\mathbf{x}_*), \ldots, \nabla h_k(\mathbf{x}_*)$ are linearly independent. Then there exists a unique vector $\boldsymbol{\mu}$ in $\mathbb{R}^k$ such that*

$$\nabla f(\mathbf{x}_*) + \boldsymbol{\mu}^t \nabla \mathbf{h}(\mathbf{x}_*) = \mathbf{0}$$

or, in component form,

$$\frac{\partial f}{\partial x_j}(\mathbf{x}_*) + \sum_{i=1}^{k} \mu_i \frac{\partial h_i(\mathbf{x}_*)}{\partial x_j} = 0, \qquad j = 1, \ldots, n.$$

The following example illustrates CQ by showing what happens when it is not fulfilled. Also, the vectors in (21) are linearly dependent in this example.

Example 5.2. Let $X_0 = \mathbb{R}^2$, let $f(\mathbf{x}) = -x_1$, and let

$$\mathbf{g}(\mathbf{x}) = \begin{pmatrix} -x_1 \\ -x_2 \\ x_2 + (x_1 - 1)^3 \end{pmatrix} \leqslant \begin{pmatrix} 0 \\ 0 \\ 0 \end{pmatrix}.$$

The feasible set X is sketched in Figure 4.2.

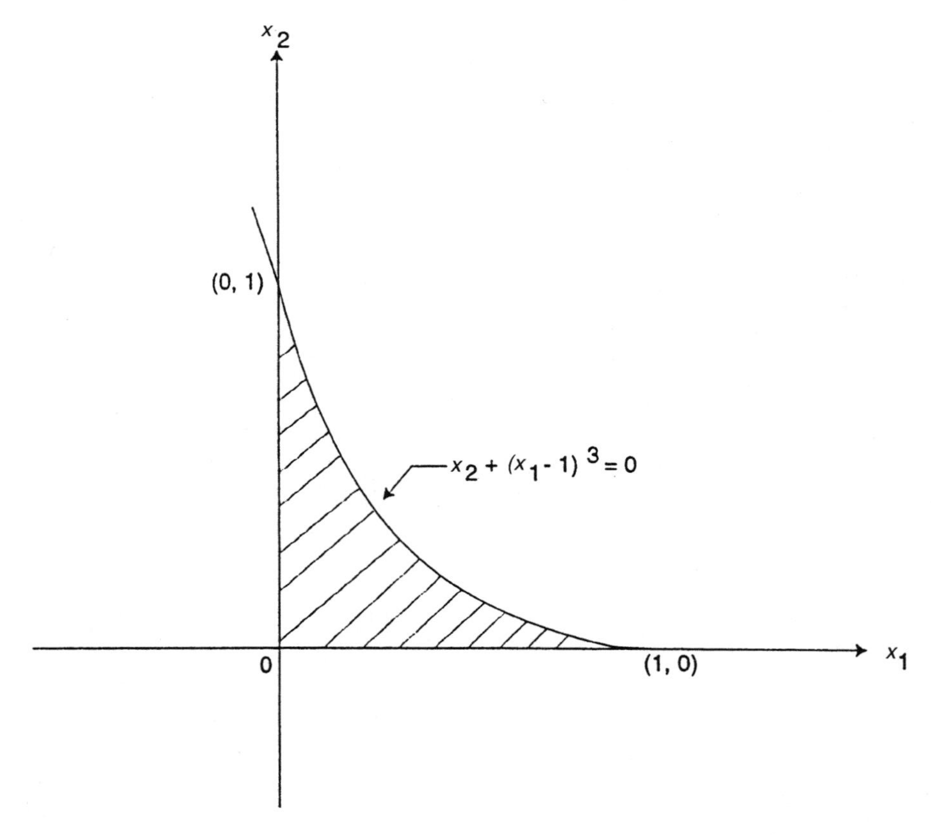

Figure 4.2.

The minimum is attained at $\mathbf{x}_* = (1, 0)$, and so

$$\mathbf{g}_E(\mathbf{x}) = \begin{pmatrix} -x_2 \\ x_2 + (x_1 - 1)^3 \end{pmatrix}.$$

Hence

$$\nabla \mathbf{g}_E(\mathbf{x}_*)\mathbf{z} = \begin{pmatrix} 0 & -1 \\ 0 & 1 \end{pmatrix}\begin{pmatrix} z_1 \\ z_2 \end{pmatrix} = \begin{pmatrix} -z_2 \\ z_2 \end{pmatrix} < \begin{pmatrix} 0 \\ 0 \end{pmatrix}$$

has no solution in $\mathbb{R}^2$. Thus neither CQ nor the linear independence of the vectors in (21) holds at $\mathbf{x}_* = (1, 0)$.

The necessary condition with $\lambda_0 = 1$ would imply

$$\nabla f(\mathbf{x}_*) + (\lambda_2, \lambda_3)\begin{pmatrix} 0 & -1 \\ 0 & 1 \end{pmatrix} = (0, 0)$$

or

$$(-1, 0) + (0, -\lambda_2 + \lambda_3) = (0, 0),$$

which is impossible. On the other hand, if we took $\lambda_0 = 0$, then any (λ_2, λ_3) with $\lambda_2 = \lambda_3 \neq 0$ would satisfy the conditions of Theorem 5.1.

Example 5.3. In this example we consider a problem with equality constraints only in which CQ does not hold at the minimum point. Let $X_0 = \mathbb{R}^2$, let $f(\mathbf{x}) = 2x_1^2 - x_2^2$, and let $h(\mathbf{x}) = x_1^2 x_2 - x_2^3$. The constraint set X consists of the lines $x_2 = 0$ and $x_2 = \pm x_1$. The level curves of f are the two branches of the hyperbolas $2x_1^2 - x_2^2 = c$ and the asymptotes of this family, the lines $x_2 = \pm\sqrt{2}x_1$. See Figure 4.3.

From the figure we see that the point $\mathbf{x}_* = (0, 0)$ furnishes a minimum. Since $\nabla f(\mathbf{x}) = (4x_1, -2x_2)$ and $\nabla h(\mathbf{x}) = (2x_1x_2, x_1^2 - 3x_2^2)$, we get that

$$\nabla f(\mathbf{x}_*) = \nabla f(0, 0) = (0, 0), \qquad \nabla h(\mathbf{x}_*) = (0, 0).$$

Since $\nabla h(\mathbf{x}_*) = \mathbf{0}$, the constraint qualification CQ does not hold at $\mathbf{x}_*$. Any (λ_0, μ) of the form $(0, \mu)$, μ arbitrary, will serve as a Lagrange multiplier, as will any $(1, \mu)$, μ arbitrary.

Example 5.1 (Continued). In this example CQ holds at a point $\mathbf{x}_0$ if $g(\mathbf{x}_0) = 0$ and there exists a solution $\mathbf{z}$ to the equation

$$\langle \nabla g(\mathbf{x}_0), \mathbf{z} \rangle = \langle (2k, -2x_{02}), (z_1, z_2) \rangle < 0.$$

Clearly, $\mathbf{z} = (-k, 0)$ is a solution. Thus CQ holds at all boundary points of the constraint set. This is consistent with our being able to rule out $\lambda_0 = 0$ at the outset in our first treatment of this example.

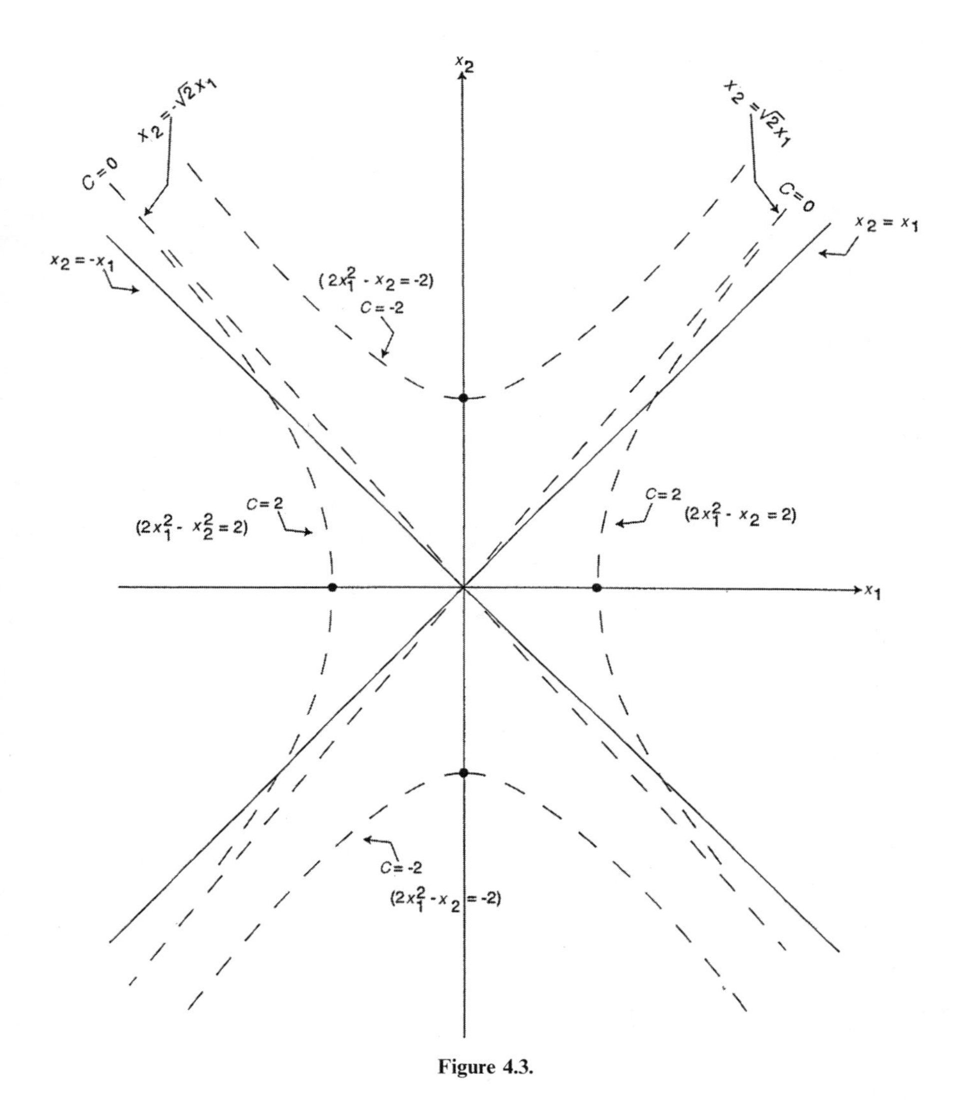

Figure 4.3.

Our next theorem is a sufficiency theorem which states that if X_0, f, and $\mathbf{g}$ are convex and if the KKT necessary conditions hold, then $\mathbf{x}_*$ is a solution.

THEOREM 5.3. *Let f and $\mathbf{g}$ be as in the statement of problem PI and let X_0, f, and $\mathbf{g}$ be convex. Let $\mathbf{x}_* \in \mathbb{R}^n$ and $\boldsymbol{\lambda} \in \mathbb{R}^m$ be such that*

(*i*) $\mathbf{g}(\mathbf{x}_*) \leqslant \mathbf{0}$,
(*ii*) $\boldsymbol{\lambda} \geqslant 0$,

(*iii*) $\langle \boldsymbol{\lambda}, \mathbf{g}(\mathbf{x}_*) \rangle = 0$, *and*
(*iv*) $\nabla f(\mathbf{x}_*) + \boldsymbol{\lambda}^t \nabla \mathbf{g}(\mathbf{x}_*) = \mathbf{0}$.

Then $\mathbf{x}_*$ *is a solution of problem PI.*

Remark 5.8. The sufficiency theorem can also be applied to problems with equality constraints $\mathbf{h}(\mathbf{x}) = \mathbf{0}$ provided the function $\mathbf{h}$ is affine, that is, has the form $\mathbf{h} = A\mathbf{x} + \mathbf{b}$. For then we can write the equality constraint $\mathbf{h}(\mathbf{x}) = \mathbf{0}$ as a pair of inequality constraints $\mathbf{h}(\mathbf{x}) \leqslant \mathbf{0}$ and $-\mathbf{h}(\mathbf{x}) \leqslant \mathbf{0}$, with both $\mathbf{h}$ and $-\mathbf{h}$ convex. The function $\mathbf{g}$ in the theorem is to be interpreted as incorporating this pair of inequality constraints.

Remark 5.9. Under the assumptions of the theorem the feasible set X is convex. To see this, let $X_i = \{\mathbf{x} : \mathbf{x} \in X_0, g_i(\mathbf{x}) \leqslant 0\}$. Since each g_i is convex, each set X_i is convex. Since $X = \cap_{i=1}^m X_i$, we get that X is convex.

Proof. Condition (i) states that $\mathbf{x}_* \in X$. Since f is convex,

$$f(\mathbf{x}) - f(\mathbf{x}_*) \geqslant \langle \nabla f(\mathbf{x}_*), \mathbf{x} - \mathbf{x}_* \rangle, \qquad \mathbf{x} \in X.$$

Substituting (iv) into this inequality gives

$$\begin{aligned} f(\mathbf{x}) - f(\mathbf{x}_*) &\geqslant \langle -\boldsymbol{\lambda}^t \nabla \mathbf{g}(\mathbf{x}_*), \mathbf{x} - \mathbf{x}_* \rangle \\ &= -\langle \boldsymbol{\lambda}, \nabla \mathbf{g}(\mathbf{x}_*)(\mathbf{x} - \mathbf{x}_*) \rangle, \qquad \mathbf{x} \in X. \end{aligned} \tag{22}$$

Since $\mathbf{g}$ is convex,

$$\mathbf{g}(\mathbf{x}) - \mathbf{g}(\mathbf{x}_*) \geqslant \nabla \mathbf{g}(\mathbf{x}_*)(\mathbf{x} - \mathbf{x}_*).$$

Therefore, since $\boldsymbol{\lambda} \geqslant \mathbf{0}$,

$$\langle \boldsymbol{\lambda}, \mathbf{g}(\mathbf{x}) - \mathbf{g}(\mathbf{x}_*) \rangle \geqslant \langle \boldsymbol{\lambda}, \nabla \mathbf{g}(\mathbf{x}_*)(\mathbf{x} - \mathbf{x}_*) \rangle.$$

Substituting this inequality into (22) and then using (iii) give

$$f(\mathbf{x}) - f(\mathbf{x}_*) \geqslant \langle \boldsymbol{\lambda}, \mathbf{g}(\mathbf{x}_*) - \mathbf{g}(\mathbf{x}) \rangle = -\langle \boldsymbol{\lambda}, \mathbf{g}(\mathbf{x}) \rangle.$$

Since $\boldsymbol{\lambda} \geqslant \mathbf{0}$ and since for $\mathbf{x}$ in X we have $\mathbf{g}(\mathbf{x}) \leqslant \mathbf{0}$, it follows that $-\langle \boldsymbol{\lambda}, \mathbf{g}(\mathbf{x}) \rangle \geqslant 0$. Hence $f(\mathbf{x}) \geqslant f(\mathbf{x}_*)$ for all $\mathbf{x}$ in X, and so $\mathbf{x}_*$ is a solution to problem PI.

The function g in Example 5.1 is not convex, as a calculation of the Hessian matrix of g shows. Therefore Theorem 5.3 cannot be used to analyze Example 5.1.

Exercise 5.1. State and prove the form of Theorem 5.3 that results when affine equality constraints $\mathbf{h}(\mathbf{x}) = \mathbf{0}$ are also present.

Exercise 5.2. Use the following to complete the exercise:

(a) Sketch the feasible set.
(b) Use an appropriate theorem to show that a solution exists.
(c) Find all critical points and associated multipliers.
(d) Without using the second-order criteria of the next section determine, if possible, whether a given critical point furnishes a solution.

(i) Minimize $f(\mathbf{x}) = x_1^2 + (x_2 + 1)^2$ subject to

$$x_1^2 + x_2^2 \leqslant 4, \qquad x_2^2 \leqslant x_1 - 1, \qquad -2x_2 \leqslant x_1 - 1.$$

(ii) Minimize $f(\mathbf{x}) = x_1^2 + x_2^2$ subject to

$$x_1^2 + x_2^2 \leqslant 4, \qquad x_2^2 \geqslant x_1, \qquad -2x_2 \leqslant x_1 - 1, \qquad -2x_2 \geqslant x_1 - 2.$$

(iii) Minimize $(x_1 - 5)^2 + x_2^2$ subject to

$$x_1^2 + x_2^2 \leqslant 4, \qquad x_1^2 - 2x_1 + x_2^2 \leqslant 0, \qquad x_1 + x_2^2 \leqslant 2.$$

(iv) Minimize $f(\mathbf{x}) = \|\mathbf{x}\|^2$, where $\mathbf{x} \in \mathbb{R}^2$, subject to

$$x_1 x_2 \geqslant 1, \qquad x_1 \geqslant 0, \qquad x_2 \geqslant 0.$$

(v) *Maximize* $f(\mathbf{x}) = x_1^2 + x_2^2 + a x_1 x_2$, $a > 0$, subject to $x_1^2 + x_2^2 = 1$. What can you say about minimizing $f(\mathbf{x})$ subject to $x_1^2 + x_2^2 = 1$. [In the minimization problem you need not answer questions (c) and (d)].

Exercise 5.3. Minimize $f(\mathbf{x}) = (x_1 - 2)^2 + (x_2 - 1)^2$ subject to

$$x_1^2 - x_2 \leqslant 0, \qquad x_1 + x_2 - 4 \leqslant 0, \qquad -x_1 \leqslant 0, \qquad -x_2 \leqslant 0.$$

Give a full justification of your answer. You may use geometric considerations to assist you in finding critical points.

Exercise 5.4. Maximize $f(\mathbf{x}) = x_1^2 + x_1 x_2 + x_2^2$ subject to

$$-3x_1 - 2x_2 + 6 \leqslant 0, \qquad -x_1 + x_2 - 3 \leqslant 0, \qquad 4(x_1 - 2)^2 + 4(x_2 - 3)^2 \leqslant 25.$$

Hint: Compare with Exercise 3.2.

Exercise 5.5. Minimize $f(\mathbf{x}) = e^{-(x_1+x_2)}$ subject to $e^{x_1} + e^{x_2} \leqslant 20$.

Exercise 5.6. Minimize $f(\mathbf{x}) = \sum_{i=1}^{n} c_i/x_i$ subject to

$$\sum_{i=1}^{n} a_i x_i = b, \qquad x_i > 0, \qquad i = 1, \dots, n,$$

where c_i, a_i, $i = 1, \dots, n$, and b are positive constants.

Exercise 5.7. Let A be a positive-definite $n \times n$ symmetric matrix and let $\mathbf{y}$ be a fixed vector in $\mathbb{R}^n$. Show that the maximum value of $f(\mathbf{x}) = \langle \mathbf{y}, \mathbf{x} \rangle$ subject to the constraint $\langle \mathbf{x}, A\mathbf{x} \rangle \leqslant 1$ is $\langle \mathbf{y}, A^{-1}\mathbf{y} \rangle^{1/2}$. Use this result to establish the generalization of the Cauchy–Schwarz inequality

$$|\langle \mathbf{x}, \mathbf{y} \rangle|^2 \leqslant \langle \mathbf{x}, A\mathbf{x} \rangle \langle \mathbf{y}, A^{-1}\mathbf{y} \rangle.$$

Exercise 5.8. Use the Lagrange multiplier rule to show that the distance $\rho(\mathbf{x}_0)$ from the point $\mathbf{x}_0$ to the hyperplane $H_{\mathbf{a}}^{\alpha} = \{\mathbf{x} : \langle \mathbf{a}, \mathbf{x} \rangle = \alpha\}$ is given by

$$\rho(\mathbf{x}_0) = \left| \frac{\langle \mathbf{a}, \mathbf{x}_0 \rangle - \alpha}{\|\mathbf{a}\|} \right|.$$

(*Hint:* In problems involving the minimization of a norm or of a distance it is often simpler to minimize the square of the norm or distance.)

Exercise 5.9. In this problem we will use the Lagrange multiplier rule to deduce important properties of eigenvectors and eigenvalues of a real $n \times n$ symmetric matrix A.

(a) Show that the maximum of $f(\mathbf{x}) = \langle \mathbf{x}, A\mathbf{x} \rangle$ subject to $\|\mathbf{x}\| = 1$ is attained at a unit eigenvector $\mathbf{x}_1$ and that $f(\mathbf{x}_1)$ is the largest eigenvalue λ_1 of A.

(b) Suppose $m < n$ mutually orthogonal unit eigenvectors $\mathbf{x}_1, \dots, \mathrm{x}_m$ of A have been determined. Show that the maximum of $f(\mathbf{x}) = \langle \mathbf{x}, A\mathbf{x} \rangle$ subject to $\|\mathbf{x}\| = 1$ and $\langle \mathbf{x}, \mathbf{x}_j \rangle = 0$, $j = 1, \dots, m$, is attained a point $\mathbf{x}_{m+1}$ that is an eigenvector of A and that $f(\mathbf{x}_{m+1})$ is the corresponding eigenvalue.

(c) Use (a) and (b) to show that a real $n \times n$ symmetric matrix has a set of n orthonormal eigenvectors $\mathbf{x}_1, \dots, \mathbf{x}_n$ with corresponding eigenvalues $\lambda_1 \geqslant \lambda_2 \geqslant \cdots \geqslant \lambda_n$.

Exercise 5.10. In this exercise we will use Farkas's lemma and a constraint qualification introduced by Kuhn and Tucker [1951] that is different from CQ to establish the conclusion of Theorem 5.2 without using Theorem 5.1. The *Kuhn–Tucker constraint qualification* KTQ is said to hold at a point $\mathbf{x}_0$ if there exists a vector $\mathbf{z}$ such that

$$\nabla \mathbf{g}_E(\mathbf{x}_0)\mathbf{z} \leqslant \mathbf{0}, \qquad \nabla \mathbf{h}(\mathbf{x}_0)\mathbf{z} = \mathbf{0}$$

and if for *each such* $\mathbf{z}$ there exists a $C^{(1)}$ function $\xi(\cdot)$ defined on an interval $[0, \tau)$ such that

$$\begin{aligned} &\xi(0) = \mathbf{x}_0, && \xi'_+(0) = \mathbf{z}, \\ &\mathbf{g}(\xi(t)) \leqslant \mathbf{0}, && \mathbf{h}(\xi(t)) = \mathbf{0} \quad \text{for all } t \text{ in } [0, \tau). \end{aligned}$$

The Kuhn–Tucker constraint qualification is difficult to check. In Lemma 6.1 we give a verifiable condition that implies KTQ.

Let $\mathbf{x}_*$ be a solution of problem PI and let KTQ hold at $\mathbf{x}_*$. Show that the conclusion of Theorem 5.2 holds without invoking Theorem 5.1. *Hint:* What can you say about the right-hand derivatives at $t = 0$ of the mappings $t \to f(\xi(t))$, $t \to \mathbf{g}_E(\xi(t))$, and $t \to \mathbf{h}(\xi(t))$ in relation to Farkas's theorem?

6. SECOND-ORDER CONDITIONS

In this section we first obtain second-order necessary conditions. Points that satisfy the first-order necessary conditions but fail to satisfy the second-order conditions cannot solve the programming problem. We conclude this section with a presentation of some second-order sufficiency conditions. Points that satisfy the first-order necessary conditions and these sufficiency conditions are solutions to the programming problem.

Since second-order conditions involve second derivatives, in this section we assume that the functions f, $\mathbf{g}$, and $\mathbf{h}$ that occur in the statement of problem PI are of class $C^{(2)}$ on X_0.

The derivation of our second-order necessary condition will require an application of the implicit function theorem, which we now discuss.

Consider the system of equations

$$\begin{aligned} u_1(x_1, \ldots, x_n, y_1, \ldots, y_m) &= 0, \\ &\vdots \\ u_m(x_1, \ldots, x_n, y_1, \ldots, y_m) &= 0, \end{aligned} \tag{1}$$

where the real-valued functions u_i, $i = 1, \ldots, m$, are defined on an open set $D = D_1 \times D_2$, where D_1 is open in the $\mathbf{x}$-space $\mathbb{R}^n$ and D_2 is open in the $\mathbf{y}$-space $\mathbb{R}^m$. Equation (1) asks us to find all $(\mathbf{x}, \mathbf{y})$ in D such that (1) is satisfied. For a given $\mathbf{x}$ in D_1, the system (1) can therefore be interpreted as a system of m equations for the m unknowns $(y_1, \ldots, y_m)$. If for each $\mathbf{x}$ in some subset S of D_1 we can find a unique solution $\mathbf{y}(\mathbf{x}) = (y_1(\mathbf{x}), \ldots, y_m(\mathbf{x}))$, then in this fashion we obtain a function $\mathbf{y}(\cdot)$ with domain S. In this event, we say that (1) *defines* $\mathbf{y}$ *as a function of* $\mathbf{x}$ *implicitly*.

Let $(\mathbf{x}_0, \mathbf{y}_0)$ be a point in D that satisfies (1). The implicit function theorem gives conditions ensuring that (1) has a unique solution for points $\mathbf{x}$ sufficiently close to $\mathbf{x}_0$. The resulting function $\mathbf{y}(\cdot)$ will inherit the regularity properties of the functions $u_1, \ldots, u_m$.

To illustrate these ideas, take $D_1 = \mathbb{R}^2$, $D_2 = \mathbb{R}^1$, and $D = \mathbb{R}^3 = \mathbb{R}^2 \times \mathbb{R}^1$. Consider

$$u(x_1, x_2, y) = x_1^2 + x_2^2 + y^2 - 1 = 0. \tag{2}$$

The set of points $(\mathbf{x}, y) = (x_1, x_2, y)$ in $\mathbb{R}^3$ that satisfy (2) is the surface of the ball with center at the origin and radius 1.

Given any point $(\mathbf{x}_0, y_0) = (x_{01}, x_{02}, y_0)$ *with* $y_0 > 0$ that satisfies (2), we can find an $\alpha > 0$ and a $\beta > 0$ such that

$$y = \sqrt{1 - x_1^2 - x_2^2}$$

is the unique solution of (2) for $\mathbf{x}$ in $B(\mathbf{x}_0, \alpha)$ having range in $B(y_0, \beta)$ The point $(\mathbf{x}_0, -y_0)$ also satisfies (2), and

$$y = -\sqrt{1 - x_1^2 - x_2^2}$$

is the unique solution of (2) for $\mathbf{x}$ in $B(\mathbf{x}_0, \alpha)$ having range in $B(-y_0, \beta)$.

Although all points of $\mathscr{E} = \{(\mathbf{x}, y) : x_1^2 + x_2^2 = 1,\ y = 0\}$ are solutions of (2), equation (2) does not define y as a function of $\mathbf{x}$ on any ball about a point $\mathbf{x}_0$ such that $\|\mathbf{x}_0\| = 1$. Since $\partial f/\partial y = 2y$, we see that at any point $(\mathbf{x}_0, y_0)$ that satisfies (2) we have $\partial f/\partial y \neq 0$ if $y_0 \neq 0$ and $\partial f/\partial y = 0$ if $y_0 = 0$. Thus at all points of $\mathscr{E}$, $\partial f/\partial y = 0$. The reader should return to this paragraph after reading Theorem 6.1.

Equation (1) in vector form is

$$\mathbf{u}(\mathbf{x}, \mathbf{y}) = \mathbf{0}.$$

THEOREM 6.1 (IMPLICIT FUNCTION THEOREM). *Let D_1 be an open set in $\mathbb{R}^n$ and let D_2 be an open set in $\mathbb{R}^m$. Let $D = D_1 \times D_2$. Let*

$$\mathbf{u} : (\mathbf{x}, \mathbf{y}) \to \mathbf{u}(\mathbf{x}, \mathbf{y}), \qquad \mathbf{x} \in D_1,\ \mathbf{y} \in D_2,$$

be a mapping that is of class $C^{(p)}$ on D and with range in $\mathbb{R}^m$. Let $(\mathbf{x}_0, \mathbf{y}_0)$, $\mathbf{x}_0 \in D_1$, $\mathbf{y}_0 \in D_2$, be a solution of

$$\mathbf{u}(\mathbf{x}, \mathbf{y}) = \mathbf{0}$$

and suppose that the $m \times m$ *matrix*

$$J(\mathbf{x}_0, \mathbf{y}_0) = \left(\frac{\partial u_i}{\partial y_j} (\mathbf{x}_0, \mathbf{y}_0) \right), \qquad i = 1, \ldots, m, \qquad j = 1, \ldots, m, \tag{3}$$

is nonsingular. Then there exist an $\alpha > 0$, *a* $\beta > 0$, *and a function* $\mathbf{y}(*)$ *defined on* $B(\mathbf{x}_0, \alpha) \subset D_1$ *with range in* $B(\mathbf{y}_0, \beta) \subset D_2$ *such that*

$$\mathbf{y}(\cdot) \textit{ is of class } C^p, \tag{i}$$

$$\mathbf{y}(\mathbf{x}, \mathbf{y}(\mathbf{x})) = \mathbf{0} \quad \textit{for all } \mathbf{x} \in B(\mathbf{x}_0, \alpha). \tag{ii}$$

Moreover, for $\mathbf{x}$ *in* $B(\mathbf{x}_0, \alpha)$, $\mathbf{y}(\mathbf{x})$ *is the only solution of* $\mathbf{u}(\mathbf{x}, \mathbf{y}) = \mathbf{0}$ *such that* $\mathbf{y}(\mathbf{x}) \in B(\mathbf{y}_0, \beta)$.

For a proof of this theorem we refer the reader to Rudin [1976] or Fleming [1977].

Definition 6.1. Let $\mathbf{x}_0$ be a point of X_0. The vector $\mathbf{z} \neq \mathbf{0}$ in $\mathbb{R}^n$ is said to satisfy the *tangential constraints* at $\mathbf{x}_0$ if

$$\nabla \mathbf{g}_E(\mathbf{x}_0)\mathbf{z} \leqslant \mathbf{0}, \qquad \nabla \mathbf{h}(\mathbf{x}_0)\mathbf{z} = \mathbf{0}, \tag{4}$$

where $E = \{i : g_i(\mathbf{x}_0) = \mathbf{0}\}$.

For definiteness we shall henceforth suppose that $E = \{1, \ldots, r\}$. Let $I = \{r + 1, \ldots, m\}$. Since either E or I can be empty, $0 \leqslant r \leqslant m$.

Let $\mathbf{z}$ be a vector that satisfies the tangential constraints. Let

$$E_1 = \{i : i \in E, \langle \nabla g_i(\mathbf{x}_0), \mathbf{z} \rangle = 0\},$$

$$E_2 = \{i : i \in E, \langle \nabla g_i(\mathbf{x}_0), \mathbf{z} \rangle < 0\}.$$

Then $E = E_1 \cup E_2$,

$$\nabla \mathbf{g}_{E_1}(\mathbf{x}_0)\mathbf{z} = \mathbf{0}, \qquad \nabla \mathbf{h}(\mathbf{x}_0)\mathbf{z} = \mathbf{0}, \tag{5}$$

and

$$\nabla \mathbf{g}_{E_2}(\mathbf{x}_0)\mathbf{z} < \mathbf{0}. \tag{6}$$

For the sake of definiteness suppose that $E_1 = \{1, \ldots, q\}$. Since E_1 may be empty, $0 \leqslant q \leqslant r$. If $q = 0$, statements involving E_1 in (5) in the ensuing discussion should be deleted. The reader should keep in mind that *the sets* E_1 *and* E_2 *depend on* $\mathbf{z}$ *and* $\mathbf{x}_0$.

The next lemma gives a condition under which, given a vector $\mathbf{z}$ that satisfies the tangential constraints, we can construct a curve $\xi(\cdot)$ emanating from $\mathbf{x}_0$ and going into the feasible set in the direction given by $\mathbf{z}$.

LEMMA 6.1. *Let* $\mathbf{g}$ *and* $\mathbf{h}$ *be of class* $C^{(p)}$ *on* X_0, $p \geqslant 1$. *Let* $\mathbf{x}_0$ *be feasible and let* $\mathbf{z}$ *satisfy the tangential constraints at* $\mathbf{x}_0$. *Suppose that the vectors*

$$\nabla g_1(\mathbf{x}_0), \ldots, \nabla g_q(\mathbf{x}_0), \nabla h_1(\mathbf{x}_0), \ldots, \nabla h_k(\mathbf{x}_0) \tag{7}$$

are linearly independent. Then there exists a $\tau > 0$ *and a* $C^{(p)}$ *mapping* $\xi(\cdot)$ *from* $(-\tau, \tau)$ *to* $\mathbb{R}^n$ *such that*

$$\begin{aligned} &\xi(0) = \mathbf{x}_0, && \xi'(0) = \mathbf{z}, && \\ &\mathbf{g}_{E_1}(\xi(t)) = \mathbf{0}, && \mathbf{h}(\xi(t)) = \mathbf{0}, && \mathbf{g}_I(\xi(t)) < \mathbf{0} \\ &\mathbf{g}_{E_2}(\xi(t)) < \mathbf{0} && \text{for } 0 < t < \tau. \end{aligned}$$

Proof. Let $\rho = q + k$. In view of (7), $\rho \leqslant n$. If $\rho = n$, then again in view of (7) the only solution of (5) is $\mathbf{z} = \mathbf{0}$. Hence, $E_2 = \varnothing$ and $E_1 = E$. The mapping $\xi(\cdot)$ is then the constant map $\xi(t) = \mathbf{x}_0$.

We now suppose that $\rho < n$. Let Γ denote the $\rho \times n$ matrix whose rows are the vectors in (7). Thus

$$\Gamma = \begin{pmatrix} \nabla \mathbf{g}_{E_1}(\mathbf{x}_0) \\ \nabla \mathbf{h}(\mathbf{x}_0) \end{pmatrix},$$

and Γ has rank ρ. By relabeling the coordinates corresponding to ρ linearly independent columns of Γ, we may suppose without loss of generality that the first ρ columns of the matrix Γ are linearly independent. Thus the $\rho \times \rho$ matrix

$$\Gamma_\rho = \begin{pmatrix} \dfrac{\partial g_i(\mathbf{x}_0)}{\partial x_j} \\ \dfrac{\partial h_s(\mathbf{x}_0)}{\partial x_j} \end{pmatrix} \quad i = 1, \ldots, q, \qquad s = 1, \ldots, k, \qquad j = 1, \ldots, \rho,$$

is nonsingular.

The system of equations

$$\begin{aligned} \mathbf{g}_{E_1}(\mathbf{x}) &= 0, \\ \mathbf{h}(\mathbf{x}) &= 0, \\ x_{\rho+1} - x_{0,\rho+1} - tz_{\rho+1} &= 0, \\ &\vdots \\ x_n - x_{0,n} - tz_n &= 0, \end{aligned} \tag{8}$$

where $z_{\rho+1}, \ldots, z_n$ are the last $n - \rho$ components of the vector $\mathbf{z}$, has a solution $(\mathbf{x}, t) = (\mathbf{x}_0, 0)$. We shall use the implicit function theorem to show that the system (8) determines $\mathbf{x}$ as a function of t near $t = 0$.

For the system (8), viewed as defining $\mathbf{x}$ implicitly as a function of t, the matrix J defined in (3) becomes

$$J = \begin{pmatrix} \Gamma_\rho & \Gamma_{n-\rho} \\ 0_{n-\rho,\rho} & I_{n-\rho} \end{pmatrix},$$

where $\Gamma_{n-\rho}$ is the $\rho \times (n - \rho)$ matrix whose columns are the last $n - \rho$ columns of Γ, the matrix $0_{n-\rho,\rho}$ is the $(n - \rho) \times \rho$ zero matrix, and $I_{n-\rho}$ is the $(n - \rho) \times (n - \rho)$ identity matrix. Since Γ_ρ is nonsingular, so is J. Hence by Theorem 6.1 there exists a $\tau > 0$ and a $C^{(p)}$ function $\xi(\cdot)$ defined on the interval $(-\tau, \tau)$ and with range in $\mathbb{R}^n$ such that

$$\begin{aligned} &\xi(0) = \mathbf{x}_0, \\ &\mathbf{g}_{E_1}(\xi(t)) = \mathbf{0}, \qquad \mathbf{h}(\xi(t)) = \mathbf{0}, \\ &\xi_i(t) = x_{0,i} + tz_i, \qquad i = \rho + 1, \ldots, n. \end{aligned} \tag{9}$$

Differentiation of the equations in (9) with respect to t and then setting $t = 0$ give

$$\begin{pmatrix} & \Gamma & \\ 0_{n-\rho,\rho} & & I_{n-\rho} \end{pmatrix} \xi'(0) = \begin{pmatrix} \mathbf{0} \\ \hat{\mathbf{z}} \end{pmatrix},$$

where $\hat{\mathbf{z}} = (z_{\rho+1}, \ldots, z_n)$. The matrix on the left is the matrix J for the system (8) and is nonsingular. Therefore the system

$$\begin{pmatrix} & \Gamma & \\ 0_{n-\rho,\rho} & & I_{n-\rho} \end{pmatrix} \mathbf{w} = \begin{pmatrix} \mathbf{0} \\ \hat{\mathbf{z}} \end{pmatrix}$$

has a unique solution. Since $\xi'(0)$ and $\mathbf{z}$ are both solutions, we get that $\xi'(0) = \mathbf{z}$.

Since $\mathbf{g}_I(\mathbf{x}_0) < \mathbf{0}$ and $\xi(0) = \mathbf{x}_0$, it follows from the continuity of $\mathbf{g}$ and ξ that for τ sufficiently small we have $\mathbf{g}_I(\xi(t)) < 0$ for $|t| \leqslant \tau$.

Since

$$\frac{d\mathbf{g}_{E_2}(\xi(t))}{dt} = \nabla \mathbf{g}_{E_2}(\xi(t))\xi'(t)$$

and since $\xi(0) = \mathbf{x}_0$ and $\xi'(0) = \mathbf{z}$, it follows from (6) that $d\mathbf{g}_{E_2}(\xi(0))/dt < \mathbf{0}$. It then follows from continuity that by taking τ to be smaller, if necessary, there is an interval $[0, \tau)$ on which all of the previous conclusions hold and on which

$d\mathbf{g}_{E_2}(\boldsymbol{\xi}(t))/dt < \mathbf{0}$. Since $\mathbf{g}_{E_2}(\boldsymbol{\xi}(0)) = \mathbf{g}_{E_2}(\mathbf{x}_0) = \mathbf{0}$, we have $\mathbf{g}_{E_2}(\boldsymbol{\xi}(t)) < \mathbf{0}$ on $(0, \tau)$. This completes the proof.

COROLLARY 1. *Let CQ hold at a point $\mathbf{x}_0$. Then for every $\mathbf{z}$ satisfying relation (18) in Section 5 there exists a $C^{(p)}$ function $\boldsymbol{\xi}(\ )$ such that $\boldsymbol{\xi}(0) = \mathbf{x}_0$, $\boldsymbol{\xi}'(0) = \mathbf{z}$ and $\boldsymbol{\xi}(t)$ is feasible for all t in some interval $[0, \tau)$. Moreover, $\mathbf{g}(\boldsymbol{\xi}(t)) < \mathbf{0}$ on $(0, \tau)$.*

Proof. A vector $\mathbf{z}$ satisfying relation (18) in Section 5 satisfies the tangential constraints, and the set of indices E_1 corresponding to $\mathbf{z}$ is empty.

Example 5.2 (Continued). At the point $\mathbf{x}_* = (1, 0)$

$$\nabla \mathbf{g}_E(\mathbf{x}_*)\mathbf{z} = \begin{pmatrix} 0 & -1 \\ 0 & 1 \end{pmatrix}\begin{pmatrix} z_1 \\ z_2 \end{pmatrix} = \begin{pmatrix} -z_2 \\ z_2 \end{pmatrix}.$$

Therefore, since there are no equality constraints, every vector of the form $(z_1, 0)$ satisfies the tangential constraints. Clearly, there exists no curve $\boldsymbol{\xi}(\cdot)$ emanating from $\mathbf{x}_* = (1, 0)$ with tangent vector $(1, 0)$ at $\mathbf{x}_*$ such that $\boldsymbol{\xi}(t)$ is feasible for t in some interval $[0, \tau)$. Lemma 6.1 is not contradicted, since we have $E_1 = E$ and the rows of $\nabla \mathbf{g}_E(\mathbf{x}_*)$ are linearly dependent. Also, the conclusion of the corollary is not contradicted since CQ does not hold at $(1, 0)$.

We now present a second-order necessary condition.

Definition 6.2. Let $(\mathbf{x}_*, \boldsymbol{\lambda}, \boldsymbol{\mu})$ be as in Theorem 5.2. Let $\mathbf{z}$ be a vector that satisfies the tangential constraints (4) at $\mathbf{x}_*$. Let

$$E_1 = E_1(\mathbf{z}) = \{i : i \in E,\ \langle \nabla g_i(\mathbf{x}_*), \mathbf{z} \rangle = 0\},$$
$$E_2 = E_2(\mathbf{z}) = \{i : i \in E,\ \langle \nabla g_i(\mathbf{x}_*), \mathbf{z} \rangle < 0\}.$$

The vector $\mathbf{z}$ will be called a *second-order test vector* if (i) $\lambda_i = 0$ for $i \in E_2$ and (ii) the rows of the matrix

$$\begin{pmatrix} \nabla \mathbf{g}_{E_1}(\mathbf{x}_*) \\ \nabla \mathbf{h}(\mathbf{x}_*) \end{pmatrix}$$

are linearly independent.

THEOREM 6.2. *Let f, $\mathbf{g}$, and $\mathbf{h}$ be of class $C^{(2)}$ on X_0. Let $\mathbf{x}_*$ be a solution to problem PI. Let $\boldsymbol{\lambda}$ and $\boldsymbol{\mu}$ be KKT multipliers as in Theorem 5.2 and let $F(\ , \boldsymbol{\lambda}, \boldsymbol{\mu})$ be defined by*

$$F(\mathbf{x}, \boldsymbol{\lambda}, \boldsymbol{\mu}) = f(\mathbf{x}) + \langle \boldsymbol{\lambda}, \mathbf{g}(\mathbf{x}) \rangle + \langle \boldsymbol{\mu}, \mathbf{h}(\mathbf{x}) \rangle.$$

Then for every second-order test vector $\mathbf{z}$,

$$\langle \mathbf{z}, F_{\mathbf{xx}}(\mathbf{x}_*, \boldsymbol{\lambda}, \boldsymbol{\mu})\mathbf{z} \rangle \geqslant 0, \tag{10}$$

where

$$F_{\mathbf{xx}} = \left(\frac{\partial^2 F}{\partial x_j \, \partial x_i} \right), \qquad i = 1, \ldots, n, \qquad j = 1, \ldots, n.$$

We first establish an elementary result.

LEMMA 6.2. *Let ϕ be a $C^{(2)}$ function defined on an interval $(-\tau, \tau)$ such that $\phi'(0) = 0$ and $\phi(t) \geqslant \phi(0)$ for all t in $[0, \tau)$. Then $\phi''(0) \geqslant 0$.*

Proof. By Taylor's theorem, for $0 < t < \tau$

$$\phi(t) - \phi(0) = \tfrac{1}{2}\phi''(\theta t)t^2,$$

where $0 < \theta < 1$. If $\phi''(0)$ were negative, then by continuity there would exist an interval $[0, \tau')$ on which ϕ'' would be negative. Hence $\phi(t) < \phi(0)$ on this interval, contradicting the hypothesis.

Proof (Theorem 6.2). Let $\mathbf{z}$ be a second-order test vector. Since $\lambda_i = 0$ for $i \in E_2$ and

$$\boldsymbol{\lambda} = (\boldsymbol{\lambda}_E, \boldsymbol{\lambda}_I) = (\boldsymbol{\lambda}_{E_1}, \boldsymbol{\lambda}_{E_2}, \boldsymbol{\lambda}_I) = (\boldsymbol{\lambda}_{E_1}, \mathbf{0}, \mathbf{0}),$$

we have

$$F(\mathbf{x}, \boldsymbol{\lambda}, \boldsymbol{\mu}) = f(\mathbf{x}) + \langle \boldsymbol{\lambda}_{E_1}, \mathbf{g}_{E_1}(\mathbf{x}) \rangle + \langle \boldsymbol{\mu}, \mathbf{h}(\mathbf{x}) \rangle.$$

Let $\boldsymbol{\xi}(\cdot)$ be the function corresponding to $\mathbf{z}$ as in Lemma 6.1, with $\mathbf{x}_0 = \mathbf{x}_*$. Since $\mathbf{g}_{E_1}(\boldsymbol{\xi}(t)) = \mathbf{0}$ and $\mathbf{h}(\boldsymbol{\xi}(t)) = \mathbf{0}$ for $|t| < \tau$, we have

$$f(\boldsymbol{\xi}(t)) = F(\boldsymbol{\xi}(t), \boldsymbol{\lambda}, \boldsymbol{\mu}) \quad \text{for } |t| < \tau.$$

Since for $0 \leqslant t < \tau$, all points $\boldsymbol{\xi}(t)$ are feasible, and since $\boldsymbol{\xi}(0) = \mathbf{x}_*$, the mapping $t \to f(\boldsymbol{\xi}(t))$, where $0 \leqslant t < \tau$ has a minimum at $t = 0$. Hence, so does the mapping $\phi(\cdot)$ defined on $[0, \tau)$ by

$$\phi(t) = F(\boldsymbol{\xi}(t), \boldsymbol{\lambda}, \boldsymbol{\mu}).$$

Straightforward calculation gives

$$\phi'(t) = \langle \nabla F(\boldsymbol{\xi}(t), \boldsymbol{\lambda}, \boldsymbol{\mu}), \boldsymbol{\xi}'(t) \rangle$$
$$\phi''(t) = \langle \boldsymbol{\xi}'(t), F_{\mathbf{xx}}(\boldsymbol{\xi}(t), \boldsymbol{\lambda}, \boldsymbol{\mu})\boldsymbol{\xi}'(t) \rangle + \langle \nabla F(\boldsymbol{\xi}(t), \boldsymbol{\lambda}, \boldsymbol{\mu}), \boldsymbol{\xi}''(t) \rangle.$$

Setting $t = 0$ in the first equation and using the relations $\boldsymbol{\xi}(0) = \mathbf{x}_*$ and $\nabla F(\mathbf{x}_*, \boldsymbol{\lambda}, \boldsymbol{\mu}) = \mathbf{0}$ give $\phi'(0) = 0$. Hence by Lemma 6.2 we have $\phi''(0) \geqslant 0$. Setting $t = 0$ in the second equation and using $\boldsymbol{\xi}(0) = \mathbf{x}_*$, $\nabla F(\mathbf{x}_*, \boldsymbol{\lambda}, \boldsymbol{\mu}) = \mathbf{0}$, and $\boldsymbol{\xi}'(0) = \mathbf{z}$ give the conclusion of the theorem.

The following corollary requires that (10) holds for fewer $\mathbf{z}$ than the theorem does and thus is more restrictive than the theorem but easier to apply.

COROLLARY 2. *Let the functions f, $\mathbf{g}$, and $\mathbf{h}$ be as in Theorem 6.2 and let $\mathbf{x}_*$ be a solution to problem PI. Suppose that the vectors*

$$\nabla g_1(\mathbf{x}_*), \ldots, \nabla g_r(\mathbf{x}_*), \nabla h_1(\mathbf{x}_*), \ldots, \nabla h_k(\mathbf{x}_*)$$

are linearly independent. Then there exist KKT multipliers $\boldsymbol{\lambda}$, $\boldsymbol{\mu}$ as in Theorem 5.2, and (10) holds for every vector $\mathbf{z}$ satisfying

$$\nabla \mathbf{g}_E(\mathbf{x}_*)\mathbf{z} = \mathbf{0}, \qquad \nabla \mathbf{h}(\mathbf{x}_*)\mathbf{z} = \mathbf{0}. \tag{11}$$

Proof. The existence of KKT multipliers follows from Corollary 1 of Theorem 5.2. The second conclusion follows from the theorem by taking $E_1 = E$, in which case $E = \{1, \ldots, r\}$ and $E_2 = \varnothing$.

If the inequality constraints are absent, then the linear independence hypothesis of Theorem 6.2 and Corollary 2 becomes "the vectors $\nabla h_1(\mathbf{x}_*), \ldots, \nabla h_k(\mathbf{x}_*)$ are linearly independent," which is also the constraint qualification CQ for this problem. Thus, we get the following corollary.

COROLLARY 3. *Let f and $\mathbf{h}$ be as in Theorem 6.2, let the inequality constraints be absent, let $\mathbf{x}_*$ be a solution of problem PI and let CQ hold at $\mathbf{x}_*$. Then there exists a unique vector $\boldsymbol{\mu}$ in $\mathbb{R}^k$ such that the function $H(\ ,\boldsymbol{\mu})$ defined by*

$$H(\mathbf{x}, \boldsymbol{\mu}) = f(\mathbf{x}) + \langle \boldsymbol{\mu}, \mathbf{h}(\mathbf{x}) \rangle$$

satisfies

$$(i) \quad \nabla f(\mathbf{x}_*) + \boldsymbol{\mu}^t \nabla \mathbf{h}(\mathbf{x}_*) = \mathbf{0}$$

and

$$(ii) \quad \langle \mathbf{z}, H_{\mathbf{xx}}(\mathbf{x}_*, \boldsymbol{\mu})\mathbf{z} \rangle \geqslant 0$$

for all $\mathbf{z}$ *in* $\mathbb{R}^n$ *satisfying*

$$\nabla \mathbf{h}(\mathbf{x}_*)\mathbf{z} = \mathbf{0}.$$

Corollary 3 is the classical second-order necessary condition for equality-constrained problems.

Example 5.1 (Continued). The reader will recall that for $0 < k < 1$ the first-order necessary conditions were satisfied by three points and corresponding multipliers given by

$$\begin{array}{lll} x_1 = 1 - k, & x_2 = \pm\sqrt{2k(1-k)}, & \lambda = 1 \\ x_1 = 0, & x_2 = 0, & \lambda = 1/k. \end{array}$$

We will use the second-order necessary condition as given in Corollary 2 to rule out the origin as a candidate for optimality when $0 < k < 1$.

Recall that for any $\mathbf{x}$ we have $\nabla g(\mathbf{x}) = (2k, \ -2x_2)$. At $\mathbf{x}_* = (0, 0)$, $\nabla g(\mathbf{x}_*) \neq \mathbf{0}$, so the linear independence requirement of Corollary 2 holds. Thus for $\mathbf{z}$ to satisfy (11) we require that

$$\langle \nabla g(\mathbf{x}_*), \mathbf{z} \rangle = \langle (2k, 0), (z_1, z_2) \rangle = 0.$$

This is satisfied by all vectors (z_1, z_2) of the form

$$z_1 = 0, \quad z_2 \text{ arbitrary.} \tag{12}$$

The function F with $\lambda = 1/k$ becomes

$$F = (x_1 - 1)^2 + x_2^2 + \frac{1}{k}(2kx_1 - x_2^2).$$

Condition (10) with $\mathbf{x}_* = (0, 0)$ becomes

$$(z_1, z_2)\begin{pmatrix} 2 & 0 \\ 0 & 2(1 - \frac{1}{k}) \end{pmatrix}\begin{pmatrix} z_1 \\ z_2 \end{pmatrix} = 2z_1^2 + 2z_2^2\left(1 - \frac{1}{k}\right) \geqslant 0$$

for all (z_1, z_2) satisfying (12). Since $z_1 = 0$ and for $0 < k < 1$ the coefficient of $2z_2^2$ is negative; the last relation only holds for $z_2 = 0$. Thus (10) fails to hold for all (z_1, z_2) satisfying (12), and so $\mathbf{x}_* = (0, 0)$ is not optimal if $0 < k < 1$.

For $k \geqslant 1$ the only critical point is $\mathbf{x}_* = (0, 0)$ with $\lambda = 1/k$. We now have $\langle \mathbf{z}, F_{\mathbf{xx}}\mathbf{z} \rangle \geqslant 0$ for all $\mathbf{z}$, so the second-order conditions of Theorem 6.2 and Corollary 2 both hold at $(0, 0)$.

We now show that for $0 < k < 1$ the second-order condition (10) holds at the two points $(1 - k, \pm\sqrt{2k(1 - k)})$. At these points $\lambda = 1$. Thus, for a second-order test vector $\mathbf{z}$ we have $E_2 = \varnothing$ and $E_1 = E$. Hence (5) and (6) reduce to $\nabla g_E(\mathbf{x}_*)\mathbf{z} = 0$ at these points. Thus a test vector $\mathbf{z}$ satisfies

$$\langle (2k, \mp 2\sqrt{2k(1 - k)}), (z_1, z_2) \rangle = 2kz_1 \mp 2z_2\sqrt{k(1 - k)} = 0. \tag{13}$$

The function F with $\lambda = 1$ becomes

$$F = (x_1 - 1)^2 + x_2^2 + (2kx_1 - x_2^2)$$

and condition (10) becomes

$$(z_1, z_2) \begin{pmatrix} 2 & 0 \\ 0 & 0 \end{pmatrix} \begin{pmatrix} z_1 \\ z_2 \end{pmatrix} = 2(z_1)^2 \geqslant 0.$$

This inequality is satisfied for all $\mathbf{z}$ and hence for all $\mathbf{z}$ satisfying (13).

We now state and prove a second-order sufficiency theorem [Han and Mangasarian, 1979]. This theorem is more useful and easier to apply than the second-order necessary conditions.

THEOREM 6.3. *Let* $(\mathbf{x}_*, \lambda_0, \boldsymbol{\lambda}, \boldsymbol{\mu})$ *satisfy the FJ necessary conditions of Theorem* 5.1. *Suppose that every* $\mathbf{z} \neq \mathbf{0}$ *that satisfies the tangential constraints* (4) *at* $\mathbf{x}_0 = \mathbf{x}_*$ *and the inequality* $\langle \nabla f(\mathbf{x}_*), \mathbf{z} \rangle \leqslant 0$ *also satisfies*

$$\langle \mathbf{z}, F^0_{\mathbf{xx}}(\mathbf{x}_*, \boldsymbol{\lambda}, \boldsymbol{\mu})\mathbf{z} \rangle > 0, \tag{14}$$

where

$$F^0(\mathbf{x}, \boldsymbol{\lambda}, \boldsymbol{\mu}) = \lambda_0 f(\mathbf{x}) + \langle \boldsymbol{\lambda}, \mathbf{g}(\mathbf{x}) \rangle + \langle \boldsymbol{\mu}, \mathbf{h}(\mathbf{x}) \rangle.$$

Then f *attains a strict local minimum for problem PI at* $\mathbf{x}_*$.

Note that strict inequality is required in (14), while equality is allowed in the necessary condition (10).

Proof. To simplify notation, we assume that $\mathbf{x}_* = \mathbf{0}$ and that $f(\mathbf{0}) = 0$. For any real-valued function $\gamma : X_0 \to \mathbb{R}$, the symbol $\gamma_{\mathbf{xx}}$ will denote the Hessian matrix whose $(i - j)$th entry is $\partial^2\gamma/\partial x_j \partial x_i$.

If f did not attain a strict local minimum at $\mathbf{0}$, then for every positive integer q there would exist a feasible point $\mathbf{v}_q \neq \mathbf{0}$ such that $\mathbf{v}_q \in B(\mathbf{0}, 1/q)$ and $f(\mathbf{v}_q) \leqslant 0$.

Thus there would exist a sequence of points $\{\mathbf{v}_q\}$ in X such that

$$\mathbf{v}_q \neq \mathbf{0}, \qquad \lim_{q\to\infty} \mathbf{v}_q = \mathbf{0}, \qquad f(\mathbf{v}_q) \leqslant 0, \qquad \mathbf{g}(\mathbf{v}_q) \leqslant \mathbf{0}, \qquad \mathbf{h}(\mathbf{v}_q) = \mathbf{0}.$$

Let $E = \{1, \ldots, r\}$. It then follows from Taylor's theorem (Corollary 1, Theorem I.11.2) and our assumption that $\mathbf{x}_* = \mathbf{0}$ and $f(\mathbf{0}) = 0$ that

$$f(\mathbf{v}_q) = \langle \nabla f(v_0 \mathbf{v}_q), \mathbf{v}_q \rangle \leqslant 0 \tag{15i}$$

and that for $i = 1, \ldots, r$ and $j = 1, \ldots, k$

$$\begin{aligned} g_i(\mathbf{v}_q) &= \langle \nabla g_i(v_i \mathbf{v}_q), \mathbf{v}_q \rangle \leqslant 0, \\ h_j(\mathbf{v}_q) &= \langle \nabla h_j(v_j \mathbf{v}_q), \mathbf{v}_q \rangle = 0, \end{aligned} \tag{15ii}$$

where the v_α are numbers in the open interval (0, 1).

It also follows from Taylor's theorem that

$$f(\mathbf{v}_q) = \langle \nabla f(\mathbf{0}), \mathbf{v}_q \rangle + \tfrac{1}{2}\langle \mathbf{v}_q, f_{\mathbf{xx}}(\theta_0 \mathbf{v}_q)\mathbf{v}_q \rangle \leqslant 0 \tag{16i}$$

and that for $i = 1, \ldots, r$ and $j = 1, \ldots, k$

$$\begin{aligned} g_i(\mathbf{v}_q) &= \langle \nabla g_i(\mathbf{0}), \mathbf{v}_q \rangle + \tfrac{1}{2}\langle \mathbf{v}_q, g_{i\mathbf{xx}}(\theta_i \mathbf{v}_q)\mathbf{v}_q \rangle \leqslant 0, \\ h_j(\mathbf{v}_q) &= \langle \nabla h_j(0), \mathbf{v}_q \rangle + \tfrac{1}{2}\langle \mathbf{v}_q, h_{j\mathbf{xx}}(\theta_j \mathbf{v}_q)\mathbf{v}_q \rangle = 0, \end{aligned} \tag{16ii}$$

where the θ_α are real numbers in the open interval (0, 1).

Since $\mathbf{v}_q \neq \mathbf{0}$, the sequence $\mathbf{v}_q/\|\mathbf{v}_q\|$ is a sequence of unit vectors. Hence there exists a unit vector $\mathbf{v}$ and a subsequence that we relabel as $\{\mathbf{v}_q\}$ such that

$$\mathbf{v}_q/\|\mathbf{v}_q\| \to \mathbf{v}.$$

Now suppose that the sequence $\{\mathbf{v}_q\}$ in (15) and (16) is the subsequence. If we divide through by $\|\mathbf{v}_q\|$ in the rightmost inequalities and the equalities in (15) and then let $q \to +\infty$, we get that $\mathbf{v}$ satisfies

$$\langle \nabla f(\mathbf{0}), \mathbf{v} \rangle \leqslant 0, \qquad \nabla \mathbf{g}_E(\mathbf{0})\mathbf{v} \leqslant \mathbf{0}, \qquad \nabla \mathbf{h}(\mathbf{0})\mathbf{v} = \mathbf{0}.$$

Thus $\mathbf{v}$ satisfies the tangential constraints (4).

Next, multiply the first inequality in (16) by $\lambda_0 \geqslant 0$, the ith inequality in (16) by $\lambda_i \geqslant 0$ (the ith component of $\boldsymbol{\lambda}$), and the jth equation in (10) by μ_j (the jth component of $\boldsymbol{\mu}$) and then add the resulting equations and inequalities. Since

$\boldsymbol{\lambda} = (\boldsymbol{\lambda}_E, \boldsymbol{\lambda}_I)$ and $\boldsymbol{\lambda}_I = \mathbf{0}$, we get

$$\langle \lambda_0 \nabla f(\mathbf{0}) + \boldsymbol{\lambda}^t \nabla \mathbf{g}(\mathbf{0}) + \boldsymbol{\mu}^t \nabla \mathbf{h}(\mathbf{0}), \mathbf{v}_q \rangle + \tfrac{1}{2}\langle \mathbf{v}_q, F^0_{\mathbf{xx}}(\theta \mathbf{v}_q)\mathbf{v}_q \rangle \leqslant 0,$$

where

$$F^0_{\mathbf{xx}}(\theta \mathbf{v}_q) = \lambda_0 f_{\mathbf{xx}}(\theta_0 \mathbf{v}_q) + \sum_{i=1}^{n} \lambda_i g_{i\mathbf{xx}}(\theta_i \mathbf{v}_q) + \sum_{j=1}^{k} \mu_j h_{j\mathbf{xx}}(\theta_j \mathbf{v}_q).$$

Since $(\mathbf{0}, \lambda^0, \boldsymbol{\lambda}, \boldsymbol{\mu})$ satisfies the FJ necessary conditions of Theorem 5.1, we get that

$$\tfrac{1}{2}\langle \mathbf{v}_q, F^0_{\mathbf{xx}}(\theta \mathbf{v}_q)\mathbf{v}_q \rangle \leqslant 0.$$

If we now divide both sides of this inequality by $\|\mathbf{v}_q\|^2$ and then let $q \to \infty$, we get that $\langle \mathbf{v}, F^0_{\mathbf{xx}}(\mathbf{0})\mathbf{v} \rangle \leqslant 0$ for a nonzero vector that satisfies the tangential constraints (4) and $\langle \nabla f(\mathbf{0}), \mathbf{v} \rangle \leqslant 0$. This contradicts the assumption that (14) holds for all such $\mathbf{z}$.

If there exists a set of multipliers at $\mathbf{x}_*$ with $\lambda_0 = 1$, we have a slightly different version of the sufficient conditions.

Corollary 4. *Let* $(\mathbf{x}_*, \boldsymbol{\lambda}, \boldsymbol{\mu})$ *satisfy the KKT necessary conditions of Theorem 5.2. Suppose that every* $\mathbf{z} \neq \mathbf{0}$ *in* $\mathbb{R}^n$ *that satisfies the tangential constraints* (4) *at* $\mathbf{x}_0 = \mathbf{x}_*$ *and the equality* $\langle \nabla f(\mathbf{x}_*), \mathbf{z} \rangle = 0$ *also satisfies*

$$\langle \mathbf{z}, F_{\mathbf{xx}}(\mathbf{x}_*, \boldsymbol{\lambda}, \boldsymbol{\mu})\mathbf{z} \rangle > 0, \tag{17}$$

where

$$F(\mathbf{x}, \boldsymbol{\lambda}, \boldsymbol{\mu}) = f(\mathbf{x}) + \langle \boldsymbol{\lambda}, \mathbf{g}(\mathbf{x}) \rangle + \langle \boldsymbol{\mu}, \mathbf{h}(\mathbf{x}) \rangle. \tag{18}$$

Then f attains a strict local minimum for problem PI at $\mathbf{x}_*$.

The corollary differs from the theorem in that the condition $\langle \nabla f(\mathbf{x}_*), \mathbf{z} \rangle \leqslant 0$ is replaced by $\langle \nabla f(\mathbf{x}_*), \mathbf{z} \rangle = 0$.

Proof. If $(\mathbf{x}_*, \boldsymbol{\lambda}, \boldsymbol{\mu})$ satisfies the conditions of Theorem 5.2, then

$$\nabla f(\mathbf{x}_*) + \boldsymbol{\lambda}^t \nabla \mathbf{g}(\mathbf{x}_*) + \boldsymbol{\mu}^t \nabla h(\mathbf{x}_*) = \mathbf{0}.$$

Since $\boldsymbol{\lambda}_I = \mathbf{0}$,

$$\nabla f(\mathbf{x}_*) = -\boldsymbol{\lambda}^t_E \nabla \mathbf{g}_E(\mathbf{x}_*) - \boldsymbol{\mu}^t \nabla \mathbf{h}(\mathbf{x}_*).$$

Hence, for any $\mathbf{z} \in \mathbb{R}^n$

$$\langle \nabla f(\mathbf{x}_*), \mathbf{z} \rangle = -\langle \boldsymbol{\lambda}_E, \nabla \mathbf{g}_E(\mathbf{x}_*)\mathbf{z} \rangle - \langle \boldsymbol{\mu}, \nabla \mathbf{h}(\mathbf{x}_*)\mathbf{z} \rangle.$$

Since $\boldsymbol{\lambda}_E \geqslant \mathbf{0}$, it follows that for any $\mathbf{z}$ satisfying the tangential constraints (4) we have $\langle \nabla f(\mathbf{x}_*), \mathbf{z} \rangle \geqslant 0$. Hence the condition $\langle \nabla f(\mathbf{x}_*), \mathbf{z} \rangle \leqslant 0$ reduces to $\langle \nabla f(\mathbf{x}_*), \mathbf{z} \rangle = 0$.

If the inequality constraints are absent, then a vector $\mathbf{z}$ satisfies the tangential constraints at a point $\mathbf{x}_0$ if and only if $\nabla \mathbf{h}(\mathbf{x}_0)\mathbf{z} = \mathbf{0}$. Thus, if at a point $\mathbf{x}_*$ the necessary conditions of Corollary 2 of Theorem 5.2 are satisfied, the proof of Corollary 4 shows that for every vector $\mathbf{z}$ such that $\nabla \mathbf{h}(\mathbf{x}_*)\mathbf{z} = \mathbf{0}$ we have $\langle \nabla f(\mathbf{x}_*), \mathbf{z} \rangle = 0$. Therefore the following corollary holds.

COROLLARY 5. *Let* $(\mathbf{x}_*, \boldsymbol{\mu})$ *satisfy the necessary condition of Corollary* 2 *of Theorem* 5.2. *Suppose that for every* $\mathbf{z}$ *such that* $\nabla \mathbf{h}(\mathbf{x}_*)\mathbf{z} = \mathbf{0}$

$$\langle \mathbf{z}, H_{\mathbf{xx}}(\mathbf{x}, \mu)\mathbf{z} \rangle > 0,$$

where

$$H(\mathbf{x}, \boldsymbol{\mu}) = f(\mathbf{x}) + \langle \boldsymbol{\mu}, \mathbf{h}(\mathbf{x}) \rangle.$$

The f *attains a strict local minimum for the problem of minimizing* f *subject to* $\mathbf{h}(\mathbf{x}) = \mathbf{0}$.

There is a sufficient condition due to Pennisi [1953] which is not as convenient to apply as Corollary 4 but is equivalent to it. We now state this condition and prove its equivalence to Corollary 4.

COROLLARY 6. *Let* $(\mathbf{x}_*, \boldsymbol{\lambda}, \boldsymbol{\mu})$ *satisfy the KKT necessary conditions of Theorem* 5.2. *Let* $E^* = \{i : i \in E \text{ and } \lambda_i > 0\}$. *Suppose that for every* $\mathbf{z} \neq \mathbf{0} \in \mathbb{R}^n$ *that satisfies*

$$\langle \nabla g_i(\mathbf{x}_*), \mathbf{z} \rangle = 0,\ i \in E^*, \qquad \langle \nabla g_i(\mathbf{x}_*), \mathbf{z} \rangle \leqslant 0, \qquad i \in E, \qquad i \notin E^*,$$
$$\nabla \mathbf{h}(\mathbf{x}_*)\mathbf{z} = 0, \tag{19}$$

the inequality

$$\langle \mathbf{z}, F_{xx}(\mathbf{x}_*, \boldsymbol{\lambda}, \boldsymbol{\mu})z \rangle > 0$$

holds, where F *is as in* (18). *Then* f *attains a strict local minimum for problem PI at* $\mathbf{x}_*$.

Proof. To prove the corollary it suffices to show that the set V_1 of vectors that satisfy (19) is the same as the set V_2 of vectors that satisfy the tangential constraints (4) and the equality $\langle \nabla f(\mathbf{x}_*), \mathbf{z} \rangle = 0$, and then invoke Corollary 4.

We first show that $V_1 \subseteq V_2$. If $V_1 = \varnothing$, there is nothing to prove, so we assume that $V_1 \neq \varnothing$. From (ii) of Theorem 5.2 we get that for $\mathbf{z}$ in $\mathbb{R}^n$

$$\langle \nabla f(\mathbf{x}_*), \mathbf{z} \rangle + \langle \boldsymbol{\lambda}, \nabla \mathbf{g}(\mathbf{x}_*)\mathbf{z} \rangle + \langle \boldsymbol{\mu}, \nabla \mathbf{h}(\mathbf{x}_*)\mathbf{z} \rangle = 0. \tag{20}$$

Now let $\mathbf{z} \in V_1$. Then $\nabla g_E(\mathbf{x}_*)\mathbf{z} \leqslant \mathbf{0}$, and $\nabla \mathbf{h}(\mathbf{x}_*)\mathbf{z} = \mathbf{0}$. Thus $\mathbf{z}$ satisfies the tangential constraints (4). Moreover, since $\lambda_i = 0$ for $i \in I$, we have that $\lambda_i = 0$ for $i \notin E^*$. Thus (20) becomes

$$\langle \nabla f(\mathbf{x}_*), \mathbf{z} \rangle + \sum_{i \in E^*} \lambda_i \langle \nabla g_i(\mathbf{x}_*), \mathbf{z} \rangle = 0$$

for all $\mathbf{z}$ in V_1. For such $\mathbf{z}$ and for $i \in E^*$ we have $\langle \nabla g_i(\mathbf{x}), \mathbf{z} \rangle = 0$. Hence $\langle \nabla f(\mathbf{x}_*), \mathbf{z} \rangle = 0$, and so $V_1 \subseteq V_2$.

We now show that $V_2 \subseteq V_1$. We assume that $V_2 \neq \varnothing$. Since $\lambda_i = 0$ for $i \notin E^*$ and $\langle \nabla f(\mathbf{x}_*), \mathbf{z} \rangle = 0$, it follows that, for $\mathbf{z} \in V_2$, (20) becomes

$$\sum_{i \in E^*} \lambda_i \langle \nabla g_i(\mathbf{x}_*), \mathbf{z} \rangle = 0.$$

Since for any vector $\mathbf{z}$ that satisfies the tangential constraints (4) we have $\langle \nabla g_i(\mathbf{x}_*), \mathbf{z} \rangle \leqslant 0$, and since $\lambda_i > 0$ for $i \in E_*$, the last equality implies that $\langle \nabla g_i(\mathbf{x}_*), \mathbf{z} \rangle = 0$ for $i \in E^*$ Since $\nabla g_E(\mathbf{x}_*)\mathbf{z} \leqslant \mathbf{0}$ for any vector satisfying the tangential constraints, we get that $V_2 \subseteq V_1$.

Example 5.1 (Concluded). We now apply Corollary 4 to show that the points given in relation (4) in Section 5 furnish local minima for problems π_k when $0 < k < 1$ and that the origin furnishes a local minimum for problems π_k when $k > 1$.

We first consider the problem π_k for $0 < k < 1$. At the point $\mathbf{x}_{1*} = (1 - k, \sqrt{2k(1-k)})$, the condition $\langle \nabla f(\mathbf{x}_*), \mathbf{z} \rangle = 0$ becomes

$$\langle (-2k, 2\sqrt{2k(1-k)}), (z_1, z_2) \rangle = 0.$$

At the point $\mathbf{x}_{2*} = (1 - k, -\sqrt{2k(1-k)})$ the condition $\langle \nabla f(\mathbf{x}_*), \mathbf{z} \rangle = 0$ becomes

$$\langle (-2k, -2\sqrt{2k(1-k)}), (z_1, z_2) \rangle = 0.$$

Thus at $\mathbf{x}_{1*}$ and $\mathbf{x}_{2*}$ we require that

$$\mathbf{x}_{1*}: z_1 = z_2 \sqrt{\frac{2(1-k)}{k}}, \qquad \mathbf{x}_{2*}: z_1 = -z_2 \sqrt{\frac{2(1-k)}{k}}. \tag{21}$$

The tangential constraints at the points $\mathbf{x}_{1*}$ and $\mathbf{x}_{2*}$ become

$$\mathbf{x}_{1*}:\langle(2k,\ -2\sqrt{2k(1-k)}),\ (z_1, z_2)\rangle \leqslant 0,$$
$$\mathbf{x}_{2*}:\langle(2k,\ 2\sqrt{2k(1-k)}),\ (z_1, z_2)\rangle \leqslant 0$$

or

$$\mathbf{x}_{1*}: z_1 \leqslant z_2\sqrt{\frac{2(1-k)}{k}}, \qquad \mathbf{x}_{2*}: z_1 \leqslant -z_2\sqrt{\frac{2(1-k)}{k}}.$$

Hence at $\mathbf{x}_{1*}$ the vectors $\mathbf{z} = (z_1, z_2)$ that satisfy $(\nabla f(\mathbf{x}_{1*}, \mathbf{z}\rangle = 0$ and the tangential constraints are given in (21). Similarly, at $\mathbf{x}_{2*}$ the vectors that satisfy $\langle\nabla f(\mathbf{x}_*), \mathbf{z}\rangle = 0$ and the tangential constraints are given in (21). Since $\lambda = 1$ at $\mathbf{x}_{1*}$ and $\mathbf{x}_{2*}$, the function F in (18) becomes

$$F = (x_1 - 1)^2 + x_2^2 + (2kx_1 - x_2^2).$$

Thus

$$\langle\mathbf{z},\ F_{\mathbf{xx}}(\mathbf{x},\ 1,\ \mu)\mathbf{z}\rangle = (z_1,\ z_2)\begin{pmatrix}2 & 0\\ 0 & 0\end{pmatrix}\begin{pmatrix}z_1\\ z_2\end{pmatrix} = 2(z_1)^2. \tag{22}$$

At $\mathbf{x}_{1*}$ and at $\mathbf{x}_{2*}$ vectors $\mathbf{z} \neq \mathbf{0}$ at which (21) holds have $z_1 \neq 0$. It therefore follows that the hypotheses of Corollary 4 hold at $\mathbf{x}_{1*}$ and $\mathbf{x}_{2*}$, and so both of these points furnish local minima. We leave it as an exercise to show that these points furnish minima for problem π_k, with $0 < k < 1$.

We now consider the problem π_k, $k > 1$. The origin is the only point that satisfies the conditions of Theorem 5.2. Since $\nabla f(\mathbf{0}) = (-2, 0)$, all vectors $\mathbf{z}$ of the form $(0, z_2)$, with z_2 arbitrary, satisfy $\langle\nabla f(\mathbf{0}), \mathbf{z}\rangle = 0$. Since $\nabla g(\mathbf{0}) = (2k, 0)$, all vectors $\mathbf{z}$ of the form $\mathbf{z} = (z_1, z_2)$, with $z_1 \leqslant 0$, and z_2 arbitrary, satisfy the tangential constraint $\langle\nabla g(\mathbf{0}), \mathbf{z}\rangle \leqslant 0$. Hence the vectors that satisfy the tangential constraints and $\langle\nabla f(0), \mathbf{z}\rangle = 0$ are vectors of the form $(0, z_2)$, where z_2 is arbitrary.

The function F with $\lambda = 1/k$ becomes

$$F = (x_1 - 1)^2 + x_2^2 + \frac{2kx_1 - x_2^2}{k}.$$

Thus (17) with $\mathbf{z} = (0,\ z_2)$ becomes

$$(0,\ z_2)\begin{pmatrix}2 & 0\\ 0 & 2(1-\frac{1}{k})\end{pmatrix}\begin{pmatrix}0\\ z_2\end{pmatrix} = 2\left(1 - \frac{1}{k}\right)(z_2)^2 > 0, \qquad z_2 \neq 0.$$

Hence the origin furnishes a strict local minimum for problem π_k, $k > 1$. Again, we leave it as an exercise for the reader to show that the origin furnishes a global minimum.

If $k = 1$, then $\mathbf{x}_* = (0, 0)$ and

$$\langle \nabla f(\mathbf{0}), \mathbf{z} \rangle = \langle (-2, 0), (z_1, z_2) \rangle = -2z_1,$$

$$\langle \nabla g(\mathbf{0}), \mathbf{z} \rangle = \langle (2, 0), (z_1, z_2) \rangle = 2z_1.$$

Hence a vector $\mathbf{z}$ satisfies the tangential constraint $\langle \nabla g(\mathbf{0}), \mathbf{z} \rangle \leqslant 0$ and the constraint $\langle \nabla f(\mathbf{0}), \mathbf{z} \rangle = 0$ if and only if it has the form $\mathbf{z} = (0, z_2)$, z_2 arbitrary. If $\mathbf{x}_* = \mathbf{0}$, then $\lambda = 1$ and (22) gives $\langle \mathbf{z}, F_{\mathbf{xx}}(\mathbf{0}, 1, \mu)\mathbf{z} \rangle = 0$. Hence the sufficient condition gives no information in this case.

Exercise 6.1. Show that the points $\mathbf{x}_{1*}$, $\mathbf{x}_{2*}$ and $\mathbf{0}$ in Example 5.1 furnish minima for the relevant problems.

Exercise 6.2. In Exercises 5.1–5.5 use an appropriate second-order sufficiency condition to determine whether the critical points furnish a local minimum.

Exercise 6.3. Given m distinct points $\mathbf{y}_1, \ldots, \mathbf{y}_m$ in $\mathbb{R}^n$, find the smallest closed ball $\overline{B(\mathbf{x}, \rho)}$ that encloses these points. This problem can be formulated analytically as follows: Minimize $f(\mathbf{x}, \rho) = \rho^2$ subject to $\|\mathbf{x} - \mathbf{y}_i\|^2 \leqslant \rho^2$.

(a) Show that a solution $(\mathbf{x}_*, \rho_*)$ exists.

(b) Show that at a critical point $(\mathbf{x}_0, \rho_0)$

  (i) $\lambda_0 = 1$,
  (ii) $\lambda_i \geqslant 0$, $i = 1, \ldots, m$,
  (iii) $\Sigma_{i=1}^m \lambda_i = 1$,
  (iv) $\mathbf{x}_0 = \Sigma_{i=1}^m \lambda_i \mathbf{y}_i$, and
  (v) $\rho_0^2 + \|\mathbf{x}_0\|^2 = \Sigma_{i=1}^m \lambda_i \|\mathbf{y}_i\|^2$.

Note that (ii), (iii), and (iv) show that $\mathbf{x}_0$ is in the simplex determined by $\mathbf{y}_1, \ldots, \mathbf{y}_m$.

(c) Show that with the KKT multipliers as in (b) the function F can be written as

$$F(\rho, \mathbf{x}, \boldsymbol{\lambda}) = \|\mathbf{x}\|^2 - 2\left\langle \mathbf{x}, \sum_{i=1}^{m} \lambda_i \mathbf{y}_i \right\rangle.$$

(d) Does the problem have a unique solution?

(e) Solve the special case $\mathbf{y}_i = \mathbf{e}_i$, $i = 1, 2, 3$, in detail.

V

CONVEX PROGRAMMING AND DUALITY

1. PROBLEM STATEMENT

In the preceding chapter we studied programming problems with differentiable data. In this chapter we shall replace the assumption that the data are differentiable by the assumption that the data are convex. This assumption will enable us to obtain much more detailed information about the behavior of solutions. In Section 1 of Chapter IV we defined a *convex programming problem* to be a programming problem in which the underlying set X_0 is convex, the functions f and $\mathbf{g}$ are convex, and there are no equality constraints.

Convex Programming Problem I (CPI)

Let X_0 be a convex set and let $f: X_0 \to \mathbb{R}$ and $\mathbf{g}: X_0 \to \mathbb{R}^m$ be convex.

$$\text{Minimize } \{f(\mathbf{x}) : \mathbf{x} \in X\},$$

where

$$X = \{\mathbf{x} : \mathbf{g}(\mathbf{x}) \leqslant \mathbf{0}\}.$$

The problem is often stated as

$$\text{Minimize} \quad f(\mathbf{x}) \quad \text{subject to } \mathbf{g}(\mathbf{x}) \leqslant \mathbf{0}.$$

Remark 1.1. A convex programming problem with the additional equality constraint $\mathbf{h}(\mathbf{x}) = \mathbf{0}$ can be cast in the format CPI if the function $\mathbf{h}$ is *affine*. One replaces the constraint $\mathbf{h}(\mathbf{x}) = \mathbf{0}$ by the two inequality constraints $\mathbf{h}(\mathbf{x}) \leqslant \mathbf{0}$ and $-\mathbf{h}(\mathbf{x}) \leqslant \mathbf{0}$. Since $\mathbf{h}$ is affine, both $\mathbf{h}$ and $-\mathbf{h}$ are convex.

In Remark IV.5.9 we pointed out that if the feasible set X in CPI is not empty, then it is convex. Thus in CPI we are minimizing a convex function defined on the convex set X. Therefore, the properties of solutions listed in Theorem IV.3.1 are valid for CPI.

If the functions f and $\mathbf{g}$ are differentiable on X_0, then CPI is a special case of the programming problem PI. Hence Theorem IV.5.1 (the Fritz John necessary conditions), Theorem IV.5.2 (the KKT necessary conditions), and the sufficiency theorem IV.5.3 are all valid.

It turns out that in many problems involving affine equality constraints $\mathbf{h}(\mathbf{x}) = \mathbf{0}$, sharper results can be obtained if we retain the equality constraints as such rather than replace them by two sets of inequality constraints. It will also be convenient to take $X_0 = \mathbb{R}^n$ when affine constraints are present. We therefore formulate the convex programming problem with affine constraints as follows.

Convex Programming Problem II (CPII)

Let $f:\mathbb{R}^n \to \mathbb{R}$ be convex, let $\mathbf{g}: \mathbb{R}^n \to \mathbb{R}^m$ be convex and let $\mathbf{h}: \mathbb{R}^n \to \mathbb{R}^k$ be affine. Minimize

$$\{f(\mathbf{x}) : \mathbf{x} \in X\},$$

where

$$X = \{\mathbf{x} : \mathbf{g}(\mathbf{x}) \leqslant \mathbf{0}\} \cap \{\mathbf{x} : \mathbf{h}(\mathbf{x}) = \mathbf{0}\}.$$

If the feasible set X is not empty, the problem is said to be *consistent.*

Remark 1.2. The components of the affine equality constraints $\mathbf{h}(\mathbf{x}) = \mathbf{0}$ can be written as

$$\langle \mathbf{a}_i, \mathbf{x} \rangle - b_i = 0, \qquad i = 1, \dots, k. \tag{1}$$

Thus, if the problem is consistent, then the system

$$A\mathbf{x} = \mathbf{b}, \tag{2}$$

where A is the $k \times n$ matrix whose ith row is $\mathbf{a}_i$ and $\mathbf{b} = (b_1, \dots, b_k)^T$ has solutions. We recall from linear algebra that for (2) to have solutions the rank of A must equal the rank of the augmented matrix $(A, \mathbf{b})$. If the system (2) has solutions and if the rank of $(A, \mathbf{b})$ were less than k, then one of the equations in (1) could be written as a linear combination of the others and thus eliminated from the system. Since we are only interested in problems that are consistent, there is no loss of generality in assuming that A has maximum possible rank, namely k. In that event the ranks of A and $(A, \mathbf{b})$ both equal k. If the feasible set consists of only one point, then the programming problem is trivial. Therefore, we further assume that $k < n$. For if $k = n$ and A has rank

$k = n$, then (2) has a unique solution and the feasible set consists of a single point. The assumptions that A has rank k and that $k < n$ will be in force henceforth.

In the next section we shall develop necessary conditions and a sufficient condition for a point $\mathbf{x}_*$ to be a solution of CPII without assuming differentiability.

2. NECESSARY CONDITIONS AND SUFFICIENT CONDITIONS

Our first necessary condition will involve a function Λ with domain $\mathbb{R}^n \times \mathbb{R} \times \mathbb{R}^m \times \mathbb{R}^k$ and range in $\mathbb{R}$ defined by

$$\Lambda(\mathbf{x}, \lambda_0, \boldsymbol{\lambda}, \boldsymbol{\mu}) = \lambda_0 f(\mathbf{x}) + \langle \boldsymbol{\lambda}, \mathbf{g}(\mathbf{x}) \rangle + \langle \boldsymbol{\mu}, \mathbf{h}(\mathbf{x}) \rangle. \tag{1}$$

The function Λ is sometimes called a generalized Lagrangian.

THEOREM 2.1. *Let f and $\mathbf{g}$ be convex on $\mathbb{R}^n$ and let $\mathbf{h}$ be affine. Let $\mathbf{x}_*$ be a solution of CPII. Then there exist a real number λ_{0*}, a vector $\boldsymbol{\lambda}_*$ in $\mathbb{R}^m$, and a vector $\boldsymbol{\mu}_*$ in $\mathbb{R}^k$ with $(\lambda_{0*}, \boldsymbol{\lambda}_*) \geqslant (0, \mathbf{0})$ and $(\lambda_{0*}, \boldsymbol{\lambda}_*, \boldsymbol{\mu}_*) \neq (0, \mathbf{0}, \mathbf{0})$ such that*

$$\langle \boldsymbol{\lambda}_*, \mathbf{g}(\mathbf{x}_*) \rangle = 0 \tag{2}$$

and such that for all $\mathbf{x}$ in $\mathbb{R}^n$, all $\boldsymbol{\lambda} \geqslant \mathbf{0}$ in $\mathbb{R}^n$, and all $\boldsymbol{\mu}$ in $\mathbb{R}^k$

$$\Lambda(\mathbf{x}_*, \lambda_{0*}, \boldsymbol{\lambda}, \boldsymbol{\mu}) \leqslant \Lambda(\mathbf{x}_*, \lambda_{0*}, \boldsymbol{\lambda}_*, \boldsymbol{\mu}_*) \leqslant \Lambda(\mathbf{x}, \lambda_{0*}, \boldsymbol{\lambda}_*, \boldsymbol{\mu}_*). \tag{3}$$

Moreover,

$$\Lambda(\mathbf{x}_*, \lambda_{0*}, \boldsymbol{\lambda}_*, \boldsymbol{\mu}_*) = \lambda_{0*} f(\mathbf{x}_*). \tag{4}$$

Proof. Since $\mathbf{x}_*$ is a solution of CPII, the system

$$f(\mathbf{x}) - f(\mathbf{x}_*) < 0,$$
$$\mathbf{g}(\mathbf{x}) < \mathbf{0}, \qquad \mathbf{h}(\mathbf{x}) = \mathbf{0},$$

has no solution in $\mathbb{R}^n$. Hence by Theorem III.4.1 there exists a vector $(\lambda_{0*}, \boldsymbol{\lambda}_*, \boldsymbol{\mu}_*) \neq (0, \mathbf{0}, \mathbf{0})$ with $(\lambda_{0*}, \boldsymbol{\lambda}_*) \geqslant \mathbf{0}$ such that

$$\lambda_{0*}(f(\mathbf{x}) - f(\mathbf{x}_*)) + \langle \boldsymbol{\lambda}_*, \mathbf{g}(\mathbf{x}) \rangle + \langle \boldsymbol{\mu}_*, \mathbf{h}(\mathbf{x}) \rangle \geqslant 0$$

for all $\mathbf{x}$ in $\mathbb{R}^n$. Hence

$$\lambda_{0*} f(\mathbf{x}_*) \leqslant \lambda_{0*} f(\mathbf{x}) + \langle \boldsymbol{\lambda}_*, \mathbf{g}(\mathbf{x}) \rangle + \langle \boldsymbol{\mu}_*, \mathbf{h}(\mathbf{x}) \rangle \tag{5}$$

for all $\mathbf{x}$ in $\mathbb{R}^n$. In particular, (5) holds for $\mathbf{x} = \mathbf{x}_*$. Since $\mathbf{h}(\mathbf{x}_*) = \mathbf{0}$, we get that $\langle \boldsymbol{\lambda}_*, \mathbf{g}(\mathbf{x}_*) \rangle \geqslant 0$. On the other hand, since $\mathbf{x}_*$ is feasible, $\mathbf{g}(\mathbf{x}_*) \leqslant \mathbf{0}$. But $\boldsymbol{\lambda}_* \geqslant \mathbf{0}$, so $\langle \boldsymbol{\lambda}_*, \mathbf{g}(\mathbf{x}_*) \rangle \leqslant 0$. Hence, $\langle \boldsymbol{\lambda}_*, \mathbf{g}(\mathbf{x}_*) \rangle = 0$, and (2) is established.

From (2), (5), and $\mathbf{h}(\mathbf{x}_*) = \mathbf{0}$ we now get

$$\begin{aligned} \lambda_{0*} f(\mathbf{x}_*) &+ \langle \boldsymbol{\lambda}_*, \mathbf{g}(\mathbf{x}_*) \rangle + \langle \boldsymbol{\mu}_*, \mathbf{h}(\mathbf{x}_*) \rangle \\ &= \lambda_{0*} f(\mathbf{x}_*) \leqslant \lambda_{0*} f(\mathbf{x}) + \langle \boldsymbol{\lambda}_*, \mathbf{g}(\mathbf{x}) \rangle + \langle \boldsymbol{\mu}_*, \mathbf{h}(\mathbf{x}) \rangle \end{aligned}$$

for all $\mathbf{x}$. This establishes (4) and the rightmost inequality in (3). From (2) and from $\mathbf{g}(\mathbf{x}_*) \leqslant \mathbf{0}$ it follows that for $\boldsymbol{\lambda} \geqslant \mathbf{0}$

$$\langle \boldsymbol{\lambda}, \mathbf{g}(\mathbf{x}_*) \rangle \leqslant 0 = \langle \boldsymbol{\lambda}_*, \mathbf{g}(\mathbf{x}_*) \rangle.$$

From this and from $\mathbf{h}(\mathbf{x}_*) = \mathbf{0}$ we get that

$$\lambda_{0*} f(\mathbf{x}_*) + \langle \boldsymbol{\lambda}, \mathbf{g}(\mathbf{x}_*) \rangle + \langle \boldsymbol{\mu}, \mathbf{h}(\mathbf{x}_*) \rangle \leqslant \lambda_{0*} f(\mathbf{x}_*) + \langle \boldsymbol{\lambda}_*, \mathbf{g}(\mathbf{x}_*) \rangle + \langle \boldsymbol{\mu}_*, \mathbf{h}(\mathbf{x}) \rangle,$$

which is the leftmost inequality in (3).

Remark 2.1. Since $\boldsymbol{\lambda}_* \geqslant \mathbf{0}$ and $\mathbf{g}(\mathbf{x}_*) \leqslant \mathbf{0}$, condition (2) is equivalent to

$$\lambda_{*i} g_i(\mathbf{x}_*) = 0, \qquad i = 1, \ldots, m. \tag{2'}$$

Remark 2.2. Although (3) is written as a saddle point condition for the function Λ, the leftmost inequality does not involve $\boldsymbol{\mu}$ since $\mathbf{h}(\mathbf{x}_*) = \mathbf{0}$.

Remark 2.3. If the functions f and $\mathbf{g}$ are differentiable, then since $\mathbf{x}_*$ minimizes the function $\Lambda(\cdot, \lambda_{0*}, \boldsymbol{\lambda}_*, \boldsymbol{\mu}_*)$ on $\mathbb{R}^n$, where

$$\Lambda(\mathbf{x}, \lambda_{0*}, \boldsymbol{\lambda}_*, \boldsymbol{\mu}_*) = \lambda_{0*} f(\mathbf{x}) + \langle \boldsymbol{\lambda}_*, \mathbf{g}(\mathbf{x}) \rangle + \langle \boldsymbol{\mu}_*, \mathbf{h}(\mathbf{x}) \rangle,$$

we have

$$\frac{\partial \Lambda}{\partial x_i}(\mathbf{x}_*, \lambda_{0*}, \boldsymbol{\lambda}_*, \boldsymbol{\mu}_*) = 0, \qquad i = 1, \ldots, n.$$

Thus, if A is the matrix in (1.2),

$$\lambda_{0*} \nabla f(\mathbf{x}_*) + \boldsymbol{\lambda}_*^t \nabla \mathbf{g}(\mathbf{x}_*) + \boldsymbol{\mu}_*^t A = \mathbf{0},$$

which together with (2) and the condition $(\lambda_{0*}, \boldsymbol{\lambda}_*, \boldsymbol{\mu}_*) \neq (0, \mathbf{0}, \mathbf{0})$ constitute the FJ necessary condition of Theorem IV.5.1.

Remark 2.4. In the proof of Theorem 2.1, the optimality of $\mathbf{x}_*$ is shown to imply the existence of a vector $(\lambda_{0*}, \boldsymbol{\lambda}_*, \boldsymbol{\mu}_*) \neq (0, \mathbf{0}, \mathbf{0})$ with $\lambda_{0*} \geqslant 0$ and $\boldsymbol{\lambda}_* \geqslant \mathbf{0}$ such that relation (5) holds. *All the conclusions of the theorem then follow from* (5) *and the feasibility of* $\mathbf{x}_*$. We shall use this observation in the sequel.

As in the study of differentiable problems, we seek a constraint qualification that guarantees that $\lambda_{0*} \neq 0$, for then we can take $\lambda_{0*} = 1$. For the convex programming problem the following simple constraint qualification suffices.

Definition 2.1. The problem CPII is said to be *strongly consistent*, or to satisfy the *Slater condition*, if there exists an $\mathbf{x}_0$ such that $\mathbf{g}(\mathbf{x}_0) < \mathbf{0}$ and $\mathbf{h}(\mathbf{x}_0) = \mathbf{0}$.

THEOREM 2.2. *Let* $\mathbf{x}_*$ *be a solution of CPII and let the problem be strongly consistent. Then there exists a* $\boldsymbol{\lambda}_* \geqslant \mathbf{0}$ *in* $\mathbb{R}^m$ *such that*

$$\langle \boldsymbol{\lambda}_*, \mathbf{g}(\mathbf{x}_*) \rangle = 0 \tag{6}$$

and a vector $\boldsymbol{\mu}_*$ *in* $\mathbb{R}^k$ *such that for all* $\mathbf{x}$, *all* $\boldsymbol{\lambda} \geqslant \mathbf{0}$ *in* $\mathbb{R}^m$, *and all* $\boldsymbol{\mu}$ *in* $\mathbb{R}^k$

$$L(\mathbf{x}_*, \boldsymbol{\lambda}, \boldsymbol{\mu}) \leqslant L(\mathbf{x}_*, \boldsymbol{\lambda}_*, \boldsymbol{\mu}_*) \leqslant L(\mathbf{x}, \boldsymbol{\lambda}_*, \boldsymbol{\mu}_*), \tag{7}$$

where

$$L\langle \mathbf{x}, \boldsymbol{\lambda}, \boldsymbol{\mu}) = \Lambda(\mathbf{x}, 1, \boldsymbol{\lambda}\boldsymbol{\mu}) = f(\mathbf{x}) + \langle \boldsymbol{\lambda}, \mathbf{g}(\mathbf{x}) \rangle + \langle \boldsymbol{\mu}, \mathbf{h}(\mathbf{x}) \rangle. \tag{8}$$

Moreover,

$$f(\mathbf{x}_*) = L(\mathbf{x}_*, \boldsymbol{\lambda}_*, \boldsymbol{\mu}_*). \tag{9}$$

Proof. Since $\mathbf{x}_*$ is a solution of CPII, the conclusion of Theorem 2.1 holds. If $\lambda_{0*} = 0$, then since (2) holds and $\mathbf{h}(\mathbf{x}_*) = \mathbf{0}$, the rightmost inequality in (3) becomes

$$0 \leqslant \langle \boldsymbol{\lambda}_*, \mathbf{g}(\mathbf{x}) \rangle + \langle \boldsymbol{\mu}_*, \mathbf{h}(\mathbf{x}) \rangle \tag{10}$$

for *all* $\mathbf{x}$. If we also have $\boldsymbol{\lambda}_* = \mathbf{0}$, then the condition $(\lambda_{0*}, \boldsymbol{\lambda}_*, \boldsymbol{\mu}_*) \neq (0, \mathbf{0}, \mathbf{0})$ implies that $\boldsymbol{\mu}_* \neq \mathbf{0}$. Since $\mathbf{h}(\mathbf{x}) = A\mathbf{x} - \mathbf{b}$, the inequality (10) and $\boldsymbol{\lambda}_* = \mathbf{0}$ imply that

$$\langle \boldsymbol{\mu}_*^t A, \mathbf{x} \rangle \geqslant \langle \boldsymbol{\mu}_*, \mathbf{b} \rangle. \tag{11}$$

for all $\mathbf{x}$. The standing hypothesis made in Section 1 that A has rank k and the condition $\boldsymbol{\mu}_* \neq \mathbf{0}$ imply that $\boldsymbol{\mu}_*^t A \neq \mathbf{0}$. Hence if we take $\mathbf{x} = y\boldsymbol{\mu}_*^t A$ and then let $y \to -\infty$, we see that (11) cannot hold for all $\mathbf{x}$. Hence $\boldsymbol{\lambda}_* \neq \mathbf{0}$.

Since the problem is strongly consistent, there exists an $\mathbf{x}_0$ such that $\mathbf{g}(\mathbf{x}_0) < \mathbf{0}$ and $\mathbf{h}(\mathbf{x}_0) = \mathbf{0}$. Thus, from (10), $\langle \boldsymbol{\lambda}_*, \mathbf{g}(\mathbf{x}_0) \rangle \geqslant 0$. On the other hand,

$\boldsymbol{\lambda}_* \geqslant \mathbf{0}$ and $\boldsymbol{\lambda}_* \neq \mathbf{0}$ imply that $\langle \boldsymbol{\lambda}_*, \mathbf{g}(\mathbf{x}_0) \rangle < 0$. This contradiction shows that $\lambda_{0*} \neq 0$. Therefore, we can divide through by $\lambda_{0*} > 0$ in (2), (3), and (4), relabel $\boldsymbol{\lambda}_*/\lambda_{0*}$ as $\boldsymbol{\lambda}_*$, and relabel $\boldsymbol{\mu}_*/\lambda_0$ as $\boldsymbol{\mu}_*$ to get the conclusion of Theorem 2.2.

Remark 2.5. The rightmost inequality in (7) and equation (9) give

$$f(\mathbf{x}_*) = L(\mathbf{x}_*, \boldsymbol{\lambda}_*, \boldsymbol{\mu}_*) \leqslant f(\mathbf{x}) + \langle \boldsymbol{\lambda}_*, \mathbf{g}(\mathbf{x}) \rangle + \langle \boldsymbol{\mu}_*, \mathbf{h}(\mathbf{x}) \rangle$$

for *all* $\mathbf{x}$. Therefore, if we assume that CPII has a solution $\mathbf{x}_*$ and *if we know the multipliers* $\boldsymbol{\lambda}_*$ and $\boldsymbol{\mu}_*$, then the solution $\mathbf{x}_*$ of the constrained problem

$$\min\{f(\mathbf{x}) : \mathbf{x} \in X\}, \qquad X \equiv \{\mathbf{x} : \mathbf{g}(\mathbf{x}) \leqslant \mathbf{0}, \mathbf{h}(\mathbf{x}) = \mathbf{0}\}$$

will be a solution of the unconstrained problem

$$\min\{f(\mathbf{x}) + \langle \boldsymbol{\lambda}_*, \mathbf{g}(\mathbf{x}) \rangle + \langle \boldsymbol{\mu}_*, \mathbf{h}(\mathbf{x}) \rangle\}. \tag{12}$$

The set S of solutions of the unconstrained problem (12) may contain points that do not satisfy the constraints. To find solutions of the constrained problem, one must determine those points of S that satisfy the constraints. It follows from the rightmost inequality in (7) that if (12) has a unique solution, then it must be a solution of the constrained problem.

One sometimes finds the statement that we may replace the constrained problem by the unconstrained problem of minimizing a Lagrangian. This statement is only true in the sense described in the preceding paragraph.

It is conceivable that there exist vectors $(\boldsymbol{\lambda}, \boldsymbol{\mu})$ with $\boldsymbol{\lambda} \geqslant \mathbf{0}$ such that the unconstrained problem (12) with $(\boldsymbol{\lambda}_*, \boldsymbol{\mu}_*) = (\boldsymbol{\lambda}, \boldsymbol{\mu})$ has the same infimum as constrained problem. Such a vector will be called a *Kuhn–Tucker vector*. The multiplier vectors $(\boldsymbol{\lambda}_*, \boldsymbol{\mu}_*)$ of Theorem 2.2 are Kuhn–Tucker vectors. In the next theorem we show that if CPII has a solution, then a Kuhn–Tucker vector for CPII is a multiplier vector as in Theorem 2.2.

THEOREM 2.3. *Let CPII have a solution* $\mathbf{x}_*$. *Then a Kuhn–Tucker vector* $(\hat{\boldsymbol{\lambda}}, \hat{\boldsymbol{\mu}})$ *for CPII is a multiplier vector* $(\boldsymbol{\lambda}_*, \boldsymbol{\mu}_*)$ *as in Theorem* 2.2.

Proof. Since $\mathbf{x}_*$ is a solution of CPII and since $(\hat{\boldsymbol{\lambda}}, \hat{\boldsymbol{\mu}})$ is a Kuhn–Tucker vector,

$$f(\mathbf{x}_*) = \inf\{f(\mathbf{x}) + \langle \hat{\boldsymbol{\lambda}}, \mathbf{g}(\mathbf{x}) \rangle + \langle \hat{\boldsymbol{\mu}}, \mathbf{h}(\mathbf{x}) \rangle : \mathbf{x} \in \mathbb{R}^n\}.$$

Thus (5) holds with $\lambda_0 = 1$. Hence by the italicized statement in Remark 2.4, the vector $(\hat{\boldsymbol{\lambda}}, \hat{\boldsymbol{\mu}})$ is a multiplier vector as in Theorem 2.2.

Theorems 2.2 and 2.3 imply the following:

COROLLARY 1. *Let CPII have a solution* $\mathbf{x}_*$. *Let* $\mathfrak{M}$ *denote the set of multipliers associated with* $\mathbf{x}_*$ *as in Theorem 2.2 and let* $\mathscr{K}$ *denote the set of Kuhn–Tucker vectors for CPII. Then* $\mathfrak{M} = \mathscr{K}$.

Remark 2.6. The hypothesis of strong consistency in Theorem 2.2 guaranteed that the set $\mathfrak{M}$ is not empty. Thus if CPII is strongly consistent and has a solution, then $\mathscr{K}$ is also nonempty.

We now give an example of a problem, taken from Rockafellar [1970], for which no Kuhn–Tucker vector, and hence no vector in $\mathfrak{M}$, exists.

Example 2.1. Consider $\mathbb{R}^2$, and let

$$f(x_1, x_2) = x_1, \qquad g_1(x_1, x_2) = x_2, \qquad g_2(x_1, x_2) = x_1^2 - x_2.$$

Then the only point in $\mathbb{R}^2$ satisfying $\mathbf{g}(x_1, x_2) \leqslant \mathbf{0}$ is $(0, 0)$, and the value of the minimum is zero. Thus, a Kuhn–Tucker vector $\boldsymbol{\lambda} = (\lambda_1, \lambda_2)$ would have to satisfy $\lambda_1 \geqslant 0$, $\lambda_2 \geqslant 0$ and

$$0 \leqslant x_1 + \lambda_1 x_2 + \lambda_2(x_1^2 - x_2) = x_1 + (\lambda_1 - \lambda_2)x_2 + \lambda_2 x_1^2$$

for all $\mathbf{x}$. We leave it to the reader to check that this is impossible.

Note that the problem in this example is not strongly consistent.

Condition (3) of Theorem 2.1 and condition (7) of Theorem 2.2 are called *saddle point conditions.* The points $(\mathbf{x}_*, \lambda_{0*}, \boldsymbol{\lambda}_*, \boldsymbol{\mu}_*)$ in (3) and $(\mathbf{x}_*, \boldsymbol{\lambda}_*, \boldsymbol{\mu}_*)$ in (7) are called *saddle points.* We previously encountered saddle point conditions and saddle points in our discussion of game theory in Section 5 of Chapter III. If the convex programming problem CPII has a solution and if the problem is strongly consistent, then the saddle point condition (7) gives us a method of computing $f(\mathbf{x}_*)$, the value of the minimum, without finding the optimal point $\mathbf{x}_*$, as is shown in the following theorem. We shall return to this point in our study of duality in Section 4.

THEOREM 2.4. *Let CPII be strongly consistent, let* $\mathbf{x}_*$ *be a solution, let* v *denote the value of the minimum, and let* $\boldsymbol{\lambda}_*$, $\boldsymbol{\mu}_*$ *be the multipliers associated with* $\mathbf{x}_*$ *as in Theorem 2.2. Then*

$$\begin{aligned} v = L(\mathbf{x}_*, \boldsymbol{\lambda}_*, \boldsymbol{\mu}_*) &= \min_{\mathbf{x}} L(\mathbf{x}, \boldsymbol{\lambda}_*, \boldsymbol{\mu}_*) \\ &= \sup_{(\boldsymbol{\lambda},\boldsymbol{\mu})} \inf_{\mathbf{x}} L(\mathbf{x}, \boldsymbol{\lambda}, \boldsymbol{\mu}) = \inf_{\mathbf{x}} \sup_{(\boldsymbol{\lambda},\boldsymbol{\mu})} L(\mathbf{x}, \boldsymbol{\lambda}, \boldsymbol{\mu}), \end{aligned} \tag{13}$$

where $\boldsymbol{\lambda} \in \mathbb{R}^m$, $\boldsymbol{\lambda} \geqslant \mathbf{0}$, $\boldsymbol{\mu} \in \mathbb{R}^k$, $\mathbf{x} \in \mathbb{R}^n$.

Proof. By a slight modification of the argument in Lemma III.5.1 to take into account the possible value of $+\infty$ for a supremum and $-\infty$ for an infimum, we have that

$$\sup_{(\boldsymbol{\lambda},\boldsymbol{\mu})} \inf_{\mathbf{x}} L(\mathbf{x}, \boldsymbol{\lambda}, \boldsymbol{\mu}) \leqslant \inf_{\mathbf{x}} \sup_{(\boldsymbol{\lambda},\boldsymbol{\mu})} L(\mathbf{x}, \boldsymbol{\lambda}, \boldsymbol{\mu}). \tag{14}$$

From (7) of Theorem 2.2 we get

$$\sup_{(\boldsymbol{\lambda},\boldsymbol{\mu})} L(\mathbf{x}_*, \boldsymbol{\lambda}, \boldsymbol{\mu}) = L(\mathbf{x}_*, \boldsymbol{\lambda}_*, \boldsymbol{\mu}_*) = \inf_{\mathbf{x}} L(\mathbf{x}, \boldsymbol{\lambda}_*, \boldsymbol{\mu}_*). \tag{15}$$

Hence

$$\inf_{\mathbf{x}} \sup_{(\boldsymbol{\lambda},\boldsymbol{\mu})} L(\mathbf{x}, \boldsymbol{\lambda}, \boldsymbol{\mu}) \leqslant L(\mathbf{x}_*, \boldsymbol{\lambda}_*, \boldsymbol{\mu}_*) \leqslant \sup_{(\boldsymbol{\lambda},\boldsymbol{\mu})} \inf_{\mathbf{x}} L(\mathbf{x}, \boldsymbol{\lambda}, \boldsymbol{\mu}). \tag{16}$$

We also note that by (7) of Theorem 2.2 we can replace the infimum in (15) by a minimum. This observation and (14) and (16) establish all the equalities in (13) except the first, which follows from (9) of Theorem 2.2.

Remark 2.7. It follows from Remark 2.6 and the proof of Theorem 2.4 that the hypothesis of strong consistency in the theorem can be replaced by the hypothesis that $\mathfrak{M}$ is not empty.

Theorem 2.2 states that if CPII is strongly consistent and $\mathbf{x}_*$ is a solution, then we may take $\lambda_0 = 1$. Hence, if the functions f and $\mathbf{g}$ are differentiable, we may take $\lambda_0 = 1$ in Remark 2.3. Theorem IV.5.3 states that, for differentiable convex problems, if the KKT conditions hold at a feasible point $\mathbf{x}_*$, then $\mathbf{x}_*$ minimizes. Thus, we have the following theorem.

THEOREM 2.5. *Let f and $\mathbf{g}$ be convex and differentiable and let $\mathbf{h}$ be affine. Let CPII be strongly consistent. A necessary and sufficient condition that a feasible point $\mathbf{x}_*$ be a solution to CPII is that there exist multipliers $\boldsymbol{\lambda}_* \in \mathbb{R}^m$, $\boldsymbol{\lambda}_* \geqslant \mathbf{0}$, and $\boldsymbol{\mu}_*$ in $\mathbb{R}^k$ such that*

$$\nabla f(\mathbf{x}_*) + \boldsymbol{\lambda}_*^t \nabla g(\mathbf{x}_*) + \boldsymbol{\mu}_*^t A = \mathbf{0},$$
$$\langle \boldsymbol{\lambda}_*, \mathbf{g}(\mathbf{x}_*) \rangle = 0.$$

In the next theorem we shall show that, with *no restrictions placed on f and* $\mathbf{g}$, the saddle point condition (7) is a sufficient condition that $\mathbf{x}_*$ be a solution to the programming problem P: Minimize f over the set X, where $X = \{\mathbf{x} : \mathbf{x} \in X_0, \mathbf{g}(\mathbf{x}) \leqslant \mathbf{0}\}$.

Note that since we place no conditions on the constraints, equality constraints can be replaced by inequality constraints. We assume that this has been done, and so $\mathbf{h} = \mathbf{0}$.

THEOREM 2.6. *Let X_0 be a set in $\mathbb{R}^n$, let f be a real-valued function defined on X_0, and let $\mathbf{g}$ be a function from X_0 to $\mathbb{R}^m$. Let $\mathbf{x}_*$ be a point in X_0 and let $\boldsymbol{\lambda}_* \geqslant \mathbf{0}$ be a vector in $\mathbb{R}^m$ such that the saddle point condition* (7) *holds for all $\mathbf{x}$ in X_0 and all $\boldsymbol{\lambda} \leqslant \mathbf{0}$ in $\mathbb{R}^m$. Then $\mathbf{x}_*$ is a solution to problem P.*

Proof. We first show that $\mathbf{x}_*$ is feasible. The leftmost inequality in (7) is

$$f(\mathbf{x}_*) + \langle \boldsymbol{\lambda}, \mathbf{g}(\mathbf{x}_*) \rangle \leqslant f(\mathbf{x}_*) + \langle \boldsymbol{\lambda}_*, \mathbf{g}(\mathbf{x}_*) \rangle, \qquad \boldsymbol{\lambda} \geqslant \mathbf{0}.$$

Hence

$$\langle \boldsymbol{\lambda} - \boldsymbol{\lambda}_*, \mathbf{g}(\mathbf{x}_*) \rangle \leqslant 0. \tag{17}$$

Now fix a value j in the set $\{1, \ldots, m\}$. Let

$$\boldsymbol{\lambda}_j = (\lambda_{*1}, \ldots, \lambda_{*j-1}, \lambda_{*j} + 1, \lambda_{*j+1}, \ldots, \lambda_{*m}).$$

If for each $j = 1, \ldots, m$ we take $\boldsymbol{\lambda} = \boldsymbol{\lambda}_j$ in (17), we get

$$g_j(\mathbf{x}_*) \leqslant 0, \qquad j = 1, \ldots, m.$$

Thus $\mathbf{x}_*$ is feasible.

We next show that (6) holds. If we take $\boldsymbol{\lambda} = \mathbf{0}$ in (17), we get that $\langle \boldsymbol{\lambda}_*, \mathbf{g}(\mathbf{x}_*) \rangle \geqslant 0$. On the other hand, $\mathbf{g}(\mathbf{x}_*) \leqslant \mathbf{0}$ and $\boldsymbol{\lambda}_* \geqslant \mathbf{0}$ imply that $\langle \boldsymbol{\lambda}_*, \mathbf{g}(\mathbf{x}_*) \rangle \leqslant 0$. Hence $\langle \boldsymbol{\lambda}_*, \mathbf{g}(\mathbf{x}_*) \rangle = 0$, which is (6).

From the rightmost inequality in (7) and from (6) we get that

$$f(\mathbf{x}_*) \leqslant f(\mathbf{x}) + \langle \boldsymbol{\lambda}_*, \mathbf{g}(\mathbf{x}) \rangle, \qquad \mathbf{x} \in X_0.$$

For $\mathbf{x}$ feasible, $\mathbf{g}(\mathbf{x}) \leqslant \mathbf{0}$. Hence for $\mathbf{x}$ feasible

$$f(\mathbf{x}_*) \leqslant f(\mathbf{x}),$$

and so $f(\mathbf{x}_*) = \min\{f(\mathbf{x}) : \mathbf{x} \in X\}$.

Remark 2.8. Note that the proof of Theorem 2.6 only involves very elementary arguments.

The next theorem emphasizes some of the points implicit in the material presented in this section.

THEOREM 2.7. *A necessary and sufficient condition for a point $(\mathbf{x}_*, \boldsymbol{\lambda}_*, \boldsymbol{\mu}_*)$ to be a saddle point for the Lagrangian L defined in* (8) *on the domain*

$$D = \{(\mathbf{x}, \boldsymbol{\lambda}, \boldsymbol{\mu}) : \mathbf{x} \in \mathbb{R}^n, \boldsymbol{\lambda} \in \mathbb{R}^m, \boldsymbol{\lambda} \geqslant \mathbf{0}, \boldsymbol{\mu} \in \mathbb{R}^k\}$$

is that

$$\begin{aligned} &(i) \quad L(\mathbf{x}_*, \boldsymbol{\lambda}_*, \boldsymbol{\mu}_*) = \min\{L(\mathbf{x}, \boldsymbol{\lambda}_*, \boldsymbol{\mu}_*) : \mathbf{x} \in \mathbb{R}^n\}; \\ &(ii) \quad \mathbf{g}(\mathbf{x}_*) \leqslant \mathbf{0}, \qquad \mathbf{h}(\mathbf{x}_*) = \mathbf{0}; \\ &(iii) \quad \langle \boldsymbol{\lambda}_*, \mathbf{g}(\mathbf{x}_*) \rangle = 0. \end{aligned} \tag{18}$$

Moreover, if (18*ii*) *and* (18*iii*) *hold, then*

$$f(\mathbf{x}_*) = L(\mathbf{x}_*, \boldsymbol{\lambda}_*, \boldsymbol{\mu}_*). \tag{19}$$

Proof. It follows from the definition of L that if (18ii) and (18iii) hold, then (19) holds.

Let $(\mathbf{x}_*, \boldsymbol{\lambda}_*, \boldsymbol{\mu}_*)$ be a saddle point for L. Then (18i) follows from the definition of saddle point. The inequality in (18ii) and the equation (18iii) were established in the course of proving Theorem 2.6. The equality in (18ii) also follows from Theorem 2.6, since there we converted the equality constraints to two inequality constraints, $\mathbf{h}(\mathbf{x}) \leqslant \mathbf{0}$ and $-\mathbf{h}(\mathbf{x}) \leqslant \mathbf{0}$, and thus $\mathbf{h}(\mathbf{x}_*) \leqslant \mathbf{0}$ and $-\mathbf{h}(\mathbf{x}_*) \leqslant \mathbf{0}$. Now suppose that (18) holds. Condition (18ii) states that $\mathbf{x}_*$ is feasible. Conditions (18i) and (19) imply that (5) holds. It now follows from Remark 2.4 that $(\mathbf{x}_*, \boldsymbol{\lambda}_*, \boldsymbol{\mu}_*)$ is a saddle point for $L(\mathbf{x}, \boldsymbol{\lambda}, \boldsymbol{\mu})$.

3. PERTURBATION THEORY

In problems arising in applications the values of various parameters are usually not known precisely but are known to lie in certain ranges of values. Thus one might assign a value to the parameters and then study the effect of perturbing the parameters. We shall assume that this is modeled by perturbing the constraints and shall study the effect of perturbations on the value of the infimum of the problem. Given the uncertainty in the parameter values, one would hope for a certain stability of the problem in the sense that small perturbations in the constraints will not lead to large changes in the value of the infimum. We shall show that if the unperturbed convex programming problem is strongly consistent and has a solution, then the change in the value of the infimum is bounded below by a linear function of the perturbations. In the preceding section we showed that under reasonable assumptions the set $\mathfrak{M}$ of multipliers and the set $\mathscr{K}$ of Kuhn–Tucker vectors are nonempty and equal. In this section we shall show that the vectors in $\mathscr{K}$ and $\mathfrak{M}$ are the coefficient vectors of the aforementioned linear function of the perturbations. Moreover, the sets $\mathscr{K}$ and $\mathfrak{M}$ are the negative subdifferential of the infimum as a function of the perturbation, evaluated at the unperturbed value. Thus, in a sense, the multipliers measure the rate of change of the value of the infimum as we perturb the constraints. We shall also give an economic interpretation of the multipliers. These statements will be made precise in this section.

Throughout this section we assume that the problem CPII is consistent, that is, it has a nonempty feasible set. For each $\mathbf{z} = (\mathbf{z}_1, \mathbf{z}_2)$ with $\mathbf{z}_1 \in \mathbb{R}^m$ and $\mathbf{z}_2 \in \mathbb{R}^k$ we define a perturbed programming problem CPII($\mathbf{z}$) as follows.

Problem CPII(z)

Minimize $f(\mathbf{x})$ subject to $\mathbf{x}$ in $X(\mathbf{z})$, where

$$X(\mathbf{z}) = \{\mathbf{x} : \mathbf{g}(\mathbf{x}) \leqslant \mathbf{z}_1,\ \mathbf{h}(\mathbf{x}) = \mathbf{z}_2\},$$

the function f is a convex function from $\mathbb{R}^n$ to $\mathbb{R}$, the function $\mathbf{g}$ is a convex function from $\mathbb{R}^n$ to $\mathbb{R}^m$, and the function $\mathbf{h}$ is an affine function from $\mathbb{R}^n$ to $\mathbb{R}^k$.

The constraint set $X(\mathbf{z})$ can also be described by

$$X(\mathbf{z}) = \{\mathbf{x} : \mathbf{g}(\mathbf{x}) - \mathbf{z}_1 \leqslant \mathbf{0},\ \mathbf{h}(\mathbf{x}) - \mathbf{z}_2 = \mathbf{0}\}.$$

For each fixed $\mathbf{z}$ the function $\mathbf{g}(\mathbf{x}) - \mathbf{z}_1$ is convex and the function $\mathbf{h}(\mathbf{x}) - \mathbf{z}_2$ is affine. Thus, each problem CPII ($\mathbf{z}$) is a convex programming problem. Note that the problem CPII is the problem CPII($\mathbf{0}$).

For each $\mathbf{z}$ in $\mathbb{R}^{m+k}$ the feasible set $X(\mathbf{z})$ is either empty or convex. For each $\mathbf{z}$ such that $X(\mathbf{z}) \neq \varnothing$, let

$$\omega(\mathbf{z}) = \inf\{f(\mathbf{x}) : \mathbf{x} \in X(\mathbf{z})\}.$$

Thus we have a function ω whose domain is the set of $\mathbf{z}$ such that $X(\mathbf{z})$ is not empty and which can assume the value $-\infty$. We shall designate the domain of ω by dom(ω) and shall call ω the *value function* of the family of perturbed problems CPII($\mathbf{z}$), or simply the value function. For $\mathbf{z}$ in dom(ω) we shall call $\omega(\mathbf{z})$ the value of CPII($\mathbf{z}$). We shall denote the infimum of the unperturbed problem CPII by v. Thus

$$v = \inf\{f(\mathbf{x}) : \mathbf{g}(\mathbf{x}) \leqslant \mathbf{0},\ \mathbf{h}(\mathbf{x}) = \mathbf{0}\},$$

so $v = \omega(\mathbf{0})$.

Since $\omega(\mathbf{z}) = -\infty$ is possible, we shall adjoin $-\infty$ to the reals $\mathbb{R}$. Although we shall not need to do so in this section, we shall also adjoin $+\infty$ to $\mathbb{R}$. To work with $-\infty$ and $+\infty$ in a consistent way, we define the following operations:

$$\begin{aligned}
(+\infty) + (+\infty) &= +\infty, \\
(-\infty) + (-\infty) &= -\infty, \\
(-\infty) + x &= -\infty, && x \text{ real}, \\
x(-\infty) &= -\infty, && x \text{ real } x > 0, \\
x(-\infty) &= +\infty, && x \text{ real } x < 0, \\
-\infty < x &< +\infty, && x \text{ real}.
\end{aligned}$$

The operation $(-\infty) + (+\infty)$ is not defined.

For functions f defined on a convex set C and taking on either real values or the value of $-\infty$, the definition of convexity is unchanged. Namely, f is said to be convex on C if for every $\mathbf{x}_1$, $\mathbf{x}_2$ in C and every (α, β) in $\mathbb{P}_2$

$$f(\alpha\mathbf{x}_1 + \beta\mathbf{x}_2) \leqslant \alpha f(\mathbf{x}_1) + \beta f(\mathbf{x}_2),$$

where the values of $-\infty$ are to be operated on as in (1). As in Chapter 3, an equivalent definition of f convex is that the set

$$\text{epi}(f) = \{(\mathbf{x}, y) : \mathbf{x} \in C,\ y \in \mathbb{R},\ y \geqslant f(\mathbf{x})\}$$

is convex.

LEMMA 3.1. *The domain of the value function ω is a convex set and ω is a convex function on dom(ω). If the problem CPII is strongly consistent, then $\mathbf{0}$ is an interior point of* dom(ω).

Proof. We first show that dom(ω) is a convex set. We shall suppose that the affine equality constraints have been written as pairs of inequality constraints. Since we assume that CPII is consistent, $\mathbf{0}$ is in dom(ω) and thus dom(ω) $\neq \varnothing$. We must show that if $\mathbf{w}_1$ and $\mathbf{w}_2$ are two points in dom(ω), then so is $\alpha\mathbf{w}_1 + \beta\mathbf{w}_2$ for all (α, β) in $\mathbb{P}_2$. Since $\mathbf{w}_1$ and $\mathbf{w}_2$ are in dom(ω), there exist points $\mathbf{x}_1$ and $\mathbf{x}_2$ in $\mathbb{R}^n$ such that

$$\mathbf{g}(\mathbf{x}_1) \leqslant \mathbf{w}_1 \quad \text{and} \quad \mathbf{g}(\mathbf{x}_2) \leqslant \mathbf{w}_2.$$

Since $\mathbf{g}$ is convex, for any (α, β) in $\mathbb{P}_2$

$$\mathbf{g}(\alpha\mathbf{x}_1 + \beta\mathbf{x}_2) \leqslant \alpha\mathbf{g}(\mathbf{x}_1) + \beta\mathbf{g}(\mathbf{x}_2) \leqslant \alpha\mathbf{w}_1 + \beta\mathbf{w}_2.$$

Thus, $\alpha\mathbf{x}_1 + \beta\mathbf{x}_2$ is in $X(\alpha\mathbf{w}_1 + \beta\mathbf{w}_2)$ and $\alpha\mathbf{w}_1 + \beta\mathbf{w}_2$ is in dom(ω).

To show that ω is convex, again let $\mathbf{w}_1$ and $\mathbf{w}_2$ be two points in dom(ω) and let (α, β) be in $\mathbb{P}_2$. Let $\mathbf{x}_1 \in X(\mathbf{w}_1)$ and let $\mathbf{x}_2 \in X(\mathbf{w}_2)$. Then as in the preceding paragraph $\alpha\mathbf{x}_1 + \beta\mathbf{x}_2$ is in $X(\alpha\mathbf{w}_1 + \beta\mathbf{w}_2)$. Hence

$$\begin{aligned}
\omega(\alpha\mathbf{w}_1 + \beta\mathbf{w}_2) &= \inf\{f(\mathbf{x}) : \mathbf{g}(\mathbf{x}) \leqslant \alpha\mathbf{w}_1 + \beta\mathbf{w}_2)\} \\
&\leqslant \inf\{f(\mathbf{x}) : \mathbf{x} = \alpha\mathbf{x}_1 + \beta\mathbf{x}_2,\ \mathbf{g}(\mathbf{x}_1) \leqslant \mathbf{w}_1,\ \mathbf{g}(\mathbf{x}_2) \leqslant \mathbf{w}_2\} \\
&= \inf\{f(\alpha\mathbf{x}_1 + \beta\mathbf{x}_2) : \mathbf{x}_1 \in X(\mathbf{w}_1),\ \mathbf{x}_2 \in X(\mathbf{w}_2)\}. \\
&\underset{f\ \text{convex}}{\leqslant} \inf\{\alpha f(\mathbf{x}_1) + \beta f(\mathbf{x}_2) : \mathbf{x}_1 \in X(\mathbf{w}_1),\ \mathbf{x}_2 \in X(\mathbf{w}_2)\}. \qquad (1)
\end{aligned}$$

If $\omega(\mathbf{w}_1)$ and $\omega(\mathbf{w}_2)$ are both finite, then by Exercise I.6.4 the rightmost expression in (1) is equal to

$$\inf\{\alpha f(\mathbf{x}_1) : \mathbf{x}_1 \in X(\mathbf{w}_1)\} + \inf\{\beta f(\mathbf{x}_2) : \mathbf{x}_2 \in X(\mathbf{w}_2)\} = \alpha\omega(\mathbf{w}_1) + \beta\boldsymbol{\omega}(\mathbf{w}_2).$$

If either $\omega(\mathbf{w}_1) = -\infty$ or $\omega(\mathbf{w}_2) = -\infty$, then the rightmost expression in (1) is equal to $-\infty$. Thus,

$$\omega(\alpha\mathbf{w}_1 + \beta\mathbf{w}_2) \leqslant \alpha\omega(\mathbf{w}_1) + \beta\omega(\mathbf{w}_2), \tag{2}$$

and the convexity of ω is established.

We now suppose that the problem CPII is strongly consistent and shall show that $\mathbf{0}$ is an interior point of dom(ω). We shall not incorporate the affine equality constraints into the inequality constraints. Recall that we are writing the vector $\mathbf{z}$ which is the perturbation of the constraints as $\mathbf{z} = (\mathbf{z}_1, \mathbf{z}_2)$, where $\mathbf{z}_1 \in \mathbb{R}^m$ is the perturbation of the inequality constraints and $\mathbf{z}_2 \in \mathbb{R}^k$ is the perturbation of the equality constraints.

Since CPII is strongly consistent, there exists an $\mathbf{x}_0$ such that

$$\mathbf{g}(\mathbf{x}_0) < \mathbf{0} \quad \text{and} \quad \mathbf{h}(\mathbf{x}_0) = \mathbf{0}. \tag{3}$$

Let $\mathbf{z}_1 = (z_{11}, \ldots, z_{1m})$. For each $i = 1, \ldots, m$, the function G_i defined on $\mathbb{R}^n \times \mathbb{R}$ by

$$G_i(\mathbf{x}, z_{1i}) = g_i(\mathbf{x}) - z_{1i}$$

is continuous. This holds because G_i is convex and a convex function is continuous on the interior of its domain of definition, which in this case is $\mathbb{R}^n \times \mathbb{R}$. From (3) we get that $G_i(\mathbf{x}_0, 0) < 0$. Hence, by the continuity of G_i, there exist a $\rho_i > 0$ and a $\delta_i > 0$ such that whenever $|z_{1i}| < \delta$ and $\|\mathbf{x} - \mathbf{x}_0\| < \rho_i$

$$g_i(\mathbf{x}) - z_{1i} < 0.$$

Let $\rho_0 = \min\{\rho_i : i = 1, \ldots, m\}$ and let $\delta_0 = \min\{\delta_1, \ldots, \delta_m\}$. Then

$$\mathbf{g}(\mathbf{x}) < \mathbf{z}_1 \tag{4}$$

when $\|\mathbf{z}_1\| < \delta_0$ and $\|\mathbf{x} - \mathbf{x}_0\| < \rho_0$.

We now consider the perturbed equality constraints, which in accordance with (1.2) we write as

$$A\mathbf{x} = \mathbf{b} + \mathbf{z}_2. \tag{5}$$

Recall that in Remark 1.1 we assumed that A has rank k. We assume that the square matrix A_k consisting of the first k columns of A is nonsingular. There is no loss of generality in this assumption since we can relabel the components to achieve this. Let A_{n-k} be the matrix consisting of the last $n - k$ columns of A. Let $\mathbf{x} = (\mathbf{x}_k, \mathbf{x}_{n-k})$, where $\mathbf{x}_k$ is the vector consisting of the first k components of $\mathbf{x}$ and $\mathbf{x}_{n-k}$ is the vector consisting of the last $n - k$ components of $\mathbf{x}$. Then we may write (5) as

$$A_k\mathbf{x}_k + A_{n-k}\mathbf{x}_{n-k} = \mathbf{b} + \mathbf{z}_2.$$

From this equation and the equation $A\mathbf{x}_0 = \mathbf{b}$ we get that

$$(\mathbf{x}_k - \mathbf{x}_{0,k}) = A_k^{-1}[-A_{n-k}(\mathbf{x}_{n-k} - \mathbf{x}_{0,n-k}) + \mathbf{z}_2], \tag{6}$$

where $\mathbf{x}_{0,k} = (x_{0,1}, \ldots, x_{0,k})$ and $\mathbf{x}_{0,n-k} = (x_{0,k+1}, \ldots, x_{0,n})$. It follows from (6) that there exist a $\delta' > 0$ and a $0 < \rho < \rho_0$ such that if $\|\mathbf{x}_{n-k} - \mathbf{x}_{0,n-k}\| < \frac{1}{2}\rho$ and $\|\mathbf{z}_2\| < \delta'$, then $\|\mathbf{x}_k - \mathbf{x}_{0,k}\| < \frac{1}{2}\rho_0$. Since for any $\mathbf{x}$ we have $\|\mathbf{x}\|^2 = \|\mathbf{x}_k\|^2 + \|\mathbf{x}_{n-k}\|^2$, it follows that for every vector $\mathbf{x}_{n-k}$ satisfying $\|\mathbf{x}_{n-k} - \mathbf{x}_{0,n-k}\| < \frac{1}{2}\rho$ and for every vector $\mathbf{z}_2$ satisfying $\|\mathbf{z}_2\| < \delta'$ the vector $\mathbf{x} = (\mathbf{x}_k, \mathbf{x}_{n-k})$, where $\mathbf{x}_k$ is given by (6), is such that $\|\mathbf{x} - \mathbf{x}_0\| < \rho_0/\sqrt{2}$ and (5) is satisfied. The last statement together with (4) show that for every $\mathbf{z} = (\mathbf{z}_1, \mathbf{z}_2)$ with $\|\mathbf{z}\| < \min(\delta_0, \delta')$ there exists a vector $\mathbf{x}_\mathbf{z}$ that is feasible. Thus $\mathbf{0}$ is an interior point of $\text{dom}(\omega)$.

Lemma 3.1 guarantees that dom (ω) is a convex set and ω is a convex function. It also guarantees that if CPII is strongly consistent, $\text{dom}(\omega)$ has nonempty interior. There is no guarantee, however, that in general $\text{dom}(\omega)$ will have a nonempty interior. To state two important corollaries to Lemma 3.1, it will therefore be necessary to consider the notions of relative interior and relative boundary, which were introduced in Section 7 of Chapter II.

COROLLARY 1. *If* $\omega(\mathbf{w}_1) = -\infty$ *at some point* $\mathbf{w}_1$ *in* $\text{dom}(\omega)$, *then* $\omega(\mathbf{z}) = -\infty$ *at all points of* $\text{dom}(\omega)$ *with the possible exception of relative boundary points.*

Proof. It follows from (2) that if $\omega(\mathbf{w}_1) = -\infty$ at some point $\mathbf{w}_1$ in $\text{dom}(\omega)$, then $\omega(\mathbf{z}) = -\infty$ at all points of the half open line segment $[\mathbf{w}_1, \mathbf{w}_2)$, where $\mathbf{w}_2 \neq \mathbf{w}_1$ is another point of $\text{dom}(\omega)$.

Let $\mathbf{w}_0$ be a point of $\text{dom}(\omega)$ that is not a relative boundary point. Then $\mathbf{w}_0$ is a relative interior point of $\text{dom}(\omega)$. Hence the line segment $[\mathbf{w}_1, \mathbf{w}_0]$, which lies in $\text{dom}(\omega)$, can be extended to a line segment $[\mathbf{w}_1, \mathbf{w}_2]$ that lies in $\text{dom}(\omega)$. The point $\mathbf{w}_0$ is interior to $[\mathbf{w}_1, \mathbf{w}_2]$, and so $\omega(\mathbf{w}_0) = -\infty$.

COROLLARY 2. *If* ω *is finite at a relative interior point of its domain, then* ω *is finite over its entire domain.*

This follows from Corollary 1, for if ω were not finite at some point, it would have to be infinite at all relative interior points of its domain.

Remark 3.1. The only property of ω that was used in the proof of Corollary 1 was the convexity of ω. Thus Corollaries 1 and 2 are valid for all convex functions admitting $-\infty$ as a value.

LEMMA 3.2. *Let problem* CPII *be strongly consistent and let* $v \equiv \omega(\mathbf{0})$ *be finite. Then* ω *is finite on* $\text{dom}(\omega)$, *the subdifferential* $\partial\omega(\mathbf{0}) \neq \varnothing$ *and each* $\boldsymbol{\eta}$ *in* $-\partial\omega(\mathbf{0})$ *can be written as* $\boldsymbol{\eta} = (\boldsymbol{\lambda}, \boldsymbol{\mu})$, *where* $\boldsymbol{\lambda} \in \mathbb{R}^m$, $\boldsymbol{\lambda} \geqslant 0$, $\boldsymbol{\mu} \in \mathbb{R}^k$ *and such that for all* $\mathbf{z} = (\mathbf{z}_1, \mathbf{z}_2)$ *in* $\text{dom}(\omega)$

$$\omega(\mathbf{z}) \geqslant \omega(\mathbf{0}) - \langle \boldsymbol{\eta}, \mathbf{z} \rangle = \omega(\mathbf{0}) - \langle \boldsymbol{\lambda}, \mathbf{z}_1 \rangle - \langle \boldsymbol{\mu}, \mathbf{z}_2 \rangle. \tag{7}$$

Proof. It follows from Lemma 3.1 that $\mathbf{0}$ is an interior point of $\text{dom}(\omega)$ and that ω is a convex function on the convex set $\text{dom}(\omega)$. Since $\omega(\mathbf{0})$ is finite, it follows from Corollary 2 to Lemma 3.1 that ω is finite at all points of $\text{dom}(\omega)$. Therefore, by Theorem III.2.1, the set of subgradients of ω at $\mathbf{0}$ is not empty. Relation (7) then follows by taking $\boldsymbol{\eta} = -\boldsymbol{\xi}$, where $\boldsymbol{\xi}$ is any element in $\partial\omega(\mathbf{0})$.

We now show that $\boldsymbol{\lambda} \geqslant \mathbf{0}$. Since the problem is consistent, $\{\mathbf{x} : \mathbf{g}(\mathbf{x}) \leqslant \mathbf{0}, \mathbf{h}(\mathbf{x}) = \mathbf{0}\}$ is nonempty. Hence the set $X(\mathbf{z})$ is nonempty for any vector $\mathbf{z} = (\mathbf{z}_1, \mathbf{z}_2)$ with $\mathbf{z}_1 \geqslant 0$ and $\mathbf{z}_2 = 0$. Thus, all such vectors are in $\text{dom}(\omega)$.

If $\boldsymbol{\lambda}$ were not nonnegative, then at least one component of $\boldsymbol{\lambda}$, say λ_i, would be negative. Let $\mathbf{z}_i = (\mathbf{z}_{1i}, \mathbf{0}_{2i})$ where $\mathbf{z}_{1i} = \mathbf{e}_i$, the vector in $\mathbb{R}^m$ all of whose components except the ith are zero and whose ith component is 1. Then $\mathbf{z}_i$ is in $\text{dom}(\omega)$ and

$$\omega(\mathbf{z}_i) \geqslant \omega(\mathbf{0}) - \langle \boldsymbol{\lambda}, \mathbf{z}_{1i} \rangle - \langle \boldsymbol{\mu}, \mathbf{0} \rangle = \omega(\mathbf{0}) - \lambda_i > \omega(\mathbf{0}), \tag{8}$$

the last inequality holding because $\lambda_i < 0$.

Since $\mathbf{z}_{1i} \geqslant \mathbf{0}$ and $\mathbf{z}_2 = \mathbf{0}$, we have $X(\mathbf{0}) \subseteq X(\mathbf{z}_i)$. Therefore,

$$\omega(\mathbf{z}_i) = \inf\{f(\mathbf{x}) : \mathbf{x} \in X(\mathbf{z}_i)\} \leqslant \inf\{f(\mathbf{x}) : \mathbf{x} \in X(\mathbf{0})\} = \omega(\mathbf{0}),$$

which contradicts (8). Thus $\boldsymbol{\lambda} \geqslant \mathbf{0}$.

Remark 3.2. The vector $-\boldsymbol{\eta}$ is any vector in $\partial\omega(\mathbf{0})$ and so need not be unique. By Theorem III.3.1 the vector $-\boldsymbol{\eta}$ is unique if and only if ω is differentiable at $\mathbf{0}$, in which case $-\boldsymbol{\eta} = \nabla\omega(\mathbf{0})$. Thus, $-\boldsymbol{\eta}$ gives the direction of maximum rate of increase at $\mathbf{0}$ of the value function ω.

If ω is not differentiable, then by Exercise III.2.3,

$$\omega'(\mathbf{0}, \mathbf{v}) \geqslant \langle -\boldsymbol{\eta}, \mathbf{v} \rangle \quad \text{for all } \mathbf{v} \text{ in } \mathbb{R}^{m+n},$$

where $\omega'(\mathbf{0}, \mathbf{v})$ is the directional derivative of ω at $\mathbf{0}$ in the direction $\mathbf{v}$. Thus, $-\boldsymbol{\eta}$ again in a sense measures the changes in ω as we move away from the origin. The vector $\boldsymbol{\eta}$ is therefore called a *sensitivity vector*.

We noted in Remark 2.5 that if we know the multipliers $(\boldsymbol{\lambda}_*, \boldsymbol{\mu}_*)$ of Theorem 2.2, then we may replace the constrained problem by an unconstrained problem. In other words, the multipliers $(\boldsymbol{\lambda}_*, \boldsymbol{\mu}_*)$ of Theorem 2.2 are Kuhn–Tucker vectors. By Corollary 1 to Theorem 2.3, if CPII has a solution, then the set $\mathscr{K}$ of Kuhn–Tucker vectors is equal to the set $\mathfrak{M}$ of multiplier vectors.

In the next lemma we shall show that if $\omega(\mathbf{0})$ is finite, then every vector in $-\partial\omega(\mathbf{0})$ is a Kuhn–Tucker vector.

LEMMA 3.3. *Let*

$$\omega(\mathbf{0}) \equiv \inf\{f(\mathbf{x}) : \mathbf{x} \in X\},$$

where $X = \{\mathbf{x} : \mathbf{g}(\mathbf{x}) \leqslant \mathbf{0}, \mathbf{h}(\mathbf{x}) = \mathbf{0}\}$, *be finite and let* $\partial\omega(\mathbf{0}) \neq \varnothing$. *Then for each* $\boldsymbol{\eta} = (\boldsymbol{\lambda}, \boldsymbol{\mu})$ *such that* $-\boldsymbol{\eta} \in \partial\omega(\mathbf{0})$ *the vector* $\boldsymbol{\lambda}$ *is nonnegative and*

$$\omega(\mathbf{0}) = \inf\{f(\mathbf{x}) + \langle\boldsymbol{\lambda}, \mathbf{g}(\mathbf{x})\rangle + \langle\boldsymbol{\mu}, \mathbf{h}(\mathbf{x})\rangle : \mathbf{x} \in \mathbb{R}^n\}.$$

Proof. Since $-\boldsymbol{\eta} \in \partial\omega(\mathbf{0})$, for each $\mathbf{z}$ in $\operatorname{dom}(\omega)$

$$\omega(\mathbf{z}) \geqslant \omega(\mathbf{0}) - \langle\boldsymbol{\eta}, \mathbf{z}\rangle. \tag{9}$$

From the proof of Lemma 3.2 we get that $\boldsymbol{\lambda} \geqslant \mathbf{0}$. For all $\mathbf{x}$ in $\mathbb{R}^n$, $\mathbf{g}(\mathbf{x}) \leqslant \mathbf{g}(\mathbf{x})$ and $\mathbf{h}(\mathbf{x}) = \mathbf{h}(\mathbf{x})$. Therefore $\mathbf{z} = (\mathbf{g}(\mathbf{x}), \mathbf{h}(\mathbf{x}))$ is in $\operatorname{dom}(\omega)$. Taking $\mathbf{z} = (\mathbf{g}(\mathbf{x}), \mathbf{h}(\mathbf{x}))$ in (9) gives

$$\omega(\mathbf{g}(\mathbf{x}), \mathbf{h}(\mathbf{x})) + \langle\boldsymbol{\lambda}, \mathbf{g}(\mathbf{x})\rangle + \langle\boldsymbol{\mu}, \mathbf{h}(\mathbf{x})\rangle \geqslant \omega(\mathbf{0}) \tag{10}$$

for all $\mathbf{x}$ in $\mathbb{R}^n$. Now

$$\omega(\mathbf{g}(\mathbf{x}), \mathbf{h}(\mathbf{x})) = \inf\{f(\mathbf{y}) : \mathbf{y} \in X(\mathbf{g}(\mathbf{x}), \mathbf{h}(\mathbf{x}))\},$$

where $X(\mathbf{g}(\mathbf{x}), \mathbf{h}(\mathbf{x})) = \{\mathbf{y} : \mathbf{g}(\mathbf{y}) \leqslant \mathbf{g}(\mathbf{x}), \mathbf{h}(\mathbf{y}) = \mathbf{h}(\mathbf{x})\}$. But $\mathbf{x} \in X(\mathbf{g}(\mathbf{x}), \mathbf{h}(\mathbf{x}))$. Hence

$$f(\mathbf{x}) \geqslant \omega(\mathbf{g}(\mathbf{x}), \mathbf{h}(\mathbf{x})).$$

Substituting this into (10) gives

$$f(\mathbf{x}) + \langle\boldsymbol{\lambda}, \mathbf{g}(\mathbf{x})\rangle + \langle\boldsymbol{\mu}, \mathbf{h}(\mathbf{x})\rangle \geqslant \omega(\mathbf{0})$$

for all $\mathbf{x}$. Hence

$$\omega(\mathbf{0}) \leqslant \inf\{f(\mathbf{x}) + \langle\boldsymbol{\lambda}, \mathbf{g}(\mathbf{x})\rangle + \langle\boldsymbol{\mu}, \mathbf{h}(\mathbf{x})\rangle\}.$$

We now establish the reverse inequality and thus prove the lemma. We have

$$\begin{aligned}&\inf\{f(\mathbf{x}) + \langle\boldsymbol{\lambda}, \mathbf{g}(\mathbf{x})\rangle + \langle\boldsymbol{\mu}, \mathbf{h}(\mathbf{x})\rangle : \mathbf{x} \in \mathbb{R}^n\} \\ &\qquad \leqslant \inf\{f(\mathbf{x}) + \langle\boldsymbol{\lambda}, \mathbf{g}(\mathbf{x})\rangle + \langle\boldsymbol{\mu}, \mathbf{h}(\mathbf{x})\rangle : \mathbf{g}(\mathbf{x}) \leqslant \mathbf{0}, \mathbf{h}(\mathbf{x}) = \mathbf{0}\}.\end{aligned}$$

Since $\boldsymbol{\lambda} \geqslant \mathbf{0}$, $\mathbf{g}(\mathbf{x}) \leqslant \mathbf{0}$, and $\mathbf{h}(\mathbf{x}) = \mathbf{0}$, the right-hand side of this inequality is less than or equal to

$$\inf\{f(\mathbf{x}) : \mathbf{g}(\mathbf{x}) \leqslant \mathbf{0},\ \mathbf{h}(\mathbf{x}) = \mathbf{0}\} = \omega(\mathbf{0}).$$

Remark 3.3. Because of relationships (7) and (9), which bound the change in ω from below by a linear function, problems with $\partial\omega(\mathbf{0}) \neq \varnothing$ are said to be *stable.*

The next theorem is an immediate consequence of Lemmas 3.2 and 3.3.

THEOREM 3.1. *Let the programming problem CPII be strongly consistent and let $\omega(\mathbf{0})$ be finite. Then $\partial\omega(\mathbf{0})$ is nonempty and for every $\boldsymbol{\eta} = (\boldsymbol{\lambda}, \boldsymbol{\mu})$ such that $-\boldsymbol{\eta} \in \partial\omega(\mathbf{0})$ the vector $\boldsymbol{\lambda}$ is nonnegative and*

$$\omega(\mathbf{0}) = \inf\{f(\mathbf{x}) + \langle \boldsymbol{\lambda}, \mathbf{g}(\mathbf{x})\rangle + \langle \boldsymbol{\mu}, \mathbf{h}(\mathbf{x})) : \mathbf{x} \in \mathbb{R}^n\}.$$

The next theorem summarizes the relationships among the set $\partial\omega(\mathbf{0})$, the multiplier vectors, and the Kuhn–Tucker vectors.

THEOREM 3.2. *Let the programming problem CPII be strongly consistent and let it have a solution. Then the set $\mathfrak{M}$ of multiplier vectors, the set $\mathscr{K}$ of Kuhn–Tucker vectors, and the set $-\partial\omega(\mathbf{0})$ are all nonempty and equal.*

Proof. That $\mathfrak{M}$ and $\mathscr{K}$ are nonempty and equal was noted in Corollary 1 of Theorem 2.3 and Remark 2.6. Theorem 3.1 states that if CPII is strongly consistent and $\omega(\mathbf{0})$ is finite, then $\partial\omega(\mathbf{0})$ is nonempty and every vector in $-\partial\omega(\mathbf{0})$ is a Kuhn–Tucker vector.

To complete the proof of the theorem, we must show that each Kuhn–Tucker vector is an element of $-\partial\omega(\mathbf{0})$. Let $\boldsymbol{\eta}_* = (\boldsymbol{\lambda}_*, \boldsymbol{\mu}_*)$ be a Kuhn–Tucker vector. Since CPII is consistent, $\mathbf{0} \in \operatorname{dom}(\omega)$ and hence $\operatorname{dom}(\omega) \neq \varnothing$. For any $\mathbf{z}$ in $\operatorname{dom}(\omega)$

$$\begin{aligned}\omega(\mathbf{0}) &= \inf\{f(\mathbf{x}) + \langle \boldsymbol{\lambda}_*, \mathbf{g}(\mathbf{x})\rangle + \langle \boldsymbol{\mu}_*, \mathbf{h}(\mathbf{x})\rangle : \mathbf{x} \text{ in } \mathbb{R}^n\} \\ &\leqslant \inf\{f(\mathbf{x}) + \langle \boldsymbol{\lambda}_*, \mathbf{g}(\mathbf{x})\rangle + \langle \boldsymbol{\mu}_*, \mathbf{h}(\mathbf{x})\rangle : \mathbf{x} \in X(\mathbf{z})\}.\end{aligned}$$

Since $\mathbf{g}(\mathbf{x}) \leqslant \mathbf{z}_1$ and $\mathbf{h}(\mathbf{x}) = \mathbf{z}_2$ for $\mathbf{x}$ in $X(\mathbf{z})$ and since $\boldsymbol{\lambda}_* \geqslant 0$, we get that

$$\begin{aligned}\omega(\mathbf{0}) &\leqslant \inf\{f(\mathbf{x}) + \langle \boldsymbol{\lambda}_*, \mathbf{z}_1\rangle + \langle \boldsymbol{\mu}_*, \mathbf{z}_2\rangle : \mathbf{x} \in X(\mathbf{z})\} \\ &= \omega(\mathbf{z}) + \langle \boldsymbol{\lambda}_*, \mathbf{z}_1\rangle + \langle \boldsymbol{\mu}_*, \mathbf{z}_2\rangle.\end{aligned}$$

Hence

$$\omega(\mathbf{z}) \geqslant \omega(\mathbf{0}) - \langle \boldsymbol{\lambda}_*, \mathbf{z}_1\rangle - \langle \boldsymbol{\mu}_*, \mathbf{z}_2\rangle$$

for all $\mathbf{z}$ in $\operatorname{dom}(\omega)$. Thus $\boldsymbol{\eta}_* \in -\partial\omega(\mathbf{0})$.

The assumption of strong consistency in Theorem 3.2, as already noted, guarantees that the sets $\mathfrak{M}$ and $-\partial\omega(\mathbf{0})$ are nonempty. In any specific problem, the validity of the strong consistency assumption is easily verified a priori from the data of the problem. In the next theorem we show that if we drop the assumption of strong consistency but assume that one of the sets $\mathscr{K}$, $\mathfrak{M}$, or $-\partial\omega(\mathbf{0})$ is nonempty, then all of the sets are nonempty and are equal. This assumption, however, is not verifiable a priori.

THEOREM 3.3. *Let the programming problem CPII have a solution. Let one of the sets* $-\partial\omega(\mathbf{0})$, *the set* $\mathscr{K}$ *of Kuhn–Tucker multipliers, or the set* $\mathfrak{M}$ *of multipliers as in Theorem 2.2 be nonempty. Then all of the sets* $\mathfrak{M}$, $\mathscr{K}$, *and* $-\partial\omega(\mathbf{0})$ *are nonempty and*

$$\mathfrak{M} = \mathscr{K} = -\partial\omega(\mathbf{0}).$$

Proof. By Corollary 1 to Theorem 2.3, the sets $\mathfrak{M}$ and $\mathscr{K}$ are equal. By Lemma 3.3, if $\partial\omega(\mathbf{0})$ is not empty, then $-\partial\omega(\mathbf{0}) \subseteq \mathscr{K}$. Of course, if $\partial\omega(\mathbf{0}) = \varnothing$, this is also true. In proving Theorem 3.2, we showed that $\mathscr{K} \subseteq -\partial\omega(\mathbf{0})$. Thus, $\mathscr{K} = -\partial\omega(\mathbf{0})$, and the theorem is proved.

Remark 3.4. The relationship between perturbations and multiplier vectors has an interesting economic interpretation. We can consider the function f as giving the cost of an activity described by a vector $\mathbf{x}$, where $\mathbf{x}$ is subject to constraints $\mathbf{g}(\mathbf{x}) \leqslant \mathbf{0}$ and $\mathbf{h}(\mathbf{x}) = \mathbf{0}$. Our objective is to minimize f subject to the constraints. We perturb the constraints by a vector $\mathbf{z} = (\mathbf{z}_1, \mathbf{z}_2)$ in the hope of reducing the optimal cost $\omega(\mathbf{z})$. If we must pay a price $\boldsymbol{\pi} = (\boldsymbol{\pi}_1, \boldsymbol{\pi}_2)$ per unit change in the constraints, then the total cost of the perturbed process with constraints $\mathbf{g}(\mathbf{x}) \leqslant \mathbf{z}_1$ and $\mathbf{h}(\mathbf{x}) = \mathbf{z}_2$ is

$$\omega(\mathbf{z}) + \langle \boldsymbol{\pi}, \mathbf{z} \rangle.$$

Thus it will be advantageous to purchase a change in the constraints if and only if

$$\omega(\mathbf{z}) + \langle \boldsymbol{\pi}, \mathbf{z} \rangle < \omega(\mathbf{0}).$$

But

$$\omega(\mathbf{z}) \geqslant \omega(\mathbf{0}) - \langle \boldsymbol{\lambda}, \mathbf{z}_1 \rangle - \langle \boldsymbol{\mu}, \mathbf{z}_2 \rangle, \qquad (\boldsymbol{\lambda}, \boldsymbol{\mu}) \in \mathfrak{M}, \qquad (\mathbf{z}_1, \mathbf{z}_2) \in \mathbb{R}^{m+k},$$

and so

$$\omega(\mathbf{z}) + \langle \boldsymbol{\lambda}, \mathbf{z}_1 \rangle + (\boldsymbol{\mu}, \mathbf{z}_2) \geqslant \omega(\mathbf{0}).$$

Also, any $(\boldsymbol{\lambda}, \boldsymbol{\mu})$ for which the last inequality holds is in $-\partial\omega(\mathbf{0})$, and so is in

$\mathfrak{M}$. Thus we can consider a multiplier vector $(\boldsymbol{\lambda}, \boldsymbol{\mu})$ to be a price such that the purchase of a perturbation at that price will not result in a lowering of the cost. Since at these prices no perturbation is worth buying, We may consider the cost to be at equilibrium at these prices. Therefore, such prices are called equilibrium prices.

We now illustrate the results of this section by several examples.

Example 3.1. Minimize

$$f(\mathbf{x}) = e^{-(x_1+x_2)}$$

subject to

$$e^{x_1} + e^{x_2} - 20 \leqslant z.$$

This problem is a perturbation of the problem in Exercise IV.5.5. Since f and the constraint function $g(\mathbf{x}) = e^{x_1} + e^{x_2} - 20$ are convex, the nonperturbed problem is a convex programming problem, as are the perturbed problems. The problem is strongly consistent since $g(\mathbf{0}) < 20$.

Let

$$L(\mathbf{x}, \lambda, z) = e^{-(x_1+x_2)} + \lambda(e^{x_1} + e^{x_2} - 20 - z).$$

It follows from Theorem 2.5 that a solution $\boldsymbol{\xi} = (\xi_1, \xi_2)$ is characterized by the existence of a positive number $\hat{\lambda}$ such that $\hat{\lambda}(g(\boldsymbol{\xi}) - z) = 0$ and $\partial L/\partial x_1 = \partial L/\partial x_2 = 0$, where the partial derivatives are evaluated at $(\boldsymbol{\xi}, \hat{\lambda})$. Thus

$$-e^{-(\xi_1+\xi_2)} + \hat{\lambda}e^{\xi_1} = 0, \qquad -e^{-(\xi_1+\xi_2)} + \hat{\lambda}e^{\xi_2} = 0. \tag{11}$$

From this it follows that $\hat{\lambda} \neq 0$ and that

$$\hat{\lambda}e^{\xi_1} = \hat{\lambda}e^{\xi_2}.$$

Hence $\xi_1 = \xi_2$. From $\hat{\lambda} \neq 0$ and $\hat{\lambda}(g(\boldsymbol{\xi}) - z) = 0$ it follows that

$$e^{\xi_1} + e^{\xi_2} = 20 + z.$$

Let ξ denote the common value of ξ_1 and ξ_2. Then

$$2e^{\xi} = 20 + z, \tag{12}$$

and so $\xi = \ln(10 + \frac{1}{2}z)$. It follows from (11) and (12) that

$$\hat{\lambda} = e^{-3\xi} = (10 + \tfrac{1}{2}z)^{-3}. \tag{13}$$

It also follows from (12) and from the relation $f(\xi) = e^{2\xi}$ that

$$\omega(z) = e^{-2\xi} = (10 + \tfrac{1}{2}z)^{-2}.$$

Hence

$$\omega'(z) = -(10 + \tfrac{1}{2}z)^{-3}.$$

From (13) we get that the value of the multiplier λ_* for the unperturbed problem is 10^{-3}. Also, $\omega'(0) = -10^{-3}$ and so $\omega'(0) = -\lambda_*$, in agreement with the theory.

We now show that the solution of the constrained problem can be obtained by solving the unconstrained problem of minimizing $F(\mathbf{x})$, where

$$F(\mathbf{x}) = e^{-(x_1+x_2)} + \lambda_*(e^{x_1} + e^{x_2} - 20).$$

Substituting the value of λ_* in the preceding expression gives

$$F(\mathbf{x}) = e^{-(x_1+x_2)} + 10^{-3}(e^{x_1} + e^{x_2} - 20).$$

Since F is convex, any critical point furnishes a minimum. Setting $\partial F/\partial x_1$ and $\partial F/\partial x_2$ equal to zero gives

$$-e^{-(x_1+x_2)} + 10^{-3}e^{x_1} = 0, \qquad -e^{-(x_1+x_2)} + 10^{-3}e^{x_2} = 0.$$

Hence $x_1 = x_2$. Setting $x = x_1 = x_2$ gives

$$e^x = 10^3 e^{-2x},$$

whence $x = \ln 10$. This is in agreement with the solution of the constrained problem given by (12) with $z = 0$.

Example 3.2. Find the point in the half space defined by $x_1 + x_2 \leqslant z$ that is closest to the origin. This problem has a solution. (Why?)

We may formulate this problem analytically as follows: Minimize $\|\mathbf{x}\|^2$ subject to $x_1 + x_2 \leqslant z$. Thus the problem is a problem of type CPII(z). If we take $z = 0$, we see that CPII is strongly consistent. On sketching the feasible sets for $z \geqslant 0$, it becomes clear that for $z \geqslant 0$ the solution is the origin and that $\omega(z) = 0$. For $z < 0$ the minimum is the point of intersection of the line $x_2 = x_1$ and the line $x_1 + x_2 = z$. Thus the minimum is achieved at $(\frac{1}{2}z, \frac{1}{2}z)$ and $\omega(z) = \frac{1}{2}z^2$ for $z < 0$. The function ω is differentiable at all points and $\omega'(0) = 0$. Therefore 0 is the unique Kuhn–Tucker vector, and the constrained problem CPII can be replaced by the problem of minimizing $\|\mathbf{x}\|^2$. The conditions of Theorem 2.5 are clearly satisfied if we take $\mathbf{x}_* = (0, 0)$ and $\lambda_* = 0$.

We may also formulate the problem analytically as follows: Minimize $\|\mathbf{x}\|$ subject to $x_1 + x_2 \leqslant z$. Again, the solution is the origin if $z \geqslant 0$ and the point $(\frac{1}{2}z, \frac{1}{2}z)$ if $z < 0$. The function ω is now given by $\omega(z) = 0$ for $z \geqslant 0$ and $\omega(z) = -z/\sqrt{2}$ for $z < 0$. Thus, ω is not differentiable at the origin. It is a straightforward calculation to show that $\partial\omega(0)$ is the interval $[-1/\sqrt{2}, 0]$. Hence by Theorem 3.2 the Kuhn–Tucker vectors and multiplier vectors are all points in the interval $[0, 1/\sqrt{2}]$. We now verify this assertion independently.

For λ to be a Kuhn–Tucker vector, we must have

$$\inf\{\|\mathbf{x}\| + \lambda(x_1 + x_2)\} = 0.$$

When $\mathbf{x} = \mathbf{0}$, the quantity in brackets is clearly zero. Therefore, for λ to be a Kuhn–Tucker vector, we require that $\lambda \geqslant 0$ and

$$\sqrt{x_1^2 + x_2^2} + \lambda(x_1 + x_2) \geqslant 0 \quad \text{for all } \mathbf{x} \in \mathbb{R}^n.$$

To see what requirement this places on λ, we introduce polar coordinates. The preceding inequality is then satisfied at the origin and whenever

$$\lambda(\cos\theta + \sin\theta) \geqslant -1, \qquad 0 \leqslant \theta \leqslant 2\pi. \tag{14}$$

The function $\rho(\theta) = \cos\theta + \sin\theta$ achieve its minimum on the interval $[0, 2\pi]$ at $\theta = 5\pi/4$ and the value of the minimum is $-\sqrt{2}$. Hence we must have $\lambda \leqslant 1/\sqrt{2}$ as well as $\lambda \geqslant 0$.

We can replace the constrained problem by the unconstrained problem.

$$\min\{\|\mathbf{x}\| + \lambda(x_1 + x_2)\},$$

where λ is any point in $[0, 1/\sqrt{2}]$. The preferred choice of λ is clearly $\lambda = 0$. We leave it to the reader to verify that any λ in $[0, 1/\sqrt{2}]$ will serve as a multiplier λ_* as in Theorem 2.2.

Example 3.2 furnishes another illustration of the advantage of minimizing the square of the distance rather than the distance itself.

Example 3.3. Minimize $f(\mathbf{x}) = x_1$ subject to

$$g_1(\mathbf{x}) = x_2 \leqslant z_1, \qquad g_2(\mathbf{x}) = x_1^2 - x_2 \leqslant z_2.$$

When $\mathbf{z} = \mathbf{0}$, the problem reduces to the problem considered in Example 2.1. We saw there that when $\mathbf{z} = \mathbf{0}$, there exists no Kuhn–Tucker vector. By drawing an appropriate figure, the reader can see that $\text{dom}(\omega) = \{\mathbf{z} : z_1 + z_2 \geqslant 0\}$ and that $\omega(\mathbf{z}) = -\sqrt{z_1 + z_2}$. We leave the proof of this assertion as an exercise. Thus neither $\mathbf{0}$ nor any point on the line $z_1 + z_2 = 0$ are interior to $\text{dom}(\omega)$. It is easy to verify that $\partial\omega(\mathbf{z}) = \varnothing$ at each $\mathbf{z}$ on the boundary of $\text{dom}(\omega)$.

The next example has discontinuous value function at the origin.

Example 3.4. Minimize e^{-x_2} subject to $\sqrt{x_1^2 + x_2^2} - x_1 \leqslant z$. For $z = 0$, the feasible set is $\{\mathbf{x} = (x_1, x_2): x_1 \geqslant 0, x_2 = 0\}$. The minimum is attained at each point of the feasible set since $e^{-x_2} = e^{-0} = 1$. Hence $\omega(0) = 1$. If $z < 0$, the feasible set $X(z)$ is empty. If $z > 0$, points $\mathbf{x}$ of the feasible set satisfy

$$x_2^2 \leqslant 2x_1 z + z^2.$$

Thus the feasible set is the set of points on and "inside" the parabola $x_2^2 = 2x_1 z + z^2$. This parabola is symmetric with respect to the x_1-axis, has vertex at $(-\frac{1}{2}z, 0)$, and opens to the right. Hence $\inf\{e^{-x_2}: \mathbf{x} \in X(z)\} = 0$, and so $\omega(z) = 0$ for $z > 0$. Thus, $\text{dom}(\omega) = [0, \infty)$, and ω is discontinuous at $z = 0$. Also, $\partial\omega(0) = \varnothing$. Note that if $z > 0$, then the infimum is never attained at any point of $X(z)$. Thus the problem CP(z) has no solution if $z > 0$.

Exercise 3.1. Prove the assertions in Example 3.3 that

(i) $\text{dom}(\omega) = \{\mathbf{z}: z_1 + z_2 \geqslant 0\}$,
(ii) $\omega(\mathbf{z}) = -\sqrt{z_1 + z_2}$, and
(iii) $\partial\omega(\mathbf{z}) = \varnothing$ at each $\mathbf{z}$ on the boundary of $\text{dom}(\omega)$.

Exercise 3.2. In each of the following problems find ω, $\text{dom}(\omega)$, and $\partial\omega(\mathbf{0})$ if it is not empty:

(i) Minimize e^{-x} subject to $-x \leqslant 0$.
(ii) Minimize x_1 subject to $x_1^2 + x_2^2 < 1$.
(iii) Minimize $x_1^2 + x_2^2 + 2x_1 + 2x_2$ subject to $x_1 + x_2 \geqslant 2$.
(iv) Minimize x_2 subject to $e^{-x_1} - x_2 \leqslant 0$.

Exercise 3.3. Show directly that there are no Kuhn–Tucker vectors for the problem with $z = 0$ in Example 3.4.

4. LAGRANGIAN DUALITY

If CPII has a solution and is strongly consistent, then Theorem 2.4 suggests a method for computing the value of the minimum. Let

$$Y_0 = \{\boldsymbol{\eta}: \boldsymbol{\eta} = (\boldsymbol{\lambda}, \boldsymbol{\mu}): \boldsymbol{\lambda} \in \mathbb{R}^m,\ \boldsymbol{\lambda} \geqslant \mathbf{0},\ \boldsymbol{\mu} \in \mathbb{R}^k\}, \tag{1}$$

and for each $\eta \in Y_0$ let

$$\theta(\boldsymbol{\eta}) = \inf\{f(\mathbf{x}) + \langle \boldsymbol{\lambda}, \mathbf{g}(\mathbf{x})\rangle + \langle \boldsymbol{\mu}, \mathbf{h}(\mathbf{x})\rangle : \mathbf{x} \in \mathbb{R}^n\}. \tag{2}$$

Let

$$\delta = \sup\{\theta(\boldsymbol{\eta}) : \boldsymbol{\eta} \in Y_0\}. \tag{3}$$

Then by Theorem 2.4, $\delta = v$, where

$$v = \inf\{f(\mathbf{x}) : \mathbf{g}(\mathbf{x}) \leqslant \mathbf{0}, \mathbf{h}(\mathbf{x}) = \mathbf{0}\}. \tag{4}$$

A possible procedure for finding $\mathbf{x}_*$, a point at which the minimum is achieved, is to first find all solutions to the equation $f(\mathbf{x}_*) = v$ and then retain those solutions that satisfy the constraints.

This discussion suggests that it might be fruitful to consider the following problem, which we call the *problem dual* to CPII, or simply the *dual problem*, and denote by DCPII.

Problem DCPII

Let

$$Y = \{\boldsymbol{\eta} : \boldsymbol{\eta} \in Y_0, \theta(\boldsymbol{\eta}) > -\infty\}, \tag{5}$$

where Y_0 is defined in (1) and $\theta(\boldsymbol{\eta})$ in (2). Then

$$\text{Maximize } \{\theta(\boldsymbol{\eta}) : \boldsymbol{\eta} \in Y\}.$$

Note that in formulating the dual problem it is *not assumed* that CPII has a solution or is strongly consistent. The set Y is the feasible set for DCPII. If $Y \neq \varnothing$, then the dual problem is said to be consistent. The original problem CPII is often referred to as the *primal problem*.

The next theorem indicates how the dual problem gives information about the primal problem and vice-versa.

Theorem 4.1 (Lagrangian Duality Theorem for Convex Programming).

(i) *If $X \neq \varnothing$ and $Y \neq \varnothing$, then for each $\mathbf{x} \in X$ and $\boldsymbol{\eta} \in Y$*

$$\theta(\boldsymbol{\eta}) \leqslant f(\mathbf{x}). \tag{6}$$

Moreover, if δ and v are as in (3) and (4), then both are finite and $\delta \leqslant v$.

(ii) *Let $Y \neq \varnothing$. If $\theta(\boldsymbol{\eta})$ is unbounded above on Y, then $X = \varnothing$.*
(iii) *Let $X \neq \varnothing$. If $f(\mathbf{x})$ is unbounded below on X, then $Y = \varnothing$.*
(iv) *If there exists an $\boldsymbol{\eta}_* \in Y$ and an $\mathbf{x}_*$ in X such that $f(\mathbf{x}_*) = \theta(\boldsymbol{\eta}_*)$, then*

$$\delta = \theta(\boldsymbol{\eta}_*) = f(\mathbf{x}_*) = v.$$

Thus $\mathbf{x}_*$ is a solution of CPII and $\boldsymbol{\eta}_*$ is a solution of DCPII.

Proof. The key to the proof of the theorem is the observation that since $\boldsymbol{\lambda} \geqslant \mathbf{0}$ for $\boldsymbol{\eta} \in Y$ and since $\mathbf{g}(\mathbf{x}) \leqslant \mathbf{0}$ and $\mathbf{h}(\mathbf{x}) = \mathbf{0}$ for $\mathbf{x} \in X$, it follows that for $\mathbf{x} \in X$ and $\boldsymbol{\eta} \in Y$,

$$f(\mathbf{x}) + \langle \boldsymbol{\lambda}, \mathbf{g}(\mathbf{x}) \rangle + \langle \boldsymbol{\mu}, \mathbf{h}(\mathbf{x}) \rangle \leqslant f(\mathbf{x}). \tag{7}$$

In (2) we take the infimum over all $\mathbf{x}$ in $\mathbb{R}^n$. This infimum is less than the infimum in (2) taken over all $\mathbf{x} \in X$. From these observations and from (7) the inequalities (6) and $\delta \leqslant v$ follow. Assertion (ii) follows from (6), for if X were not empty, then $\theta(\boldsymbol{\eta})$ would be bounded above. Also, (iii) follows from (6) by a similar argument. Statement (iv) follows from the chain

$$\theta(\boldsymbol{\eta}_*) \leqslant \delta \leqslant v \leqslant f(\mathbf{x}_*) = \theta(\boldsymbol{\eta}_*).$$

We now see how the dual problem gives useful information about the primal problem and vice-versa. Statement (i) shows how a feasible point for the dual problem gives a lower bound for the value of the primal problem and how a feasible point for the primal problem gives an upper bound for the value of the dual problem. Statement (ii) gives a way of checking whether $X = \varnothing$ in the absence of more direct checks. Statement (iii) gives a condition under which the dual problem cannot be formulated.

Definition 4.1. Problems CPII and DCPII are said to exhibit a *duality gap* if $\delta < v$.

The next theorem gives a sufficient condition for the absence of a duality gap and relates the multipliers $\boldsymbol{\eta}_* = (\boldsymbol{\lambda}_*, \boldsymbol{\mu}_*)$ to the solutions of the dual problem.

THEOREM 4.2. *Let the primal problem CPII be strongly consistent and have a solution. Then:*

(i) *There is no duality gap.*
(ii) *If $\boldsymbol{\eta}_* = (\boldsymbol{\lambda}_*, \boldsymbol{\mu}_*)$ is a multiplier for CPII, then $\boldsymbol{\eta}_*$ is a solution of the dual problem.*
(iii) *If $\boldsymbol{\eta}_0 = (\boldsymbol{\lambda}_0, \boldsymbol{\mu}_0)$ is a solution of the dual problem, then $\boldsymbol{\eta}_0$ is a multiplier vector for the primal problem.*

Proof. Let $\mathbf{x}_*$ denote the point at which CPII achieves its minimum. Since the problem is strongly consistent by Theorem 2.2, there exists a multiplier vector $\boldsymbol{\eta}_* = (\boldsymbol{\lambda}_*, \boldsymbol{\mu}_*)$ for CPII. Thus,

$$f(\mathbf{x}_*) \leqslant f(\mathbf{x}) + \langle \boldsymbol{\lambda}_*, \mathbf{g}(\mathbf{x}) \rangle + \langle \boldsymbol{\mu}_*, \mathbf{h}(\mathbf{x}) \rangle \tag{8}$$

for all $\mathbf{x}$ in $\mathbb{R}^n$. Hence, from (2), $\theta(\boldsymbol{\eta}_*) \geqslant f(\mathbf{x}_*)$. By Theorem 4.1, $\theta(\boldsymbol{\eta}_*) \leqslant f(\mathbf{x}_*)$, so $\theta(\boldsymbol{\eta}_*) = f(\mathbf{x}_*)$. Conclusions (i) and (ii) now follow from (iv) of Theorem 4.1.

Now let $\boldsymbol{\eta}_0 = (\boldsymbol{\lambda}_0, \boldsymbol{\mu}_0)$ be a solution of the dual problem. From (2) and from the absence of a duality gap we get that

$$\inf\{f(\mathbf{x}) + \langle \boldsymbol{\lambda}_0, \mathbf{g}(\mathbf{x})\rangle + \langle \boldsymbol{\mu}_0, \mathbf{h}(\mathbf{x})\rangle : \mathbf{x} \in \mathbb{R}^n\} = \theta(\boldsymbol{\eta}_0) = f(\mathbf{x}_*).$$

Thus $(\boldsymbol{\lambda}_0, \boldsymbol{\mu}_0)$ is a Kuhn–Tucker vector, and so by Theorem 2.3, $(\boldsymbol{\lambda}_0, \boldsymbol{\mu}_0)$ is a multiplier vector for the primal problem.

We now present several examples to illustrate the theory developed in this section.

Example 4.1. We take the primal problem to be the problem in Exercise IV.5.5 or the problem in Example 3.1 with $z = 0$. The primal problem is

$$\text{Minimize } f(\mathbf{x}) = e^{-(x_1 + x_2)}$$

$$\text{subject to } e^{x_1} + e^{x_2} \leqslant 20.$$

Let

$$L(\mathbf{x}, \lambda) = e^{-(x_1 + x_2)} + \lambda(e^{x_1} + e^{x_2} - 20).$$

Then for each $\lambda \geqslant 0$ define

$$\theta(\lambda) = \inf\{L(\mathbf{x}, \lambda) : \mathbf{x} \in \mathbb{R}^2\}.$$

The dual problem is to maximize $\theta(\lambda)$.

We saw in Example 3.1 that the primal problem is strongly consistent and that it has a solution at $\mathbf{x}_* = (\xi, \xi)$, where $\xi = \ln 10$. The value of the minimum is 10^{-2} and the value of the multiplier λ_* is 10^{-3}. Thus we would expect the value of the maximum of θ to be 10^{-2} and for θ to assume this value at $\lambda_0 = 10^{-3}$. We now show that this is indeed so.

We first determine the function θ. For each $\lambda \geqslant 0$, the function $L(\ , \lambda)$ is readily seen to be a strictly convex function of $\mathbf{x}$. Hence a necessary and sufficient condition that $L(\ , \lambda)$ attain a unique minimum at $\boldsymbol{\xi} = (\xi_1, \xi_2)$ is that $\partial L/\partial x_1$ and $\partial L/\partial x_2$ both equal zero at $\boldsymbol{\xi}$. As in Example 3.1 we see that for $\lambda \neq 0$ this implies that $\xi_1 = \xi_2$. Denoting this common value by ξ, we get that

$$\theta(\lambda) = \inf\{\phi(\xi, \lambda) : \xi \in \mathbb{R}\}, \tag{9}$$

where

$$\phi(\xi, \lambda) = e^{-2\xi} + 2\lambda(e^{\xi} - 10).$$

For fixed positive λ, $\phi(\ , \lambda)$ is a strictly convex function of ξ. A necessary and sufficient condition that $\phi(\ , \lambda)$ attain a unique minimum at a point ξ_0 is that

$d\phi/d\xi$ vanish at ξ_0. Now

$$\frac{d\phi}{d\xi} = -2e^{-2\xi} + 2\lambda e^{\xi}.$$

Setting $d\phi/d\xi = 0$, we get that $\exp(-\xi_0) = \lambda^{1/3}$. Substituting this into (9) gives

$$\theta(\lambda) = 3\lambda^{2/3} - 20\lambda, \qquad \lambda > 0.$$

Since $\theta(\lambda) = 0$ when $\lambda = 0$, the preceding formula is valid for all $\lambda \geqslant 0$. An elementary calculus argument shows that θ is maximized at the unique point at which $\theta'(\lambda) = 0$. This value of λ is readily seen to be 10^{-3} and the corresponding value of θ is 10^{-2}. These values are as required.

Example 4.2. Let the primal problem be as follows: Minimize $\|\mathbf{x}\|$ subject to $x_1 + x_2 \leqslant 0$. This problem was analyzed in Example 3.2 with $z = 0$. This problem is strongly consistent and has $\mathbf{x} = \mathbf{0}$ as its solution. The value of the minimum is clearly zero. Any λ in the interval $[0, 1/\sqrt{2}]$ is a multiplier vector.

The dual problem is as follows: Maximize $\{\theta(\lambda) : \lambda \geqslant 0, \theta(\lambda) > -\infty\}$, where $\theta(\lambda) = \inf\{\|\mathbf{x}\| + \lambda(x_1 + x_2) : \mathbf{x} \in \mathbb{R}^2\}$. We introduce polar coordinates (r, ϕ) and obtain

$$\theta(\lambda) = \inf\{r(1 + \lambda(\cos\phi + \sin\phi)) : r \geqslant 0, 0 \leqslant \phi \leqslant 2\pi\}.$$

Let

$$\gamma(\phi, \lambda) = 1 + \lambda(\cos\phi + \sin\phi).$$

Then $\theta(\lambda) = 0$ if $\gamma(\phi, \lambda) \geqslant 0$ for all $0 \leqslant \phi \leqslant 2\pi$, and $\theta(\lambda) = -\infty$ if this is not so and $r \neq 0$. The argument used in Example 3.2 shows that $\gamma(\phi, \lambda) \geqslant 0$ for all ϕ in $[0, 2\pi]$ if and only if $0 \leqslant \lambda \leqslant 1/\sqrt{2}$. Thus $\theta(\lambda) = 0$ for λ in $[0, 1/\sqrt{2}]$ and $\theta(\lambda) = -\infty$ otherwise. Hence $\max\{\theta(\lambda) : \lambda \in Y\} = 0$ and the maximum is attained at each $\lambda \in [0, 1/\sqrt{2}]$. Comparison with the solution to the primal problem shows that there is no duality gap and the points at which the dual problem attains its maximum are the multipliers for the primal problem.

Example 4.3. Let the primal problem be the problem treated in Example 2.1 and in Example 3.3 with $z = 0$. The primal problem has the origin as the only feasible point. This point furnishes the solution and the value of the minimum is zero. This problem is clearly not strongly consistent.

The dual problem is as follows: Maximize $\{\theta(\boldsymbol{\lambda}) : \boldsymbol{\lambda} \in Y\}$, where

$$Y = \{\boldsymbol{\lambda} : \boldsymbol{\lambda} \geqslant \mathbf{0}, \theta(\boldsymbol{\lambda}) > -\infty\} \quad \text{and} \quad \theta(\boldsymbol{\lambda}) = \inf(x_1 + \lambda_1 x_2 + \lambda_2(x_1^2 - x_2) : \mathbf{x} \in \mathbb{R}\}.$$

If we write

$$x_1 + \lambda_1 x_2 + \lambda_2(x_1^2 - x_2) = x_1 + \lambda_2 x_1^2 + (\lambda_1 - \lambda_2)x_2,$$

we see that if $\lambda_1 \neq \lambda_2$, then $\theta(\boldsymbol{\lambda}) = -\infty$. For $\boldsymbol{\lambda} \neq \mathbf{0}$ with $\lambda_1 = \lambda_2$, we have $\lambda_2 \neq 0$ and

$$\theta(\boldsymbol{\lambda}) = \inf\{x_1 + \lambda_2 x_1^2 : x_1 \in \mathbb{R}\}.$$

The function inside the curly brackets attains its minimum at $x_1 = -1/2\lambda_2$ and

$$\theta(\boldsymbol{\lambda}) = -1/4\lambda_2. \tag{10}$$

For $\boldsymbol{\lambda} = \mathbf{0}$ we have

$$\theta(\mathbf{0}) = -\infty.$$

Thus $Y = \{\boldsymbol{\lambda} = (\lambda_1, \lambda_2) : \lambda_1 = \lambda_2, \boldsymbol{\lambda} > 0\}$. Since $\lambda_2 > 0$, it follows from (10) that $\sup\{\theta(\boldsymbol{\lambda}) : \boldsymbol{\lambda} \in Y\} = 0$. The supremum is not attained in the dual problem, even though there is no duality gap. Thus we have an example in which there is no duality gap; the primal problem has a solution but the dual problem has no solution. We also recall from Example 3.3 that $\partial\omega(\mathbf{0}) = \phi$ and there are no Kuhn–Tucker vectors.

Example 4.4. In this example we will exhibit a duality gap. Let the primal problem be the problem in Example 3.4 with $z = 0$. We saw there that the minimum is attained at every point of the feasible set and that the value of the minimum is 1.

The dual problem is to maximize $\{\theta(\lambda) : \lambda \geqslant 0, \theta(\lambda) > -\infty\}$, where

$$\theta(\lambda) = \inf\{e^{-x_2} + \lambda(\sqrt{x_1^2 + x_2^2} - x_1) : \mathbf{x} \in \mathbb{R}^2\}.$$

Let

$$L(\mathbf{x}, \lambda) = e^{-x_2} + \lambda(\sqrt{x_1^2 + x_2^2} - x_1).$$

Then $L(\mathbf{x}, \lambda) \geqslant 0$ for all $\mathbf{x}$ in $\mathbb{R}^2$. We shall show that $\theta(\lambda) = 0$.

Let $x_1 > 0$ and let $x_2 > 0$. Then

$$\begin{aligned} L(\mathbf{x}, \lambda) &= e^{-x_2} + \frac{\lambda x_2^2}{\sqrt{x_1^2 + x_2^2} + x_1} \\ &= e^{-x_2} + \lambda \frac{x_2^2}{x_1\left[\sqrt{1 + \left(\dfrac{x_2}{x_1}\right)^2} + 1\right]}. \end{aligned}$$

Now let $x_1 = (x_2)^3$ and then let $x_2 \to \infty$. We see that $L(\mathbf{x}, \lambda) \to 0$ along this curve. Thus $\theta(\lambda) = 0$ for all $\lambda \geqslant 0$ and $\max\{\theta(\lambda) : \lambda \geqslant 0\} = 0$. Thus $\delta = 0$, and since $v = 1$, we have a duality gap.

Theorem 4.2 states that if a solution to the primal problem exists, then strong consistency is a sufficient condition for the absence of a duality gap and the existence of a solution to the dual problem. The following example shows that strong consistency is not a necessary condition for the absence of a duality gap and for the existence of a solution to the dual problem.

Example 4.5. Minimize $f(\mathbf{x}) = x_1$ subject to

$$g_1(\mathbf{x}) = (x_1 + 1)^2 + (x_2)^2 - 1 \leqslant 0, \qquad g_2(\mathbf{x}) = -x_1 \leqslant 0.$$

The problem is clearly a convex programming problem. The feasible set consists of the origin $\mathbf{0}$ in $\mathbb{R}^2$, and the problem is not strongly consistent. The solution to the primal problem is $\mathbf{x}_* = \mathbf{0}$ and $v = 0$.

The dual objective function is given by

$$\theta(\boldsymbol{\lambda}) = \inf\{x_1 + \lambda_1[(x_1 + 1)^2 + (x_2)^2 - 1] - \lambda_2 x_1 : \mathbf{x} \in \mathbb{R}^2\},$$

where $\boldsymbol{\lambda} \geqslant 0$. If we take $\boldsymbol{\lambda} = \boldsymbol{\lambda}_* = (0, 1)$ we get that $\theta(\boldsymbol{\lambda}_*) = 0$. If then follows from (iv) of Theorem 4.1 that there is no duality gap and that $\boldsymbol{\lambda}_*$ is a solution of the dual problem.

We will now obtain two necessary and sufficient conditions for the absence of a duality gap and the existence of a solution to the dual problem.

THEOREM 4.3. *Let problem CPII have a solution* $\mathbf{x}_*$. *Then a necessary and sufficient condition for the absence of a duality gap and for* $\boldsymbol{\eta}_* = (\boldsymbol{\lambda}_*, \boldsymbol{\mu}_*)$, $\lambda_* \geqslant \mathbf{0}$, *to be a solution of the dual problem is that* $(\mathbf{x}_*, \boldsymbol{\lambda}_*, \boldsymbol{\mu}_*)$ *is a saddle point for L.*

Proof. Let $(\mathbf{x}_*, \boldsymbol{\lambda}_*, \boldsymbol{\mu}_*)$ be a saddle point for L. Then (2.18i) and (2.19) of Theorem 2.7 hold. Thus (8) of Theorem 4.2 holds. That there is no duality gap and that $(\boldsymbol{\lambda}_*, \boldsymbol{\mu}_*)$ is a solution to the dual problem follows by the same arguments used to establish (i) and (ii) in Theorem 4.2.

Now let $\boldsymbol{\eta}_* = (\boldsymbol{\lambda}_*, \boldsymbol{\mu}_*)$ be a solution of the dual problem and let there be no duality gap. Then, as in the proof of statement (iii) in Theorem 4.2 with $\boldsymbol{\lambda}_0 = \boldsymbol{\lambda}_*$ and $\boldsymbol{\mu}_0 = \boldsymbol{\mu}_*$, we get that $(\boldsymbol{\lambda}_*, \boldsymbol{\mu}_*)$ is a multiplier as in Theorem 2.2 for the primal problem. But (7) of Theorem 2.2 states that $(\mathbf{x}_*, \boldsymbol{\lambda}_*, \boldsymbol{\mu}_*)$ is a saddle point for L.

THEOREM 4.4. *Let CPII have a solution* $\mathbf{x}_*$. *Then a necessary and sufficient condition for the absence of a duality gap and the existence of a solution to the dual problem is that* $\partial\omega(\mathbf{0}) \neq \varnothing$. *In this event every vector* $\boldsymbol{\eta}$ *in* $-\partial\omega(\mathbf{0})$ *is a solution of the dual problem.*

Proof. We first prove the sufficiency of the condition $\partial\omega(\mathbf{0}) \neq \varnothing$. Since the primal problem has a solution, $\omega(\mathbf{0})$ is finite. By hypothesis, $\partial\omega(\mathbf{0}) \neq \varnothing$. Hence by Lemma 3.3 each vector $\boldsymbol{\eta} = (\boldsymbol{\lambda}, \boldsymbol{\mu})$ such that $-\boldsymbol{\eta} \in \partial\omega(\mathbf{0})$ is a Kuhn–Tucker vector, and such vectors exist. By Theorem 3.3 each such vector is in $\mathfrak{M}$, the set of multiplier vectors, as in Theorem 2.2. Thus $(\mathbf{x}_*, \boldsymbol{\lambda}, \boldsymbol{\mu})$ is a saddle point for L, and so by Theorem 4.3 there is no duality gap and $(\boldsymbol{\lambda}, \boldsymbol{\mu})$ is a solution to the dual problem. Now suppose that there is no duality gap and that the dual problem has a solution $\boldsymbol{\eta}_* = (\boldsymbol{\lambda}_*, \boldsymbol{\mu}_*)$. Then by Theorem 4.3 the point $(\mathbf{x}_*, \boldsymbol{\lambda}_*, \boldsymbol{\mu}_*)$ is a saddle point for L. By Theorem 2.7, $(\boldsymbol{\lambda}_*, \boldsymbol{\mu}_*)$ is a Kuhn–Tucker vector. It then follows from Theorem 3.3 that $(\boldsymbol{\lambda}_*, \boldsymbol{\mu}_*) \in -\partial\omega(\mathbf{0})$.

COROLLARY 1. *Let* CPII *have a solution* $\mathbf{x}_*$. *Then a necessary and sufficient condition for the absence of a duality gap and the existence of a solution to the dual problem is that one of the sets* $\partial\omega(\mathbf{0})$, $\mathscr{K}$, or $\mathfrak{M}$ *be nonempty.*

The corollary is an immediate consequence of Theorems 4.4 and 3.3.

Remark 4.1. In Remark 3.4 we indicated an economic interpretation of the multiplier vectors as equilibrium prices. In view of the preceding corollary, we see that the solution of the dual problem gives the equilibrium price.

Exercise 4.1. Formulate, and solve whenever possible, the problems dual to the problems in Exercise 3.2. Discuss your results as they relate to Theorem 4.2.

Exercise 4.2. The problem posed in Exercise IV.5.8 is a convex programming problem. Solve this problem by formulating the dual and solving the dual. *Hint:* To simplify the algebra involved, first solve the problem for $\mathbf{x}_0 = \mathbf{0}$. Then use a translation of coordinates to obtain the result for general $\mathbf{x}_0$.

Exercise 4.3. Show that the function θ defined by (2.2) is concave.

5. GEOMETRIC INTERPRETATION

In this section we present a geometric interpretation of the duality and perturbation results. We shall assume that all the constraints are inequality constraints and that we are considering CPI: Minimize $f(\mathbf{x})$ subject to $\mathbf{g}(\mathbf{x}) \leqslant \mathbf{0}$.

The mapping

$$\zeta = f(\mathbf{x}), \qquad \mathbf{z} = \mathbf{g}(\mathbf{x}),$$

is a mapping from $\mathbb{R}^n$ to $\mathbb{R}^{m+1}$. We have indicated two image sets,

$$I = \{(\mathbf{z}, \zeta), \qquad \zeta = f(\mathbf{x}), \qquad \mathbf{z} = \mathbf{g}(\mathbf{x}) : \mathbf{x} \in \mathbb{R}^n\}, \tag{1}$$

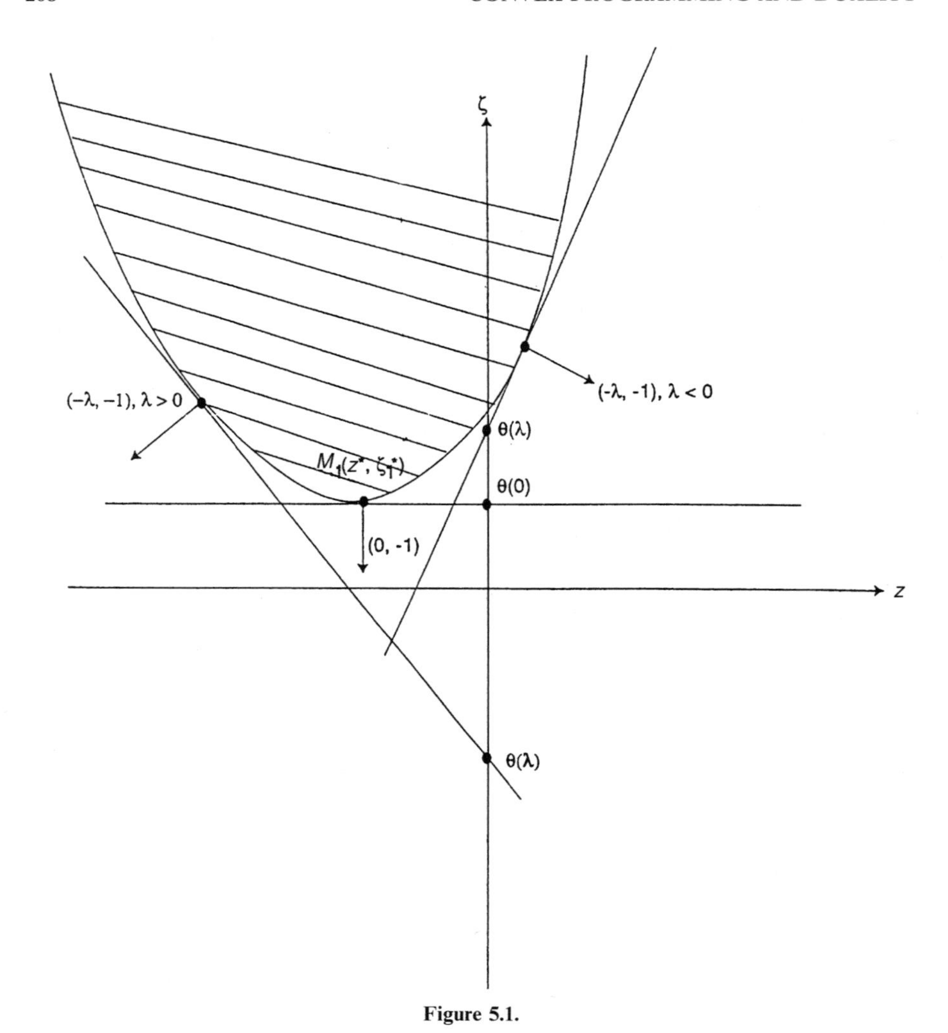

Figure 5.1.

schematically for $m = 1$ as the hatched areas in Figures 5.1 and 5.2. In geometric terms, to solve CPI, we must find a point in the intersection of I and the closed half space $\mathbf{z} \leqslant \mathbf{0}$ with smallest ζ coordinate if such a point exists. In Figure 5.1 this is the point M_1 with coordinates (z^*, ζ_1^*). In Figure 5.2 this is the point M_2 with coordinates $(0, \zeta_2^*)$.

Let $\boldsymbol{\lambda} \geqslant \mathbf{0}$. The dual problem is to maximize $\theta(\boldsymbol{\lambda})$, where θ is defined in relation (2) of Section 4. We may therefore also write

$$\begin{aligned}\theta(\boldsymbol{\lambda}) &= \inf\{\zeta + \langle \boldsymbol{\lambda}, \mathbf{z}\rangle : \zeta = f(\mathbf{x}),\ \mathbf{z} = \mathbf{g}(\mathbf{x})\} \\ &= \inf\{\alpha : \alpha = \zeta + \langle \boldsymbol{\lambda}, \mathbf{z}\rangle : (\mathbf{z}, \zeta) \in I\}.\end{aligned}$$

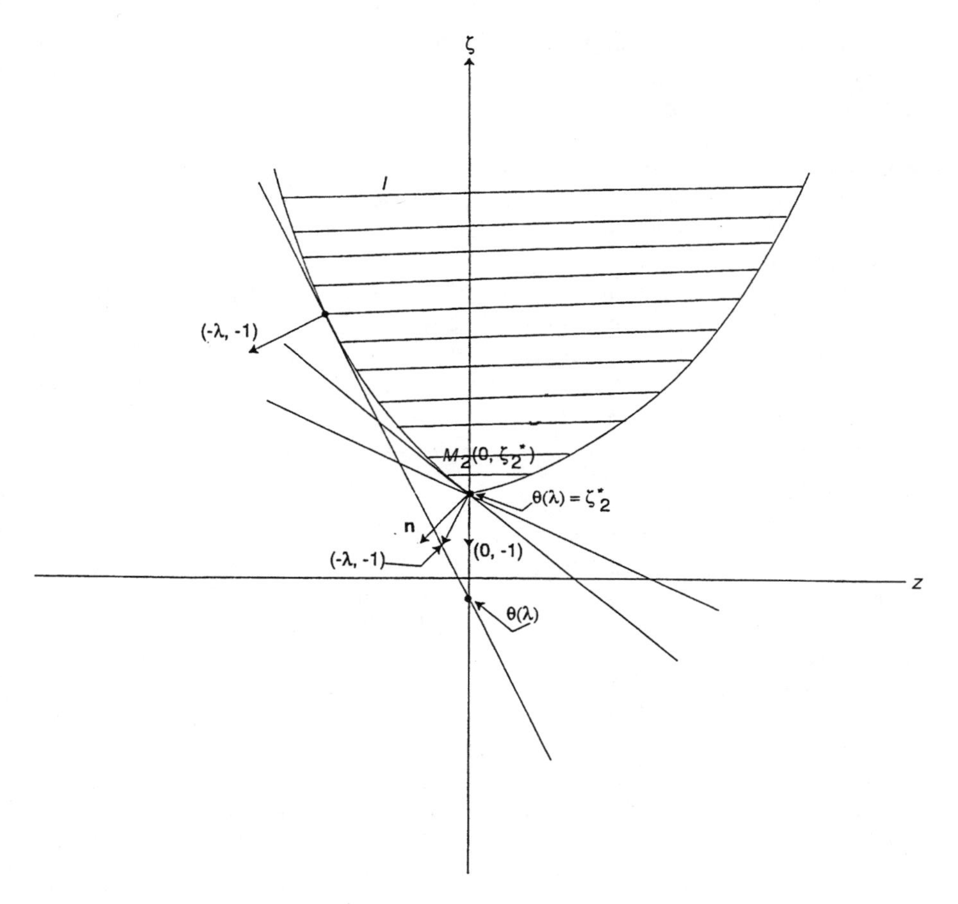

Figure 5.2.

Thus, for all $(\zeta, \mathbf{z})$ in I

$$\langle(-\boldsymbol{\lambda}, -1), (\mathbf{z}, \zeta)\rangle \leqslant -\theta(\boldsymbol{\lambda}),$$

and so the hyperplane

$$\langle(-\boldsymbol{\lambda}, -1), (\mathbf{z}, \zeta)\rangle = -\theta(\boldsymbol{\lambda}) \tag{2}$$

is a supporting hyperplane for the set I. The equation of this hyperplane can also be written as

$$\zeta = \langle -\boldsymbol{\lambda}, \mathbf{z}\rangle + \theta(\boldsymbol{\lambda}). \tag{3}$$

Hence $\theta(\boldsymbol{\lambda})$ is the ζ-intercept of the hyperplane (2). See Figures 5.1 and 5.2. The geometric interpretation of the dual problem is to find, among supporting hyperplanes (3) to I with $\boldsymbol{\lambda} \geqslant \mathbf{0}$, those whose ζ-intercept in maximized provided such hyperplanes exists.

In Figure 5.1 the supporting hyperplane to I at any point P with z-coordinate greater than z^* has $\lambda < 0$. Thus these hyperplanes do not enter into the competition. The unique hyperplane through M_1 with normal $(0, -1)$ is the required hyperplane. The ζ-intercept of this horizontal hyperplane is ζ_1^*. Thus there is no duality gap and the dual problem has a solution. The point M_1 corresponds to a point $\mathbf{x}_*$ such that $\mathbf{g}(\mathbf{x}_*) < 0$. Therefore, from the complementary slackness condition $\langle \boldsymbol{\lambda}_*, \mathbf{g}(\mathbf{x}_*) \rangle = 0$, we have that $\boldsymbol{\lambda} = 0$. This is in agreement with the geometry.

In Figure 5.2 any hyperplane through M_2 whose normal lies in the angle formed by $\mathbf{n}$ and the vector $(0, -1)$ is a supporting hyperplane with $\lambda \geqslant 0$ having maximal ζ-intercept equal to ζ_2^*. Thus, again there is no duality gap and the dual problem has a solution.

From Figure 5.1 and the definition of the function ω we see that the graph of ω is the lower boundary curve of I for $z \leqslant z_1^*$ and then the straight line $\zeta = \zeta_1^*$ for $z \geqslant z_1^*$. At M_1 the function ω is differentiable and $\omega'(0) = 0$. The dual problem has the unique solution $\lambda_* = 0$, and so $\omega'(0) = \lambda_*$.

In Figure 5.2 the graph of ω is the lower boundary curve of I for $z \leqslant 0$ and then the straight line $\zeta = \zeta_2^*$ for $z \geqslant 0$. At M_2 the function ω is not differentiable. The set of vectors $\boldsymbol{\lambda}_* \geqslant \mathbf{0}$ such that $(-\lambda_*, -1)$ lies in the angle formed by $\mathbf{n}$ and $(0, -1)$ is the set $\partial\omega(0)$. The dual problem attains its solution at any vector in $-\partial\omega(0)$.

Exercise 5.1. In Exercises 3.1(i), (ii), and (iv) sketch the sets I and carry out the geometric arguments of this section in these specific problems, whenever possible. For problems in which the arguments made in connection with Figures 5.1 and 5.2 seem to fail, explain why.

6. QUADRATIC PROGRAMMING

Quadratic Programming Problem (QP)

Let Q be a positive semidefinite $n \times n$ matrix, let $\mathbf{b}$ be a vector in $\mathbb{R}^n$, let A be an $m \times n$ matrix, and let $\mathbf{c}$ be a vector in $\mathbb{R}^m$. Minimize

$$f(\mathbf{x}) = \tfrac{1}{2}\langle \mathbf{x}, Q\mathbf{x} \rangle - \langle \mathbf{b}, \mathbf{x} \rangle$$

subject to

$$A\mathbf{x} \leqslant \mathbf{c}.$$

The formulation just given includes equality constraints of the form $H\mathbf{x} = \mathbf{d}$. This is achieved by the device already used of writing each equality constraint $\langle \mathbf{h}_i, \mathbf{x} \rangle = d_i$, where $\mathbf{h}_i$ is the ith row of H as two equality constraints.

Because Q is positive semidefinite, the function f is convex. The constraints are affine. Hence the problem is a convex programming problem. We leave it as an exercise to show that if the matrix Q is positive definite, the problem does have a solution, while the problem need not have a solution if Q is only positive semidefinite. If Q is positive definite, then f is strictly convex. Since in this case we are minimizing a strictly convex function over a convex set, the minimum is achieved at a unique point $\mathbf{x}_*$.

For the quadratic programming problem the Lagrangian L is defined by

$$L(\mathbf{x}, \boldsymbol{\lambda}) = \tfrac{1}{2}\langle \mathbf{x}, Q\mathbf{x}\rangle - \langle \mathbf{b}, \mathbf{x}\rangle + \langle \boldsymbol{\lambda}, A\mathbf{x} - \mathbf{c}\rangle. \tag{1}$$

THEOREM 6.1. *A necessary and sufficient condition for a point $\mathbf{x}_*$ to be a solution of QP is that*

$$A\mathbf{x}_* \leqslant \mathbf{c} \tag{2}$$

and there exist a vector $\boldsymbol{\lambda}_* \geqslant 0$ in $\mathbb{R}^n$ such that

$$Q\mathbf{x}_* - \mathbf{b} + A^t\boldsymbol{\lambda}_* = \mathbf{0}, \tag{3}$$

$$\langle \boldsymbol{\lambda}_*, A\mathbf{x}_* - \mathbf{c}\rangle = 0. \tag{4}$$

Proof. We first prove the necessity of the conditions. If $\mathbf{x}_*$ is a solution of QP, then $\mathbf{x}_*$ is feasible, so (2) holds. By Theorem IV.5.1 or by Remark 2.3, there exists a real number $\lambda_0 \geqslant 0$ and a vector $\boldsymbol{\lambda}_* \geqslant \mathbf{0}$ in $\mathbb{R}^n$ such that $(\lambda_0, \boldsymbol{\lambda}_*) \neq (0, \mathbf{0})$,

$$\lambda_0(Q\mathbf{x}_* - \mathbf{b}) + A^t\boldsymbol{\lambda}_* = 0, \tag{5}$$

and (4) holds. We shall show that $\lambda_0 = 0$ is not possible. Hence we may take $\lambda_0 = 1$ and get (3).

If $\lambda_0 = 0$, then $\boldsymbol{\lambda}_* \neq \mathbf{0}$, and from (5) we get that $A^t\boldsymbol{\lambda}_* = 0$. It then follows from (4) that $\langle \boldsymbol{\lambda}_*, \mathbf{c}\rangle = 0$. Thus the system

$$\begin{pmatrix} A^t \\ \mathbf{c}^t \end{pmatrix}\mathbf{y} = \mathbf{0}, \qquad \mathbf{y} \geqslant \mathbf{0}, \qquad \mathbf{y} \neq \mathbf{0}, \qquad \mathbf{y} \in \mathbb{R}^m,$$

has a solution $\mathbf{y} = \boldsymbol{\lambda}_*$. It then follows from Gordan's theorem (Theorem II.5.2) that

$$(A, -\mathbf{c})\begin{pmatrix} \mathbf{x} \\ \xi \end{pmatrix} \geqslant \mathbf{0}$$

for all $\mathbf{x} \in \mathbb{R}^n$, ξ in $\mathbb{R}$. Hence $A\mathbf{x} \geqslant \xi\mathbf{c}$ for all $\mathbf{x}$ in $\mathbb{R}^n$ and ξ in $\mathbb{R}$. In particular, this inequality holds for $\xi = 0$. Hence $A\mathbf{x} \geqslant \mathbf{0}$ and $A(-\mathbf{x}) \geqslant \mathbf{0}$ for all $\mathbf{x}$ in $\mathbb{R}^n$. Thus $A\mathbf{x} = \mathbf{0}$ for all $\mathbf{x}$ in $\mathbb{R}^n$, which is not possible since A is not the zero matrix.

The sufficiency of the conditions follows from Theorem IV.5.3 with $f(\mathbf{x}) = \frac{1}{2}\langle \mathbf{x}, Q\mathbf{x} \rangle$ and $\mathbf{g}(\mathbf{x}) = A\mathbf{x} - \mathbf{c}$.

Remark 6.1. In Theorem 6.1 we showed that in the quadratic programming problem the KKT condition is a necessary condition for a minimum *without assuming a priori that a constraint qualification* holds. The linearity of the constraints, however, played a crucial role.

Theorem 6.1 suggests the following procedure for solving QP. Solve the linear system (3) and then reject all solutions of (3) that do not satisfy (2) and (4). In actual practice, however, various computational algorithms related to the simplex method have been devised to solve quadratic programming problems. For a discussion of these see Bazzara, Shetty, and Sherali [1993].

An alternate, and possibly more fruitful, approach is through the *dual problem*, which is

$$\max\{\theta(\boldsymbol{\lambda}) : \boldsymbol{\lambda} \geqslant \mathbf{0},\ \theta(\boldsymbol{\lambda}) > -\infty\}, \tag{6}$$

where

$$\theta(\boldsymbol{\lambda}) = \inf\{L(\mathbf{x}, \boldsymbol{\lambda}) : \mathbf{x} \in \mathbb{R}^n\} \tag{7}$$

and L is defined in (1). A justification for this procedure is provided by the following theorem.

THEOREM 6.2. *Let Problem QP have a solution $\mathbf{x}_*$. Then there is no duality gap between QP and the dual problem, the dual problem has a solution, and the solution is given by the multipliers $\boldsymbol{\lambda}_*$ of Theorem* 6.1.

Proof. For fixed $\boldsymbol{\lambda} \geqslant \mathbf{0}$, the function $L(\cdot, \boldsymbol{\lambda})$ defined in (1) is convex on $\mathbb{R}^n$. Therefore, by the corollary to Theorem IV.3.2, a necessary and sufficient condition for a point $\mathbf{x}_0$ to minimize L is that $\nabla_{\mathbf{x}} L(\mathbf{x}_0, \boldsymbol{\lambda}) = \mathbf{0}$, where $\nabla_{\mathbf{x}}$ denotes the gradient with respect to $\mathbf{x}$. Since $\mathbf{x}_*$ is a solution to QP, it follows from (3) of Theorem 6.1 that $\nabla_{\mathbf{x}} L(\mathbf{x}_*, \boldsymbol{\lambda}_*) = \mathbf{0}$. Hence,

$$L(\mathbf{x}_*, \boldsymbol{\lambda}_*) = \min\{L(\mathbf{x}, \boldsymbol{\lambda}_*) : \mathbf{x} \in \mathbb{R}^n\}.$$

From this and from (2) and (4) we see that (2.18) of Theorem 2.7 holds, where $\mathbf{h}$ is the zero function. Thus, $(\mathbf{x}_*, \boldsymbol{\lambda}_*)$ is a saddle point for L. The conclusion of the theorem now follows from Theorem 4.3.

Remark 6.2. Note that we have not assumed strong consistency, but we have assumed that QP has a solution. We have already remarked that a sufficient condition for the existence of a solution is that Q be positive definite.

In determining the explicit form of the dual problem, it is useful to consider the unconstrained quadratic programming problem. Minimize

$$f(\mathbf{x}) = \tfrac{1}{2}\langle \mathbf{x}, Q\mathbf{x}\rangle - \langle \mathbf{b}, \mathbf{x}\rangle, \qquad \mathbf{x} \in \mathbb{R}^n. \tag{8}$$

Since Q is positive semidefinite, there exists an orthogonal matrix P such that PAP^t is a diagonal matrix. Under the transformation of coordinates $\mathbf{y} = P\mathbf{x}$, (8) becomes

$$F(\mathbf{y}) = f(P^t\mathbf{y}) = \tfrac{1}{2}\mathbf{y}^t PQP^t\mathbf{y} - \mathbf{b}^t P^t\mathbf{y} = \tfrac{1}{2}(\alpha_1 y_1^2 + \cdots + \alpha_k y_k^2) + \sum_{i=1}^{n} d_i y_i,$$

where the α_i are the positive eigenvalues of Q. If k is positive definite, then $k = n$; if k is positive semidefinite, then $k < n$. If a coefficient d_i with $i > k$ is different from zero, then by considering vectors $\rho\mathbf{e}_i$, where ρ is a scalar and $\mathbf{e}_i$ is the vector whose ith component is 1 and whose other components are zero, we see that $\inf\{F(\mathbf{y}) : \mathbf{y} \in \mathbb{R}^n\} = -\infty$. If $d_i = 0$ for all $i > k$, then $\nabla F(\mathbf{y}_0) = \mathbf{0}$ at any point $\mathbf{y}_0 = (y_{01}, \ldots, y_{0n})$ with

$$y_{0i} = \begin{cases} \dfrac{d_i}{\alpha_i}, & i = 1, \ldots, k, \\ \text{arbitrary}, & i = k+1, \ldots, n. \end{cases}$$

Since F is convex, it follows from the corollary to Theorem IV.3.2 that F attains a minimum at any such point $\mathbf{y}_0$. If Q is positive definite, then $k = n$ and the point $\mathbf{y}_0$ is unique. We summarize this discussion in the following lemma.

LEMMA 6.1. *The unconstrained programming problem either has infimum equal to* $-\infty$ *or attains a minimum at a finite point.*

We return to the determination of the explicit form of the dual problem. For fixed $\boldsymbol{\lambda} \geqslant \mathbf{0}$, the problem of minimizing $L(\mathbf{x}, \boldsymbol{\lambda}\rangle$ over all $\mathbf{x}$ in $\mathbb{R}^n$ is an unconstrained quadratic programming problem, since

$$\begin{aligned} L(\mathbf{x}, \boldsymbol{\lambda}) &= \tfrac{1}{2}\langle \mathbf{x}, Q\mathbf{x}\rangle - \langle \mathbf{b}, \mathbf{x}\rangle + \langle \boldsymbol{\lambda}, A\mathbf{x} - \mathbf{c}\rangle \\ &= \tfrac{1}{2}\langle \mathbf{x}, Q\mathbf{x}\rangle - \langle \mathbf{b} - A^t\boldsymbol{\lambda}, \mathbf{x}\rangle - \langle \boldsymbol{\lambda}, \mathbf{c}\rangle \\ &= \tfrac{1}{2}\langle \mathbf{x}, Q\mathbf{x}\rangle - \langle \mathbf{d}, \mathbf{x}\rangle - \langle \boldsymbol{\lambda}, \mathbf{c}\rangle, \end{aligned}$$

where $\mathbf{d} = \mathbf{b} - A^t\boldsymbol{\lambda}$. It then follows from Lemma 6.1 that if $\theta(\boldsymbol{\lambda})$ is finite, then we may replace the infimum in (7) by a minimum. Thus, for fixed $\boldsymbol{\lambda}$ such that

$\theta(\boldsymbol{\lambda}) > -\infty$, if we denote a point at which $\min L(\cdot, \boldsymbol{\lambda})$ is attained by $\mathbf{x}_\lambda$, then

$$\theta(\boldsymbol{\lambda}) = L(\mathbf{x}_\lambda, \boldsymbol{\lambda}) = \tfrac{1}{2}\langle \mathbf{x}_\lambda, Q\mathbf{x}_\lambda \rangle - \langle \mathbf{b}, \mathbf{x}_\lambda \rangle + \langle \boldsymbol{\lambda}, A\mathbf{x}_\lambda - \mathbf{c} \rangle. \tag{9}$$

Since for fixed $\boldsymbol{\lambda}$ the function $L(\cdot, \boldsymbol{\lambda})$ is convex on $\mathbb{R}^n$, it follows from the corollary to Theorem IV.3.2 that the minimum is attained at a point $\mathbf{x}_\lambda$ if and only if $\nabla_{\mathbf{x}} L(\mathbf{x}_\lambda, \boldsymbol{\lambda}) = \mathbf{0}$. A straightforward calculation shows that $\nabla_{\mathbf{x}} L(\mathbf{x}_\lambda, \boldsymbol{\lambda}) = \mathbf{0}$ if and only if

$$Q\mathbf{x}_\lambda - \mathbf{b} + A^t\boldsymbol{\lambda} = \mathbf{0}. \tag{10}$$

If we take the inner product of each side of (10) with $\mathbf{x}_\lambda$, we get

$$\langle \mathbf{x}_\lambda, Q\mathbf{x}_\lambda \rangle - \langle \mathbf{b}, \mathbf{x}_\lambda \rangle + \langle \boldsymbol{\lambda}, A\mathbf{x}_\lambda \rangle = 0.$$

Substitution of this equation into (9) gives

$$\theta(\boldsymbol{\lambda}) = -\tfrac{1}{2}\langle \mathbf{x}_\lambda, Q\mathbf{x}_\lambda \rangle - \langle \boldsymbol{\lambda}, \mathbf{c} \rangle.$$

Thus the explicit formulation of the dual problem is as follows.

Dual Quadratic Programming Problem (DQP)

Maximize

$$-\tfrac{1}{2}\langle \mathbf{x}, Q\mathbf{x} \rangle - \langle \boldsymbol{\lambda}, \mathbf{c} \rangle \tag{11}$$

subject to

$$Q\mathbf{x} + A^t\boldsymbol{\lambda} = \mathbf{b}, \qquad \boldsymbol{\lambda} \geqslant \mathbf{0}. \tag{12}$$

If Q is positive definite, then the dual problem can be simplified. From (12) we get that

$$\mathbf{x} = Q^{-1}(\mathbf{b} - A^t\boldsymbol{\lambda}). \tag{13}$$

If we substitute this into (11), then a straightforward calculation, using the fact that Q^{-1} is symmetric, gives

$$\theta(\boldsymbol{\lambda}) = -\tfrac{1}{2}\langle \boldsymbol{\lambda}, \mathbf{R}\boldsymbol{\lambda} \rangle + (\mathbf{d}, \boldsymbol{\lambda} \rangle + a, \tag{14}$$

where

$$R = AQ^{-1}A^t, \qquad \mathbf{d} = \mathbf{b}^tQ^{-1}A^t - \mathbf{c}, \qquad a = -\tfrac{1}{2}\langle \mathbf{b}, Q^{-1}\mathbf{b} \rangle.$$

Thus the dual problem becomes the following: Maximize $\theta(\boldsymbol{\lambda})$ subject to $\boldsymbol{\lambda} \geqslant \mathbf{0}$, where $\theta(\boldsymbol{\lambda})$ is given by (14).

This problem is a quadratic programming problem whose constraint set has a simple structure, namely $\{\boldsymbol{\lambda} : \boldsymbol{\lambda} \in \mathbb{R}^n, \boldsymbol{\lambda} \geqslant \mathbf{0}\}$. By Theorem 6.2 there is no duality gap and this problem has a solution $\boldsymbol{\lambda}_*$, where $\boldsymbol{\lambda}_*$ is a multiplier as in Theorem 6.1. Thus, from (3) we get that the solution of QP is

$$\mathbf{x}_* = Q^{-1}(\mathbf{b} - A^t\boldsymbol{\lambda}_*).$$

Exercise 6.1. Show that if Q is positive definite, then the quadratic programming problem has a unique solution. Does your proof apply to the unconstrained problem?

Exercise 6.2. The problems posed in Exercise IV.5.8 and Exercise 3.2(iii) are quadratic programming problems. Use the duality developed in this section to solve these problems. Compare with the duality in Exercises 4.1 and 4.2.

7. DUALITY IN LINEAR PROGRAMMING

The linear programming problem LPI formulated in Section 1 in Chapter IV can be viewed as a convex programming problem with $f(\mathbf{x}) = \langle -\mathbf{b}, \mathbf{x} \rangle$ and $\mathbf{g}(\mathbf{x}) = A\mathbf{x} - \mathbf{c}$. Thus all of the results of this chapter hold for the linear programming problem. The linear nature of LPI, however, results in a much stronger duality than that of Theorem 4.1. We proceed to determine this duality.

To maintain a symmetry in the duality, we shall *not* incorporate the constraint $\mathbf{x} \geqslant \mathbf{0}$ in the matrix A and the vector $\mathbf{c}$, as was done in Chapter IV, but shall write our primal problem (P) as

$$\text{Minimize } \langle -\mathbf{b}, \mathbf{x} \rangle \quad \text{subject to } \begin{pmatrix} A \\ -I_n \end{pmatrix} \mathbf{x} \leqslant \begin{pmatrix} \mathbf{c} \\ \mathbf{0}_n \end{pmatrix}, \tag{P}$$

where I_n is the $n \times n$ identity matrix and $\mathbf{0}_n$ is the zero vector in $\mathbb{R}^n$.

We now formulate the dual problem. Let $\boldsymbol{\eta} = (\boldsymbol{\lambda}, \boldsymbol{\mu})$, where $\boldsymbol{\lambda} \in \mathbb{R}^m$, $\boldsymbol{\lambda} \geqslant \mathbf{0}$, $\boldsymbol{\mu} \in \mathbb{R}^n$, $\boldsymbol{\mu} \geqslant \mathbf{0}$. Let

$$\begin{aligned} \theta(\boldsymbol{\eta}) &= \inf\{\langle -\mathbf{b}, \mathbf{x} \rangle + \langle \boldsymbol{\lambda}, A\mathbf{x} - \mathbf{c} \rangle - \langle \boldsymbol{\mu}, \mathbf{x} \rangle : \mathbf{x} \in \mathbb{R}^n\} \\ &= \inf\{\langle (-\mathbf{b} - \boldsymbol{\mu} + A^t\boldsymbol{\lambda}), \mathbf{x} \rangle - \langle \boldsymbol{\lambda}, \mathbf{c} \rangle : \mathbf{x} \in \mathbb{R}^n\}. \end{aligned}$$

The expression inside the curly braces is linear in $\mathbf{x}$. Therefore, for it to have a finite infimum over all $\mathbf{x}$ in $\mathbb{R}^n$, the coefficient of each component of $\mathbf{x}$ must be zero, and so

$$A^t\boldsymbol{\lambda} - \mathbf{b} - \boldsymbol{\mu} = \mathbf{0}. \tag{1}$$

Since $\boldsymbol{\mu} \geqslant \mathbf{0}$, equation (1) is equivalent to

$$A^t\boldsymbol{\lambda} - \mathbf{b} \geqslant \mathbf{0}.$$

If (1) holds, then $\theta(\boldsymbol{\eta}) = \langle -\mathbf{c}, \boldsymbol{\lambda} \rangle$. Thus the problem dual to P is

$$\text{Maximize } \langle -\mathbf{c}, \boldsymbol{\lambda} \rangle \quad \text{subject to } \begin{pmatrix} A^t \\ I_m \end{pmatrix} \boldsymbol{\lambda} \geqslant \begin{pmatrix} \mathbf{b} \\ \mathbf{0}_m \end{pmatrix}, \tag{D}$$

where I_m is the $m \times m$ identity matrix and $\mathbf{0}_m$ is the zero vector in $\mathbb{R}^m$.

The dual problem (D) can be written as a linear programming problem:

$$\text{Minimize } \langle -(-\mathbf{c}), \boldsymbol{\lambda} \rangle \quad \text{subject to } \begin{pmatrix} -A^t \\ -I_m \end{pmatrix} \boldsymbol{\lambda} \leqslant \begin{pmatrix} -\mathbf{b} \\ \mathbf{0}_m \end{pmatrix}.$$

The dual of this problem is

$$\text{Maximize } \langle \mathbf{b}, \mathbf{y} \rangle \quad \text{subject to } \begin{pmatrix} -A \\ I_n \end{pmatrix} \mathbf{y} \geqslant \begin{pmatrix} -\mathbf{c} \\ \mathbf{0}_n \end{pmatrix}.$$

If we write this problem as a minimization problem and relabel the vector $\mathbf{y} \in \mathbb{R}^n$ as $\mathbf{x}$, we see that the dual of the dual problem is the primal problem. We summarize this discussion in the following lemma.

LEMMA 7.1. *The dual of the linear programming problem*

$$\textit{Minimize } \langle -\mathbf{b}, \mathbf{x} \rangle \quad \textit{subject to } \begin{pmatrix} A \\ -I_n \end{pmatrix} \mathbf{x} \leqslant \begin{pmatrix} \mathbf{c} \\ \mathbf{0}_n \end{pmatrix} \tag{P}$$

is

$$\textit{Maximize } \langle -\mathbf{c}, \boldsymbol{\lambda} \rangle \quad \textit{subject to } \begin{pmatrix} A^t \\ I_m \end{pmatrix} \boldsymbol{\lambda} \geqslant \begin{pmatrix} \mathbf{b} \\ \mathbf{0}_m \end{pmatrix}. \tag{D}$$

The dual of the dual problem (D) is the primal problem (P).

THEOREM 7.1 (DUALITY THEOREM OF LINEAR PROGRAMMING). *Let X denote the feasible set for the primal problem and let Y denote the feasible set for the dual problem. Then:*

(*i*) *For each* $\mathbf{x}$ *in* X *and* $\boldsymbol{\lambda}$ *in* Y,

$$\langle -\mathbf{c}, \boldsymbol{\lambda} \rangle \leqslant \langle -\mathbf{b}, \mathbf{x} \rangle \quad (\textit{weak duality})$$

(*ii*) *Let Y be nonempty. Then X is empty if and only if $\langle -\mathbf{c}, \boldsymbol{\lambda} \rangle$ is unbounded above on Y.*

(*iii*) *Let X be nonempty. Then Y is empty if and only if $\langle -\mathbf{b}, \mathbf{x} \rangle$ is bounded below on X.*

(*iv*) (*Strong duality*). *Let the primal problem* (*P*) *have a solution $\mathbf{x}_*$. Then the dual problem* (*D*) *has a solution $\boldsymbol{\lambda}_*$ and $\langle \mathbf{c}, \boldsymbol{\lambda}_* \rangle = \langle \mathbf{b}, \mathbf{x}_* \rangle$.*

***Remark** 7.1.* Since problem P is also a convex programming problem, all of the conclusions of Theorem 4.1 apply to the linear programming problem. Thus, (i) is a special case of (i) of Theorem 4.1. In (ii) the statement that if $\langle -\mathbf{c}, \boldsymbol{\lambda} \rangle$ is unbounded above then X is empty is a special case of (ii) of Theorem 4.1. In the linear programming problem the stronger "if and only if" statement holds. Analogous comments hold for (iii).

Also note that the phenomenon exhibited in Example 4.3 in which there is no duality gap, yet the dual problem has no solution, cannot occur.

We now prove the theorem. We pointed out in Remark 7.1 that (i) follows from (i) of Theorem 4.1.

From the discussion of (ii) in Remark 7.1 it follows that to establish (ii) we must show that if X is empty then $\langle -\mathbf{c}, \boldsymbol{\lambda} \rangle$ is unbounded above. We first note that if X is empty, then we must have $\mathbf{c} \neq \mathbf{0}$. For if $\mathbf{c} = \mathbf{0}$, then $\mathbf{x} = \mathbf{0}$ would be feasible and X would be nonempty.

If X is empty, then the system

$$\begin{pmatrix} A \\ -I_n \end{pmatrix} \mathbf{x} \leqslant \begin{pmatrix} \mathbf{c} \\ \mathbf{0}_n \end{pmatrix}$$

has no solution. It then follows from Theorem II.5.4 that if X is empty, then there exists a vector $\mathbf{y} = (\mathbf{y}_1, \mathbf{y}_2)$, where $\mathbf{y}_1 \in \mathbb{R}^m$, $\mathbf{y}_1 \geqslant \mathbf{0}$, $\mathbf{y}_2 \in \mathbb{R}^n$, $\mathbf{y}_2 \geqslant \mathbf{0}$, such that

$$\langle (\mathbf{c}, \mathbf{0}_n), (\mathbf{y}_1, \mathbf{y}_2) \rangle = -1, \qquad (A^t, -I_n) \begin{pmatrix} \mathbf{y}_1 \\ \mathbf{y}_2 \end{pmatrix} = (\mathbf{0}_m, \mathbf{0}_n). \tag{2}$$

Hence (2) becomes

$$\langle \mathbf{c}, \mathbf{y}_1 \rangle = -1, \qquad A^t \mathbf{y}_1 = \mathbf{0}_m, \qquad \mathbf{y}_1 \geqslant \mathbf{0}_m.$$

Since $Y \neq \varnothing$, there exists a $\boldsymbol{\lambda}_0 \geqslant \mathbf{0}$ such that $A^t \boldsymbol{\lambda}_0 \geqslant \mathbf{b}$. Then for any real $\sigma > 0$, the vector $\boldsymbol{\lambda}_0 + \sigma \mathbf{y}_1$ is in Y and

$$\langle -\mathbf{c}, \boldsymbol{\lambda}_0 + \sigma y_1 \rangle = \langle -\mathbf{c}, \boldsymbol{\lambda}_0 \rangle + \sigma.$$

Thus, since σ can be taken to be arbitrarily large, $\langle -\mathbf{c}, \boldsymbol{\lambda} \rangle$ is unbounded above.

We view the primal problem as the dual to the dual and view the dual problem as primal. Then from (ii) we conclude that if X is nonempty, then Y is empty if and only if $\langle \mathbf{b}, \mathbf{x} \rangle$ is unbounded above, or equivalently that $\langle -\mathbf{b}, \mathbf{x} \rangle$ is unbounded below. Thus, (iii) is proved.

We now take up the proof of (iv). Since the primal problem has a solution, there exists a vector $\boldsymbol{\lambda}$ in $\mathbb{R}^m$ as in Theorem IV.4.1. It is readily seen that this vector is a multiplier as in the KKT conditions of Theorem IV.5.2. The set $\mathfrak{M}$ is thus nonempty. The absence of a duality gap and the existence of a solution follow from Corollary 1 to Theorem 4.4. Since there is no duality gap and both the primal and dual have a solution, $\langle -\mathbf{c}, \boldsymbol{\lambda}_* \rangle = \langle -\mathbf{b}, \mathbf{x}_* \rangle$.

COROLLARY 1. *A necessary and sufficient condition that a point $\mathbf{x}_*$ in X and a point $\boldsymbol{\lambda}_*$ in Y be solutions of the primal and dual problems respectively is that*

$$\langle \mathbf{b}, \mathbf{x}_* \rangle = \langle \mathbf{c}, \boldsymbol{\lambda}_* \rangle.$$

Proof. The necessity of the condition follow from (iv) of Theorem 7.1 and the sufficiency from (iv) of Theorem 4.1.

COROLLARY 2. *Necessary and sufficient conditions that a point $\mathbf{x}_*$ in X and a point $\boldsymbol{\lambda}_*$ in Y be solutions of the primal and dual problems respectively are*

(*i*) $\langle A\mathbf{x}_* - \mathbf{c}, \boldsymbol{\lambda}_* \rangle = 0$ *and*
(*ii*) $\langle A^t\boldsymbol{\lambda}_* - \mathbf{b}, \mathbf{x}_* \rangle = 0$.

Proof. Let $\mathbf{x}_*$ be optimal for the primal problem and let $\boldsymbol{\lambda}_*$ be optimal for the dual problem. Then $A\mathbf{x}_* \leqslant \mathbf{c}$ and $\boldsymbol{\lambda}_* \geqslant 0$. Hence

$$\langle A\mathbf{x}_* - \mathbf{c}, \boldsymbol{\lambda}_* \rangle \leqslant 0. \tag{3}$$

Since $\boldsymbol{\lambda}_*$ is feasible for the dual problem, $A^t\boldsymbol{\lambda}_* \geqslant \mathbf{b}$. Since $\mathbf{x}_* \geqslant 0$, we have

$$\langle A^t\boldsymbol{\lambda}_* - \mathbf{b}, \mathbf{x}_* \rangle \geqslant 0. \tag{4}$$

By Corollary 1, $\langle \mathbf{b}, \mathbf{x}_* \rangle = \langle \mathbf{c}, \boldsymbol{\lambda}_* \rangle$, so $\langle A^t\boldsymbol{\lambda}_*, \mathbf{x}_* \rangle - \langle \mathbf{c}, \boldsymbol{\lambda}_* \rangle \geqslant 0$. Therefore $\langle A\mathbf{x}_* - \mathbf{c}, \boldsymbol{\lambda}_* \rangle \geqslant 0$. Comparing this with (3) gives (i). Also,

$$\langle A\mathbf{x}_* - \mathbf{c}, \boldsymbol{\lambda}_* \rangle = \langle A\mathbf{x}_*, \boldsymbol{\lambda}_* \rangle - \langle \mathbf{c}, \boldsymbol{\lambda}_* \rangle = \langle A^t\boldsymbol{\lambda}_* - \mathbf{b}, \mathbf{x}_* \rangle;$$

and so from (3) we get that $\langle A^t\boldsymbol{\lambda}_* - \mathbf{b}, \mathbf{x}_* \rangle \leqslant 0$. This inequality and (4) give (ii).

To prove the sufficiency, note that (i) and (ii) give

$$\langle \mathbf{c}, \boldsymbol{\lambda}_* \rangle = \langle A\mathbf{x}_*, \boldsymbol{\lambda}_* \rangle = \langle A^t\boldsymbol{\lambda}_*, \mathbf{x}_* \rangle = \langle \mathbf{b}, \mathbf{x}_* \rangle.$$

It then follows from the equality of the outermost quantities and Corollary 1 that $\mathbf{x}_*$ is a solution of the primal problem and that $\boldsymbol{\lambda}_*$ is a solution of the dual.

The next theorem is a stronger version of the strong duality statement (iv) of Theorem (7.1) in that we obtain strong duality by merely assuming the feasible sets X and Y to be nonempty.

THEOREM 7.2. *The feasible sets X and Y for the primal and dual problems respectively are nonempty if and only if there exists a solution $\mathbf{x}_*$ to the primal problem and a solution $\boldsymbol{\lambda}_*$ to the dual problem. In this event*

$$\langle \mathbf{c}, \boldsymbol{\lambda}_* \rangle = \langle \mathbf{b}, \mathbf{x}_* \rangle. \tag{5}$$

Proof. If the primal and dual problems have solutions, then the sets X and Y are clearly nonempty.

To show that $X \neq \varnothing$ and $Y \neq \varnothing$ imply that both the dual and primal problems have solutions and that (5) holds, it suffices to show that the primal problem has a solution. For then by (iv) of Theorem 7.1, the dual problem has a solution and (5) holds.

In Remark IV.4.1 we noted that relation (2iv) of Theorem IV.4.1 is a consequence of (i), (ii), (iii), and (v). Thus in Corollary IV.4.3 the same observation holds. Hence a necessary and sufficient condition for a vector $\mathbf{x}_*$ to be a solution of P is that there exist a vector $\boldsymbol{\lambda}_0$ in $\mathbb{R}^m$ such that

$$\begin{gathered}\begin{pmatrix} A \\ -I_n \end{pmatrix} \mathbf{x}_* \leqslant \begin{pmatrix} \mathbf{c} \\ \mathbf{0} \end{pmatrix}, \qquad \begin{pmatrix} A^t \\ I_m \end{pmatrix} \boldsymbol{\lambda}_0 \geqslant \begin{pmatrix} \mathbf{b} \\ \mathbf{0} \end{pmatrix}, \\ \langle \mathbf{b}, \mathbf{x}_* \rangle = \langle \mathbf{c}, \boldsymbol{\lambda}_0 \rangle. \end{gathered} \tag{6}$$

Since the existence of vectors $\mathbf{x}_*$ in $\mathbb{R}^n$ and vector $\boldsymbol{\lambda}_0$ in $\mathbb{R}^m$ that satisfy (6) is a necessary and sufficient condition that $\mathbf{x}_*$ be a solution of P, it follows that if P has no solution, then the system

$$\begin{aligned} &\text{(i)} && \begin{pmatrix} A \\ -I_n \end{pmatrix} \mathbf{x} \leqslant \begin{pmatrix} \mathbf{c} \\ \mathbf{0} \end{pmatrix}, \\ &\text{(ii)} && \begin{pmatrix} -A^t \\ -I_m \end{pmatrix} \mathbf{u} \leqslant \begin{pmatrix} -\mathbf{b} \\ \mathbf{0} \end{pmatrix}, \\ &\text{(iii)} && \langle -\mathbf{b}, \mathbf{x} \rangle + \langle \mathbf{c}, \mathbf{u} \rangle = 0 \end{aligned} \tag{7}$$

has no *solution* $(\mathbf{x}, \mathbf{u})$, $\mathbf{x} \in \mathbb{R}^n$, $\mathbf{u} \in \mathbb{R}^m$. Condition (7i) is the condition that $\mathbf{x}$ be feasible for P and condition (7ii) is the condition that $\mathbf{u}$ be feasible for the dual problem. By assumption, there exist feasible vectors $\mathbf{x} \in X$ and $\mathbf{u} \in Y$. By (i) of Theorem 7.1, for such vectors we have $\langle -\mathbf{b}, \mathbf{x} \rangle \geqslant \langle -\mathbf{c}, \mathbf{u} \rangle$, and thus $\langle -\mathbf{b}, \mathbf{x} \rangle + \langle \mathbf{c}, \mathbf{u} \rangle \geqslant 0$. Hence for a vector $(\mathbf{x}, \mathbf{u})$ that satisfies (7i) and (7ii) to fail to satisfy (7iii), it must fail to satisfy

$$\langle -\mathbf{b}, \mathbf{x} \rangle + \langle \mathbf{c}, \mathbf{u} \rangle \leqslant 0, \tag{7iii$'$}$$

so we may replace (7iii) by (7iii′) in (7).

Now consider the system

$$\begin{gathered} \zeta > 0, \qquad \zeta \in \mathbb{R}, \\ \begin{pmatrix} A \\ -I_n \end{pmatrix} \mathbf{x} \leqslant \begin{pmatrix} \zeta \mathbf{c} \\ 0 \end{pmatrix}, \qquad \begin{pmatrix} -A^t \\ -I_m \end{pmatrix} \mathbf{u} \leqslant \begin{pmatrix} -\zeta \mathbf{b} \\ \mathbf{0} \end{pmatrix}, \\ \langle -\mathbf{b}, \mathbf{x} \rangle + \langle \mathbf{c}, \mathbf{u} \rangle \leqslant 0. \end{gathered} \tag{8}$$

If (7) has a solution $(\mathbf{x}_0, \mathbf{u}_0)$, then (8) has a solution $(\zeta_1, \mathbf{x}_1, \mathbf{u}_1) = (1, \mathbf{x}_0, \mathbf{u}_0)$. Conversely, if (8) has a solution $(\zeta_1, \mathbf{x}_1, \mathbf{u}_1)$, then $(\mathbf{x}_0, \mathbf{u}_0) = (\mathbf{x}_1/\zeta_1, \mathbf{u}_1/\zeta_1)$ is a solution of (7). Therefore (7) has no solution if and only if (8) has no solution.

System (8) can be written as

$$\begin{gathered} \langle (\mathbf{0}_n, \mathbf{0}_m, 1), (\mathbf{x}, \mathbf{u}, \zeta) \rangle > 0, \\ \begin{pmatrix} A & O & -\mathbf{c} \\ -I_n & O & \mathbf{0}_n \\ O & -A^t & \mathbf{b} \\ O & -I_m & \mathbf{0}_m \\ -\mathbf{b}^t & \mathbf{c}^t & 0 \end{pmatrix} \begin{pmatrix} \mathbf{x} \\ \mathbf{u} \\ \zeta \end{pmatrix} \leqslant \mathbf{0}_{2n+2m+1}, \end{gathered} \tag{9}$$

where O represents a zero matrix of appropriate dimension and $\mathbf{0}_i$ is the zero vector in $\mathbb{R}^i$. Since this system has no solution, it follows from Theorem II.5.1 (Farkas's lemma) that there exists a vector $(\mathbf{p}_1, \mathbf{p}_2, \mathbf{q}_1, \mathbf{q}_2, \sigma) \geqslant \mathbf{0}_{2n+2m+1}$ with $\mathbf{p}_1 \in \mathbb{R}^m$, $\mathbf{p}_2 \in \mathbb{R}^n$, $\mathbf{q}_1 \in \mathbb{R}^n$, $\mathbf{q}_2 \in \mathbb{R}^m$, and $\sigma \in \mathbb{R}$ such that

$$\begin{pmatrix} A^t & -I_n & O & O & -\mathbf{b} \\ O & O & -A & -I_m & \mathbf{c} \\ -\mathbf{c}^t & \mathbf{0}_n & \mathbf{b}^t & \mathbf{0}_m & 0 \end{pmatrix} \begin{pmatrix} \mathbf{p}_1 \\ \mathbf{p}_2 \\ \mathbf{q}_1 \\ \mathbf{q}_2 \\ \sigma \end{pmatrix} = \begin{pmatrix} \mathbf{0}_n \\ \mathbf{0}_m \\ 1 \end{pmatrix}.$$

Thus

$$\begin{gathered} A^t \mathbf{p}_1 - \mathbf{p}_2 - \sigma \mathbf{b} = \mathbf{0}_n, \\ -A\mathbf{q}_1 - \mathbf{q}_2 + \sigma \mathbf{c} = \mathbf{0}_m, \\ \langle -\mathbf{p}_1, \mathbf{c} \rangle + \langle \mathbf{q}_1, \mathbf{b} \rangle = 1. \end{gathered} \tag{10}$$

We shall show that this system cannot have a solution $(\mathbf{p}_1, \mathbf{p}_2, \mathbf{q}_1, \mathbf{q}_2, \sigma) \geqslant \mathbf{0}_{2n+2m+1}$. This contradiction will prove that the primal problem has a solution, since we arrived at (10) by assuming that it did not.

Let $\mathbf{x}_0 \in X$ and let $\mathbf{y}_0 \in Y$. Such points exist since $X \neq \varnothing$ and $Y \neq \varnothing$. Suppose first that $\sigma = 0$. Then $A^t\mathbf{p}_1 = \mathbf{p}_2$ and $A\mathbf{q}_1 = -\mathbf{q}_2$. Since $\mathbf{x}_0 \in X$, we have $A\mathbf{x}_0 \leqslant \mathbf{c}$. Thus, since $\mathbf{p}_1 \geqslant \mathbf{0}$,

$$\langle -\mathbf{p}_1, \mathbf{c} \rangle \leqslant \langle -\mathbf{p}_1, A\mathbf{x}_0 \rangle = -\langle A^t\mathbf{p}_1, \mathbf{x}_0 \rangle = -\langle \mathbf{p}_2, \mathbf{x}_0 \rangle \leqslant 0, \tag{11}$$

where the last inequality follows from $\mathbf{p}_2 \geqslant \mathbf{0}$, $\mathbf{x}_0 \geqslant \mathbf{0}$. Similarly,

$$\langle \mathbf{q}_1, \mathbf{b} \rangle \leqslant \langle \mathbf{q}_1, A^t\mathbf{y}_0 \rangle = \langle A\mathbf{q}_1, \mathbf{y}_0 \rangle = -\langle \mathbf{q}_2, \mathbf{y}_0 \rangle \leqslant 0. \tag{12}$$

From (11) and (12) we get

$$\langle -\mathbf{p}_1, \mathbf{c} \rangle + \langle \mathbf{q}_1, \mathbf{b} \rangle \leqslant 0, \tag{13}$$

which contradicts the last equation in (10).

Suppose now that $\sigma > 0$. The first equation in (10) implies that $\mathbf{p}_1/\sigma$ is feasible for the dual problem. The second equation in (10) implies that $\mathbf{q}_1/\sigma$ is feasible for the primal problem. By the weak duality, $\langle -\mathbf{b}, \mathbf{q}_1/\sigma \rangle \geqslant \langle -\mathbf{c}, \mathbf{p}_1/\sigma \rangle$. Hence we again arrive at (13), which contradicts the last equation in (10).

VI

SIMPLEX METHOD

1. INTRODUCTION

In Section 4 of Chapter IV we formulated the linear programming problem and derived a necessary and sufficient condition that a point $\mathbf{x}_*$ be a solution to a linear programming problem. In Section 7 of Chapter V we presented the duality theorems for linear programming. We remarked in Section 4 of Chapter IV that in practice the dimension of the space $\mathbb{R}^n$ precluded the use of the necessary and sufficient conditions to solve linear programming problems and that instead a method known as the simplex method is used. In this chapter we shall present the essentials of the simplex method, which is an algorithm for solving linear programming problems.

To motivate the rationale of the simplex methods, we consider several examples in $\mathbb{R}^2$.

Example 1.1. Minimize $f(\mathbf{x}) = x_1 - 2x_2$ subject to

$$\begin{aligned} -x_1 - x_2 &\leqslant -1, \\ -x_1 + x_2 &\leqslant 0, \\ x_1 + 2x_2 &\leqslant 4, \\ x_1 \geqslant 0, \qquad x_2 &\geqslant 0. \end{aligned} \tag{1}$$

The feasible set is indicated by the hatched area in Figure 6.1. The level curves $f(\mathbf{x}) = 0$ and $f(\mathbf{x}) = -1$ are indicated by dashed lines; the arrows indicate the directions of the gradient vectors of f. Since the gradient vectors indicate the direction of increasing f, we see from the figure that f attains a minimum at the extreme point $(\frac{4}{3}, \frac{4}{3})$ of the feasible set.

Example 1.2. This example differs from Example 1.1 in that we take the function f to be $f(\mathbf{x}) = x_1 + x_2$. The level curves are lines parallel to the line $x_1 + x_2 = 1$, a segment of which is part of the boundary of the feasible set. Since the gradient vector of f is $(1, 1)$ and this is the direction of increasing f,

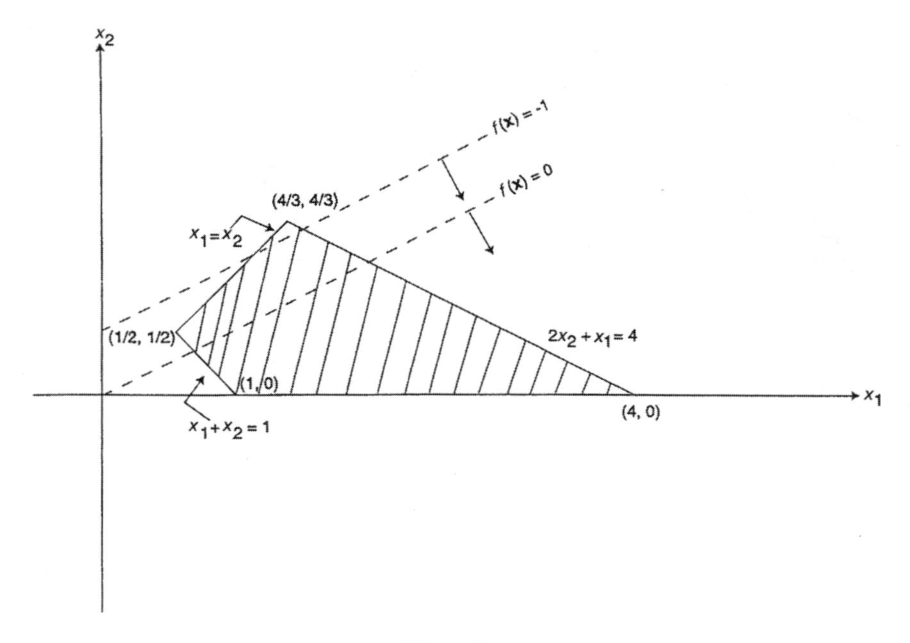

Figure 6.1.

it follows that f attains its minimum at all boundary points of the feasible set that lie on the line $x_1 + x_2 = 1$. In particular, the minimum is attained at the two extreme points $(\frac{1}{2}, \frac{1}{2})$ and $(1, 0)$.

Example 1.3. Minimize $f(\mathbf{x}) = -x_1 - x_2$ subject to

$$-x_1 + x_2 \leqslant 0,$$

$$x_1 - 2x_2 \leqslant -2,$$

$$x_1 \geqslant 0, \qquad x_2 \geqslant 0.$$

The feasible set is the hatched area in Figure 6.2. The level curve $f(\mathbf{x}) = -6$ is indicated by a dashed line. Since the gradient of f is $(-1, -1)$, it follows from the figure that this problem has no solution. The feasible set in this example is unbounded. An unbounded feasible set, however, does not imply that a linear programming problem has no solution. If we modify this example by taking $f(\mathbf{x}) = x_1 + x_2$, then the problem does have a solution at the extreme point $(2, 2)$.

In the examples that we considered either the problem had no solution or the problem had a solution and there was an extreme point of the constraint set at which the solution was attained. There is, of course, a third possibility,

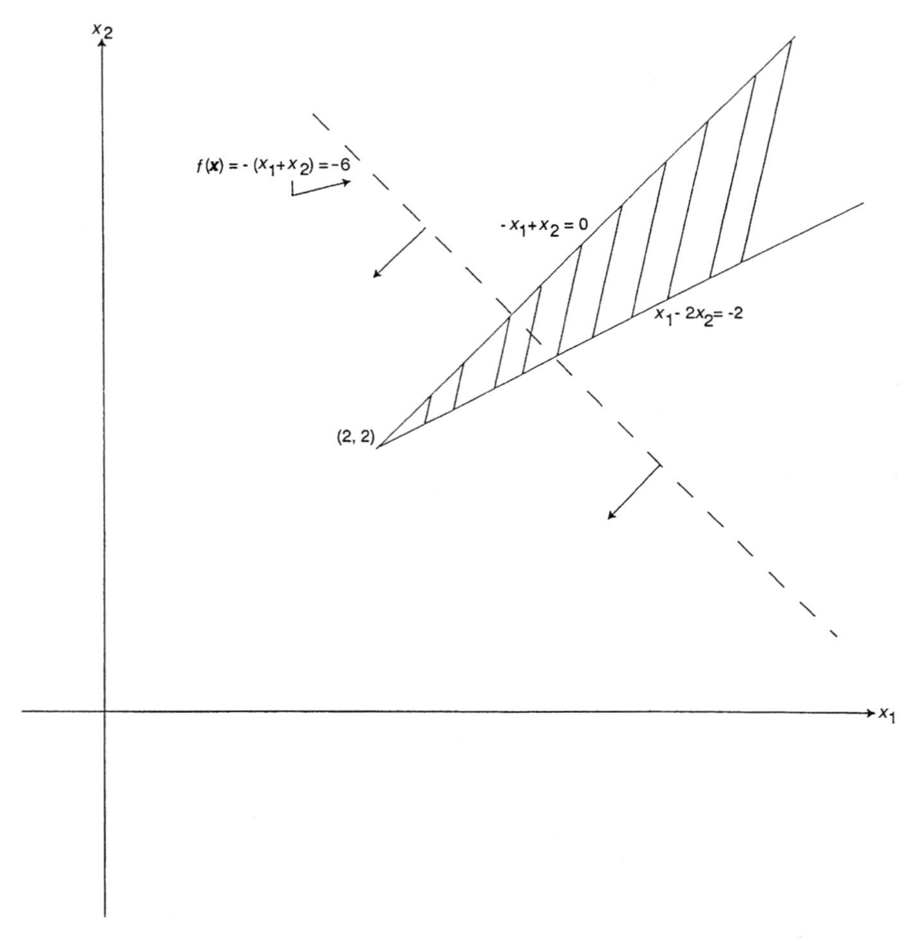

Figure 6.2.

namely that the feasible set is empty. In the next section we shall show that in a linear programming problem these are the only possibilities.

The simplex method is applied to problems in canonical form, CLP (see Section 4 of Chapter IV for definitions). To put the problem in Example 1.1 in canonical form we introduce additional slack variables x_3, x_4, x_5 and write the problem as

$$
\begin{aligned}
&\text{Minimize } f(\mathbf{x}) = x_1 - 2x_2 \\
&\text{subject to } -x_1 - x_2 + x_3 = -1, \\
&\qquad -x_1 + x_2 + x_4 = 0, \\
&\qquad x_1 + 2x_2 + x_5 = 4.
\end{aligned}
\tag{2}
$$

Since the variables x_3, x_4, x_5 do not appear in the formula for f, the problem with feasible set defined by (2) has its minimum attained at

$$x_1 = \tfrac{4}{3}, \qquad x_2 = \tfrac{4}{3} \qquad x_3 = \tfrac{5}{3}, \qquad x_4 = 0, \qquad x_5 = 0.$$

It is readily verified that this point is an extreme point of the set defined by (2). We shall leave it to the reader to show that the extreme points of the constraint set in problem SLP corresponds to the extreme points in the problem CLP obtained by introducing slack variables.

Exercise 1.1. Solve the following linear programming problem graphically. Maximize $f(\mathbf{x}) = x_1 + 2x_2$ subject to

$$\begin{aligned} x_1 + x_2 &\leqslant 5, \\ -x_1 + x_2 &\leqslant 1, \\ x_1 - x_2 &\leqslant 0, \\ -3x_1 - x_2 &\leqslant -1, \\ x_1 \geqslant 0, \qquad & x_2 \geqslant 0. \end{aligned}$$

Exercise 1.2. Let $\mathbf{x}_0$ be a feasible point for problem SLP and let $(\mathbf{x}_0, \boldsymbol{\xi}_0)$ be the corresponding feasible point for the related problem CLP; then

$$A\mathbf{x}_0 \leqslant \mathbf{c}, \qquad A\mathbf{x}_0 + I\boldsymbol{\xi}_0 = \mathbf{c},$$
$$\mathbf{x}_0 \geqslant \mathbf{0}, \qquad \boldsymbol{\xi}_0 \geqslant \mathbf{0}.$$

Show that $\mathbf{x}_0$ is an extreme point of the feasible set for problem SLP if and only if $(\mathbf{x}_0, \boldsymbol{\xi}_0)$ is an extreme point for problem CLP.

Exercise 1.3. In problem LPI, which is Minimize $\langle -\mathbf{b}, \mathbf{x} \rangle$ subject to $A\mathbf{x} \leqslant \mathbf{c}$, show that if the feasible set is compact, then the problem has a solution and there exists an extreme point of the feasible set at which the minimum is attained.

2. EXTREME POINTS OF FEASIBLE SET

In this section we shall show that if a linear programming problem has a solution, then there exists an extreme point at which the minimum is attained. We shall prove this for problems in canonical form without assuming that the feasible set is compact, as was done in Exercise 1.3. By virtue of Exercise 1.2 this also holds for problem SLP. We shall also give an analytic characterization of the extreme points of the feasible set for CLP problems. The significance of these results is that in looking for a solution to a linear programming problem we can confine ourselves to the extreme points of the feasible set, points for which we have an analytic characterization. The simplex method is an algorithm for efficiently determining whether a solution exists and, if so, for finding an extreme point which solves the problem.

We now consider the linear programming problem in canonical form:

$$\begin{aligned} &\text{Maximize } \langle \mathbf{b}, \mathbf{x} \rangle \\ &\text{subject to } A\mathbf{x} = \mathbf{c}, \qquad \mathbf{x} \geqslant \mathbf{0}. \end{aligned} \tag{1}$$

In keeping with much of the linear programming literature we have replaced the objective of minimizing $\langle -\mathbf{b}, \mathbf{x} \rangle$ by the equivalent objective of maximizing $\langle \mathbf{b}, \mathbf{x} \rangle$. We assume that $\mathbf{x}$ and $\mathbf{b}$ are in $\mathbb{R}^n$, that $\mathbf{c}$ is in $\mathbb{R}^m$, that A is $m \times n$, and that

$$\text{(i)} \quad m < n, \qquad \text{(ii)} \quad \operatorname{rank} A = m. \tag{2}$$

The rationale for the assumption (2) is precisely the same as given in Remark V.1.2 for the assumption that the $k \times n$ matrix A defining the affine constraints in the convex programming problem CPII has rank k and that $k < n$.

Before presenting the principal results of this section, we emphasize that the reader should keep in mind that the constraint $\mathbf{x} \geqslant \mathbf{0}$ implies that the components x_j of $\mathbf{x}$ are either positive or zero.

THEOREM 2.1. *A feasible point $\mathbf{x}_0$ is an extreme point of the feasible set if and only if the columns $\mathbf{A}_j$ of the matrix A that correspond to positive components of $\mathbf{x}_0$ are linearly independent.*

Proof. Suppose that the columns $\mathbf{A}_j$ of A that correspond to positive components of $\mathbf{x}_0$ are linearly independent. We shall show that $\mathbf{x}_0$ is an extreme point of the feasible set.

If $\mathbf{x}_0$ were not an extreme point of the feasible set, there would exist two distinct feasible points $\mathbf{x}_1$ and $\mathbf{x}_2$ and a scalar λ such that $0 < \lambda < 1$ and

$$\mathbf{x}_0 = \lambda \mathbf{x}_1 + (1 - \lambda)\mathbf{x}_2. \tag{3}$$

Let B denote the submatrix of A formed by the columns of A that correspond to positive components of $\mathbf{x}$ and let N denote the submatrix of A formed by the columns of A corresponding to zero components of $\mathbf{x}$. To simplify notation, let us assume that B consists of the first k columns of A and that N consist of the remaining $n - k$ columns. There is no loss of generality in making this assumption since it can be achieved by a relabeling of the coordinates. Since the columns of B are linearly independent, it follows from (2) that $k \leqslant m$. It then further follows from (2) that $n - k > 0$. For $\mathbf{x}$ in $\mathbb{R}^n$, let

$$\mathbf{x}_B = (x_1, \ldots, x_k), \qquad \mathbf{x}_N = (x_{k+1}, \ldots, x_n).$$

Then

$$\mathbf{x}_0 = (\mathbf{x}_{0B}, \mathbf{x}_{0N}) = (\mathbf{x}_{0B}, \mathbf{0}_N)$$
$$\mathbf{x}_i = (\mathbf{x}_{iB}, \mathbf{x}_{iN}), \qquad i = 1, 2,$$

where $\mathbf{x}_0, \mathbf{x}_1$, and $\mathbf{x}_2$ are as in (3) and $\mathbf{0}_N$ is the zero vector in $\mathbb{R}^{n-k}$.

Since $\lambda > 0$ and $1 - \lambda > 0$, it follows from (3) that a component of $\mathbf{x}_0$ is equal to zero if and only if the corresponding components of $\mathbf{x}_1$ and $\mathbf{x}_2$ are zero. Hence $\mathbf{x}_{0N} = \mathbf{x}_{1N} = \mathbf{x}_{2N} = \mathbf{0}_N$ and we may write

$$\mathbf{x}_i = (\mathbf{x}_{iB}, \mathbf{0}_N), \qquad i = 0, 1, 2.$$

Since $\mathbf{x}_1$ is feasible,

$$\mathbf{c} = A\mathbf{x}_1 = (B, N)\begin{pmatrix} \mathbf{x}_{1B} \\ \mathbf{0}_N \end{pmatrix} = B\mathbf{x}_{1B}.$$

Similarly, we get that $\mathbf{c} = B\mathbf{x}_{2B}$. Hence

$$B(\mathbf{x}_{1B} - \mathbf{x}_{2B}) = \mathbf{0}. \tag{4}$$

Since $\mathbf{x}_1 \neq \mathbf{x}_2$ and $\mathbf{x}_{1N} = \mathbf{x}_{2N}$, we have that $\mathbf{x}_{1B} - \mathbf{x}_{2B} \neq \mathbf{0}$. But this cannot be, since the columns of B are linearly independent. This contradiction shows that $\mathbf{x}_0$ is an extreme point.

Now let us suppose that $\mathbf{x}_0$ is an extreme point of the feasible set. To simplify notation, we suppose that the first k components of $\mathbf{x}_0$ are positive and the remaining components are zero. We shall show that the first k columns $\mathbf{A}_1, \ldots, \mathbf{A}_k$ of A are linearly independent. If these columns were not linearly independent, there would exist constants $\alpha_1, \ldots, \alpha_k$ not all zero such that

$$\alpha_1\mathbf{A}_1 + \cdots + \alpha_k\mathbf{A}_k = \mathbf{0}.$$

Let the n-vector $\mathbf{a}$ be defined by

$$\mathbf{a} = (\alpha_1, \ldots, \alpha_k, 0, \ldots, 0).$$

Then for any real ε,

$$\varepsilon A\mathbf{a} = \varepsilon \sum_{j=1}^{k} \alpha_j A_j = \mathbf{0}. \tag{5}$$

For $\varepsilon > 0$ and sufficiently small the vectors $\mathbf{x}_0 + \varepsilon\mathbf{a}$ and $\mathbf{x}_0 - \varepsilon\mathbf{a}$ have their first k components positive and the remaining $n - k$ components equal to zero. Also

$$A(\mathbf{x}_0 \pm \varepsilon\mathbf{a}) = A\mathbf{x}_0 \pm \varepsilon A\mathbf{a} = \mathbf{c},$$

where the last equality follows from (5) and the feasibility of $\mathbf{x}_0$. Thus for sufficiently small $\varepsilon > 0$ the vectors $\mathbf{x}_0 + \varepsilon\mathbf{a}$ and $\mathbf{x}_0 - \varepsilon\mathbf{a}$ are distinct and feasible. But then the relation

$$\mathbf{x}_0 = \tfrac{1}{2}(\mathbf{x}_0 + \varepsilon\mathbf{a}) + \tfrac{1}{2}(\mathbf{x}_0 - \varepsilon\mathbf{a})$$

contradicts the hypothesis that $\mathbf{x}_0$ is an extreme point. Hence the vectors $\mathbf{A}_1, \ldots, \mathbf{A}_k$ are linearly independent.

The following corollary is an immediate consequence of the theorem and the fact that A has rank m.

COROLLARY 1. *An extreme point of the feasible set has at most m positive components.*

COROLLARY 2. *The number of extreme points of the feasible set is less than or equal to*

$$\binom{n}{m} = \frac{n!}{m!(n-m)!}.$$

Proof. If $\mathbf{x}$ is an extreme point of the feasible set, then $\mathbf{x}$ has $k \leqslant m$ positive components. These k positive components correspond to k linearly independent columns of A. We assert that if $\mathbf{x}'$ is another extreme point with the same nonzero components as $\mathbf{x}$, then $\mathbf{x} = \mathbf{x}'$. To see this, let B denote the submatrix of A corresponding to the nonzero components of $\mathbf{x}$ and $\mathbf{x}'$, let $\mathbf{x} = (\mathbf{x}_B, \mathbf{0})$, and let $\mathbf{x}' = (\mathbf{x}'_B, \mathbf{0})$. Then by the argument used to establish (4), we get that $B(\mathbf{x}_B - \mathbf{x}'_B) = \mathbf{0}$. Since the columns of B are linearly independent, $\mathbf{x}_B = \mathbf{x}'_B$, and so $\mathbf{x} = \mathbf{x}'$.

If $k < m$, then since rank $A = m$, we can adjoin an additional $m - k$ columns of A to obtain a set of m linearly independent columns associated with $\mathbf{x}$. If $k = m$, then this certainly holds. Hence the number of extreme points cannot exceed the number of ways in which m columns can be selected from among n columns.

THEOREM 2.2. *Let the canonical linear programming problem as stated in* (1) *have a solution* $\mathbf{x}_0$. *Then there exists an extreme point* $\mathbf{z}$ *of the feasible set that is also a solution.*

Proof. By Theorem 2.1, if the columns of A that correspond to positive components of $\mathbf{x}_0$ are linearly independent, then $\mathbf{x}_0$ is an extreme point of the feasible set and we let $\mathbf{z} = \mathbf{x}_0$.

Let us now suppose that the columns of A that correspond to the positive components of $\mathbf{x}_0$ are linearly dependent. To simplify notation, we suppose that the first p components of $\mathbf{x}_0$ are positive and that the remaining $n - p$ components are equal to zero. Then the first p columns of A are linearly dependent and there exist scalars $\gamma_1, \ldots, \gamma_p$ not all zero such that

$$\sum_{j=1}^{p} \gamma_j \mathbf{A}_j = \mathbf{0}. \tag{6}$$

For each real number σ define a vector $\mathbf{z}(\sigma)$ as follows:

$$z_j(\sigma) = \begin{cases} x_{0j} - \sigma\gamma_j, & j = 1, \ldots, p, \\ 0, & j = p+1, \ldots, n. \end{cases} \tag{7}$$

Then, from the fact that $x_{0j} = 0$ for $j = p + 1, \ldots, n$, from the feasibility of $\mathbf{x}_0$ and from (6) we get that

$$A\mathbf{z}(\sigma) = A\mathbf{x}_0 - \sigma \sum_{j=1}^{p} \gamma_j \mathbf{A}_j = \mathbf{c}.$$

Let

$$\sigma_0 = \min\left\{\frac{x_{0j}}{|\gamma_j|} : j = 1, \ldots, p \text{ and } \gamma_j \neq 0\right\}. \tag{8}$$

It follows from (7) that $\mathbf{z}(\sigma) \geqslant \mathbf{0}$ whenever $|\sigma| \leqslant \sigma_0$. Hence $\mathbf{z}(\sigma)$ is feasible whenever $|\sigma| \leqslant \sigma_0$. Thus,

$$\langle \mathbf{b}, \mathbf{x}_0 \rangle \geqslant \langle \mathbf{b}, \mathbf{z}(\sigma) \rangle = \langle \mathbf{b}, \mathbf{x}_0 \rangle - \sigma \sum_{j=1}^{p} b_j \gamma_j$$

for *all* σ such that $|\sigma| \leqslant \sigma_0$. Hence

$$\sum_{j=1}^{p} b_j \gamma_j = 0 \quad \text{and} \quad \langle \mathbf{b}, \mathbf{x}_0 \rangle = \langle \mathbf{b}, \mathbf{z}(\sigma) \rangle.$$

Thus, $\mathbf{z}(\sigma)$ is a solution whenever $|\sigma| \leqslant \sigma_0$. Let the minimum in (8) be attained at the index r. Let $\sigma_1 = x_{0r}/\gamma_r$. Then $\sigma_1 = \sigma_0$ if $\gamma_r > 0$ and $\sigma_1 = -\sigma_1$ if $\gamma_r < 0$. Hence $\mathbf{z}(\sigma_1)$ is feasible and is a solution. Moreover $\mathbf{z}(\sigma_1)$ has fewer then p positive components. If the corresponding columns of A are linearly independent, then $\mathbf{z}(\sigma_1)$ is an extreme point. If the columns are not linearly independent, we repeat the process, taking $\mathbf{x}_0$ to be $\mathbf{z}(\sigma_1)$. In a finite number of iterations we shall obtain a solution such that the column vectors corresponding to the positive components are linearly independent. This solution will be an extreme point of the feasible set.

3. PRELIMINARIES TO SIMPLEX METHOD

In the last section we showed that if a solution to the canonical linear programming problem exists, then it exists at an extreme point of the feasible set. We also showed that there are at most $n!/m!(n - m)!$ such points. For a moderate system of 6 constraint equations and 30 variables, there can be as many as

$$\binom{30}{6} = \frac{30!}{6!24!} = 593{,}775$$

extreme points. Thus the simple-minded scheme of evaluating the objective function at all possible extreme points and then selecting the point at which the largest value of the objective function is attained is not practical. The simplex method is an algorithm for finding a solution by going from one extreme point of the feasible set to another in such a way that the objective function is not decreased.

We now introduce the notation and definitions that will be used. Consider the system

$$A\mathbf{x} = \mathbf{c} \tag{1}$$

that defines the feasible set for CLP, the canonical linear programming problem. Thus, A is $m \times n$ and of rank m. Let B denote an $m \times m$ submatrix of A obtained by taking m linearly independent columns of A and let N denote the submatrix of A consisting of the remaining $n - m$ columns. We can write a vector $\mathbf{x}$ in $\mathbb{R}^n$ as

$$\mathbf{x} = (\mathbf{x}_B, \mathbf{x}_N),$$

where $\mathbf{x}_B$ consists of those components of $\mathbf{x}$ corresponding to the columns in B and $\mathbf{x}_N$ consists of those components of $\mathbf{x}$ corresponding to the columns in N.

Definition 3.1. The columns of B are called a *basis*. The components of $\mathbf{x}_B$ are called *basic* variables. The components of $\mathbf{x}_N$ are called *nonbasic variables* or *free variables*.

We may write the system (1) as

$$(B, N)\begin{pmatrix} \mathbf{x}_B \\ \mathbf{x}_N \end{pmatrix} = \mathbf{c}. \tag{2}$$

Any vector of the form $(\xi_B, \mathbf{x}_N)$, where $\mathbf{x}_N$ is arbitrary and

$$\xi_B = B^{-1}\mathbf{c} - B^{-1}N\mathbf{x}_N,$$

will be a solution of (2). (This is why the components of $\mathbf{x}_N$ are called free variables.) There is a unique solution of (2) with $\mathbf{x}_N = \mathbf{0}_N$, namely

$$\xi = (\xi_B, \mathbf{0}_N), \qquad \xi_B = B^{-1}\mathbf{c}. \tag{3}$$

Definition 3.2. (i) A solution of (1) of the form (3) is called a *basic solution.* (ii) If $\xi_B \geqslant \mathbf{0}_B$, then ξ is feasible and is called a *basic feasible solution.* (iii) If some of the components of ξ_B are zero, then ξ is called a *degenerate basic solution.*

We emphasize the following aspects of our notation. By $\mathbf{x}_B$ we mean the vector consisting of those components of $\mathbf{x}$ that correspond to the columns of B. By $\mathbf{x}_N$ we mean the vector consisting of those components of $\mathbf{x}$ that correspond to the columns of N. By ξ_B we mean the *value* of the basic variables in a basic solution. By ξ we mean the *value* of a solution of (1) and by ξ_N we mean the value of a particular $\mathbf{x}_N$.

We also call attention to a change in terminology. Up to now we have used the term *solution* to mean a solution of the linear programming problem. We are now using it to mean a solution of (1). Henceforth we shall use the term *optimal solution* to designate a solution of the linear programming problem, that is, a *feasible point* at *which the maximum is attained.*

Remark 3.1. It follows from the preceding and from Theorem 2.1 that a basic feasible solution is an extreme point and conversely. It follows from Theorem 2.2 that if an optimal solution exists, then there exists a basic feasible solution that is an *optimal solution.* Therefore, we need only consider basic feasible solutions in our search for the optimal solution.

To apply the simplex method, we reformulate the problem stated in (1) of Section 2 by introducing a scalar variable z and writing the problem as

$$\begin{aligned} &\text{Maximize } z \\ &\text{subject to } A\mathbf{x} = \mathbf{c}, \\ &\quad \langle \mathbf{b}, \mathbf{x}\rangle - z = 0, \\ &\quad \mathbf{x} \geqslant \mathbf{0}. \end{aligned} \tag{4}$$

The simplex method consists of two phases. In phase I the method either determines a basic feasible solution or determines that the feasible set is empty. In phase II the method starts with a basic feasible solution and either determines that no optimal solution exists or determines an optimal solution. In the latter case it does so by moving from one basic feasible solution to another in such a way that the value of z is not decreased.

We conclude this section with a review of some aspects of linear algebra involving the solution of systems of linear equations and the calculation of inverse of matrices. These aspects of linear algebra arise in the simplex method.

Consider a system of linear equations

$$M\mathbf{y} = \mathbf{d}, \tag{5}$$

where M is a $k \times n$ matrix of rank k, $\mathbf{y}$ is a vector in $\mathbb{R}^n$, and $\mathbf{d}$ is a vector in $\mathbb{R}^k$. A system of linear equations $M'\mathbf{y} = \mathbf{d}'$ is said to be *equivalent* to (5) if the solution sets of both equations are equal. We define an *elementary operation* on the system (5) to be any one of the following operations.

(i) Interchange two equations.
(ii) Multiply an equation by a nonzero constant.
(iii) Add a nonzero multiple of an equation to another equation.

Any sequence of elementary operations applied to (5) will yield a system equivalent to (5).

Let R denote a $k \times k$ submatrix of M that has rank k and let S denote the submatrix of M formed by the remaining $n - k$ columns. Then we may write (5) as

$$(R, S)\begin{pmatrix} \mathbf{y}_r \\ \mathbf{y}_s \end{pmatrix} = \mathbf{d}, \tag{6}$$

where $\mathbf{y}_r$ is the vector obtained from $\mathbf{y}$ by taking the components of $\mathbf{y}$ that correspond to the columns in R and $\mathbf{y}_s$ has an analogous meaning. By an appropriate sequence of elementary operations, we can obtain a system equivalent to (6) having the form

$$(U, V)\begin{pmatrix} \mathbf{y}_r \\ \mathbf{y}_s \end{pmatrix} = \mathbf{d}', \tag{7}$$

where U is a $k \times k$ upper triangular matrix with diagonal entries equal to 1 and V is a $k \times (n - k)$ matrix. To simplify notation, assume that R consists of the first k columns of M. Then (7) in component form reads

$$\begin{aligned} y_1 + \alpha_{12}y_2 + \cdots + \alpha_{1k}y_k + \alpha_{1,k+1}y_{k+1} + \cdots + \alpha_{1n}y_n &= d'_1, \\ y_2 + \cdots + \alpha_{2k}y_k + \alpha_{2,k+1}y_{k+1} + \cdots + \alpha_{2n}y_n &= d'_2, \\ &\vdots \\ y_k + \alpha_{k,k+1}y_{k+1} + \cdots + \alpha_{kn}y_n &= d'_k. \end{aligned} \tag{8}$$

The last equation in (8) expresses y_k in terms of $y_{k+1}, \ldots, y_n$. Proceeding upward through the system (8) by backward substitution, we can express $y_1, \ldots, y_k$ in terms of $y_{k+1}, \ldots, y_n$ and thus obtain the general solution of (7) and hence of (5). The method of solving (5) just described is called *Gauss elimination.*

In solving the system (5) or (6) we need not write down the equations and the variables. We need only to write down the augmented matrices $(M, \mathbf{d})$ or $(R, S, \mathbf{d})$ and apply the elementary operations to the rows of the augmented matrices. In so doing we speak of elementary row operations. The reader will recall, or can verify, that an elementary row operation on a $k \times n$ matrix M can be effected in the following fashion. Perform the elementary operation on the $k \times k$ identity matrix to obtain a matrix E. Then premultiply the matrix M by E. The matrices E are called *elementary matrices.* In computer programs for solving systems of linear equations by Gauss elimination, elementary row operations are executed by premultiplications by elementary matrices. Such computer programs also incorporate rules for choosing the components $\mathbf{y}_r$ of $\mathbf{y}$ so as to increase accuracy and speed of computation. We shall not take up these questions here.

By an additional sequence of elementary row operations, we can obtain a system equivalent to (7) having the form

$$(I_k, W)\begin{pmatrix} \mathbf{y}_r \\ \mathbf{y}_s \end{pmatrix} = \mathbf{d}'', \tag{9}$$

where I_k is the $k \times k$ identity matrix. We can rewrite (9) as

$$I_k \mathbf{y}_r = d'' - W\mathbf{y}_s \tag{10}$$

and thus obtain $\mathbf{y}_r$ in terms of $\mathbf{y}_s$ directly, without backward substitution. The method of solving (5) by reducing it to the equivalent system (9) is known as *Gauss–Jordan elimination.* For most large systems, computer implementations of Gauss elimination are superior to those of Gauss–Jordan elimination.

We next recall an effective way of calculating the inverse of a nonsingular matrix R. We first note that an elementary matrix E is nonsingular. Since R is nonsingular, we can reduce R to the identity matrix by applying a sequence of elementary row operations. But each elementary row operation is effected by a premultiplication by an elementary matrix. Thus, we have a sequence E_p, $E_{p-1}, \ldots, E_1$ of elementary matrices such that

$$E_p \ldots E_2 E_1 R = I.$$

Multiplying this equation through on the right by R^{-1} gives

$$E_p \ldots E_1 I = R^{-1}. \tag{11}$$

This is the justification for the following method of calculating R^{-1} for matrices of low dimension. Reduce the matrix (R, I) to the form (I, T) by a sequence of elementary operations. Then $T = R^{-1}$. The so-called product form of the inverse will also be used in the simplex method.

Remark 3.2. In reducing (6) to (9), we performed a sequence of elementary operations, which in effect was a sequence of premultiplications of (6) by elementary matrices. Thus,

$$E_p \ldots E_1(R, S, \mathbf{d}) = (I_k, W, \mathbf{d}''). \tag{12}$$

Since the sequence of elementary operations reduced R to I_k, we have that $R^{-1} = E_p, \ldots, E_1 I$. Hence, from (12) we have

$$W = R^{-1}S \quad \text{and} \quad \mathbf{d}'' = R^{-1}\mathbf{d}.$$

Substituting this into (10) gives

$$\mathbf{y}_r = R^{-1}\mathbf{d} - R^{-1}S\mathbf{y}_s.$$

4. PHASE II OF SIMPLEX METHOD

Our aim in this and the next section is to explain the essentials of the simplex method; it is not to present the most current computer implementations of the simplex method. In Section 7, where we discuss the revised simplex method, we shall go into greater detail concerning the computations in current computer codes. For more complete discussions of computational questions see Dantzig [1963], Chvátal [1983], and Gass [1985].

We suppose that the problem is formulated as in relation (4) of Section 3. Phase I of the simplex, which we will describe in Section 6, either determines that the feasible set is empty or determines a basic feasible solution. We begin our discussion of phase II with the assumption that by means of phase I, or otherwise, we have obtained a basic feasible solution $\boldsymbol{\xi} = (\boldsymbol{\xi}_B, \mathbf{0}_N)$, where $\mathbf{0}_N$ is the zero vector in $\mathbb{R}^{n-m}$. We denote the value of the objective function at $\boldsymbol{\xi}$ by ζ. Thus, $\zeta = \langle \mathbf{b}_B, \boldsymbol{\xi}_B \rangle$. We call $\boldsymbol{\xi} = (\boldsymbol{\xi}_B, \mathbf{0}_N)$ the *current basic feasible solution*, the matrix B associated with $\boldsymbol{\xi}_B$ the *current basis*, and ζ the *current value* of the objective function. We abuse the precept for good terminology and also call the components of $\boldsymbol{\xi}_B$ the *current basis*. The simplex method consists of a sequence of iterations. In a given iteration we determine either that the current basic feasible solution is optimal or that there is no optimal solution or we determine a new basic feasible solution that does not decrease the value of the objective function. In the last event, we repeat the iteration, with the new basic feasible solution as the current one. Each iteration consists of four steps. We shall illustrate the process using Example 1.1.

Step 1. Express z and the basic variables $\mathbf{x}_B$ in terms of the nonbasic variables $\mathbf{x}_N$.

The equation in (4) of Section 3 can be written as

$$\begin{pmatrix} B & N & \mathbf{0}_m \\ \mathbf{b}_B^t & \mathbf{b}_N^t & -1 \end{pmatrix} \begin{pmatrix} \mathbf{x}_B \\ \mathbf{x}_N \\ z \end{pmatrix} = \begin{pmatrix} \mathbf{c} \\ 0 \end{pmatrix}, \tag{1}$$

where $\mathbf{0}_m$ is the zero vector in $\mathbb{R}^m$. If we multiply the augmented matrix corresponding to the first m equations in (1) by B^{-1} on the left, we see that (1) is equivalent to

$$\begin{pmatrix} I_m & B^{-1}N & \mathbf{0}_m \\ \mathbf{b}_B^t & \mathbf{b}_N^t & -1 \end{pmatrix} \begin{pmatrix} \mathbf{x}_B \\ \mathbf{x}_N \\ z \end{pmatrix} = \begin{pmatrix} B^{-1}\mathbf{c} \\ 0 \end{pmatrix}, \tag{2}$$

where I_m is the $m \times m$ identity, Setting $\mathbf{x}_B = \xi_B$ and $\mathbf{x}_N = \mathbf{0}_N$ in (2) gives

$$\xi_B = B^{-1}\mathbf{c}. \tag{3}$$

From (2) and (3) we get that

$$\mathbf{x}_B = \xi_B - B^{-1}N\mathbf{x}_N. \tag{4}$$

From (2) and (4) we get that

$$z = \langle \mathbf{b}, \xi_B \rangle + \langle \mathbf{b}_N - \mathbf{b}_B^t B^{-1}N, \mathbf{x}_N \rangle.$$

Setting $\mathbf{x}_N = \mathbf{0}_N$ gives

$$\zeta = \langle \mathbf{b}_B, \xi_B \rangle \tag{5}$$

and so

$$\mathbf{z} = \zeta + \langle \mathbf{b}_N - \mathbf{b}_B^t B^{-1}N, \mathbf{x}_N \rangle. \tag{6}$$

Equations (4) and (6) express $\mathbf{x}_B$ and z in terms of $\mathbf{x}_N$.

Let

$$\mathbf{d}_N = \mathbf{b}_N - \mathbf{b}_B^t B^{-1}N = (d_{j_{m+1}}, \ldots, d_{j_n}), \tag{7}$$

where $j_{m+1}, \ldots, j_n$ are the indices of the components of $\mathbf{x}$ that constitute $\mathbf{x}_N$. Then we may write (6) as

$$z = \zeta + \langle \mathbf{d}_N, \mathbf{x}_N \rangle = \zeta + \sum_{p=m+1}^{n} d_{i_p} x_{i_p}. \tag{8}$$

Remark 4.1. Note that by virtue of (3) the right-hand side of (2) can be written as $(\xi, 0)^t$.

The difficult calculation in (4) and (6) is the calculation of B^{-1}. The other calculations involve straightforward matrix and vector–matrix multiplications. We now present a second, seemingly different, method of obtaining $\mathbf{x}_B$ and z in terms of $\mathbf{x}_N$. Computer implementation of the second method turns out in essence to be the same as the computer implementation of the previous method. Our use of the second method is didactic. For small-scale problems solved by hand this method is advantageous and will be the one used in our illustrative example. The *tableau method* is a method that systematically records the sequence of steps and calculations that we shall describe.

We again start with the augmented matrix of (1). By a sequence of elementary operations applied to the first m rows of (1), we can transform this matrix to an equivalent matrix

$$\begin{pmatrix} I_m & D & \mathbf{0}_m & \mathbf{c}' \\ \mathbf{b}_B^t & \mathbf{b}_N^t & -1 & 0 \end{pmatrix}. \tag{9}$$

Each of these operations can be effected by a premultiplication by an elementary matrix E_i. Since the sequence of operations transforms B to I, the product of the elementary matrices associated with these transformations is equal to B^{-1}, as was shown in the preceding section. Thus, the matrix (9) can also be obtained by a premultiplication of the first m rows of the augmented coefficient matrix (1) by the matrix B^{-1}. Hence, $D = B^{-1}N$ and $\mathbf{c}' = B^{-1}\mathbf{c}$, and we have, as before, that (1) is equivalent to a system with augmented matrix

$$\begin{pmatrix} I_m & B^{-1}N & \mathbf{0}_m & \mathbf{c}' \\ \mathbf{b}_B^t & \mathbf{b}_N^t & -1 & 0 \end{pmatrix}, \tag{10}$$

where $\mathbf{c}' = B^{-1}\mathbf{c}$. From (10) we again get (3) and (4). Let us now suppose that the columns of B are the columns $j_1, \ldots, j_m$ of A. If we successively multiply row 1 of (10) by $-b_{j_1}$ and add the result to the last row, multiply row 2 of (10) by $-b_{j_2}$ and add the result to the last row, and so on, then we transform (10) into

$$\begin{pmatrix} I_m & B^{-1}N & \mathbf{0}_m & \mathbf{c}' \\ \mathbf{0}_B & \mathbf{b}_N^t - \mathbf{b}_B^t B^{-1}N & -1 & -\langle \mathbf{b}_B, \mathbf{c}' \rangle \end{pmatrix}. \tag{11}$$

From the last row of this matrix and from the relation $\mathbf{c}' = B^{-1}\mathbf{c}$, we immediately obtain

$$z = \langle \mathbf{b}_B, B^{-1}\mathbf{c} \rangle + \langle \mathbf{b}_N^t - \mathbf{b}_B^t B^{-1}N, \mathbf{x}_N \rangle.$$

Remark 4.2. From (11) and $\mathbf{c}' = B^{-1}\mathbf{c}$ we again get (3) and (5). Consequently, the right-hand side of (11) can be written as $(\xi, -\zeta)^t$. From this relation and from (3) and (5) we again get (6).

We illustrate the preceding description of Step 1 using the second method.

Example 4.1. We first introduce slack variabless x_3, x_4, x_5 and write the problem as in (2) of Section 1. We then write the problem as a *maximization problem* in the format of (4) of Section 3.

Maximize z subject to

$$\begin{aligned} -x_1 - x_2 + x_3 \qquad\qquad &= -1, \\ -x_1 + x_2 \qquad + x_4 \qquad &= 0, \\ x_1 + 2x_2 \qquad\qquad + x_5 &= 4, \\ -x_1 + 2x_2 \qquad\qquad\qquad - z &= 0. \end{aligned} \tag{12}$$

It is readily verified that $\xi = (1, 0, 0, 1, 3)$, which corresponds to the point $(1, 0)$ in Figure 6.1, is a basic feasible solution. The value of z at this point is -1. Thus, x_1, x_4, x_5 are the basic variables and x_2, x_3 are the nonbasic variables. The augmented coefficient matrix of (12) is

$$\begin{matrix} \\ x_1 \\ x_4 \\ x_5 \\ \\ \end{matrix} \begin{pmatrix} x_1 & x_2 & x_3 & x_4 & x_5 & z & \mathbf{c} \\ -1 & -1 & 1 & 0 & 0 & 0 & -1 \\ -1 & 1 & 0 & 1 & 0 & 0 & 0 \\ 1 & 2 & 0 & 0 & 1 & 0 & 4 \\ -1 & 2 & 0 & 0 & 0 & -1 & 0 \end{pmatrix}. \tag{13}$$

Above each column of the matrix we have indicated the variable corresponding to the column. At the left of the matrix we have listed the current basic variables. The last column is the column of right-hand sides of (12). In the tableau description of the simplex method the matrix (13) is the initial tableau. The matrix B in (13) is the submatrix formed by the first three rows and columns 1, 4, and 5. Thus

$$B = \begin{pmatrix} -1 & 0 & 0 \\ -1 & 1 & 0 \\ 1 & 0 & 1 \end{pmatrix}$$

To obtain the matrix corresponding to (11), we must first transform B to the identity matrix by a sequence of elementary row operations on (13). From the form of B it is clear that the most effective sequence of operations is the following. (i) Multiply row 1 by -1. (ii) Add the transformed row 1 to row 2.

(iii) Add the negative of the transformed row 1 to row 3. To obtain the row corresponding to the last row of (11), add the transformed row 1 to row 4. We thus obtain

$$\begin{array}{c} \\ x_1 \\ x_4 \\ x_5 \\ \\ \end{array} \begin{pmatrix} x_1 & x_2 & x_3 & x_4 & x_5 & z & \\ 1 & 1 & -1 & 0 & 0 & 0 & 1 \\ 0 & 2 & -1 & 1 & 0 & 0 & 1 \\ 0 & 1 & 1 & 0 & 1 & 0 & 3 \\ 0 & 3 & -1 & 0 & 0 & -1 & 1 \end{pmatrix}. \tag{14}$$

From this we read off

$$\begin{aligned} x_1 &= 1 - x_2 + x_3, \\ x_4 &= 1 - 2x_2 + x_3, \\ x_5 &= 3 - x_2 - x_3, \end{aligned} \tag{15}$$

and

$$z = -1 + 3x_2 - x_3, \tag{16}$$

which are (4) and (6) for this example. Note that the last column of (14) gives the current values of ξ_1, ξ_4, $\xi_5 - \zeta$ in agreement with (4) and (6) and Remark 4.2.

Step 2. Check the current basic feasible solution for optimality.

If every component of the vector $\mathbf{d}_N$ defined in (7) is nonpositive (i.e., $\leqslant 0$), then the current basic feasible solution is optimal and the process terminates.

To see that this is so, recall that by Remark 3.1 we need only consider basic feasible solutions in our search for the optimal solution. The current basic feasible solution is the unique basic feasible solution with $\mathbf{x}_N = \mathbf{0}_N$. Thus any other basic feasible solution will have at least one of the components of $\mathbf{x}_N$ positive. But since $d_N \leqslant \mathbf{0}$, we have from (8) that the value of the objective function will either decrease or be unchanged. Thus the current basic feasible solution is optimal.

In our example, we see from (16) that the current solution is not optimal.

Step 3. If the current basic feasible solution is not optimal, either determine a new basis and corresponding basic feasible solution that does not decrease the current value of z or determine that z is unbounded and that therefore the problem has no solution.

We first motivate Step 3 using Example 4.1.

From (16) we see that if we keep the value of x_3 at zero and increase the value of x_2, then we shall increase the value of z. The variables x_1, x_2, x_3, x_4, x_5 must satisfy (15) and the condition $x_i \geqslant 0$, $i = 1, \ldots, 5$, for feasibility. Setting $x_3 = 0$ in (15) and taking into account the requirement that x_1, x_4, and x_5 must be nonnegative give the following inequalities that x_2 must satisfy:

$$1 - x_2 \geqslant 0, \qquad 1 - 2x_2 \geqslant 0, \qquad 3 - x_2 \geqslant 0.$$

For all three of these inequalities to hold we can only increase the value of x_2 to $\frac{1}{2}$. If $x_2 = \frac{1}{2}$ and $x_3 = 0$, then from (15) we get that $x_1 = \frac{1}{2}$, $x_4 = 0$, and $x_5 = \frac{5}{2}$. We now have another feasible solution $(\frac{1}{2}, \frac{1}{2}, 0, 0, \frac{5}{2})$. This corresponds to the point $(\frac{1}{2}, \frac{1}{2})$ in Figure 6.1. The value of z is increased from -1 to $\frac{1}{2}$.

We next show that this solution is basic. The matrix corresponding to the system (15) is the submatrix of (14) consisting of the first three rows of (14). The matrix B corresponding to $(\frac{1}{2}, \frac{1}{2}, 0, 0, \frac{5}{2})$ is the matrix formed by taking columns 1, 2, and 5 of this submatrix. Thus

$$B = \begin{pmatrix} 1 & 1 & 0 \\ 0 & 2 & 0 \\ 0 & 1 & 1 \end{pmatrix}.$$

Since the columns of B are linearly independent, the point $(\frac{1}{2}, \frac{1}{2}, 0, 0, \frac{5}{2})$ is a basic feasible solution.

In summary, the original basic variable x_4 whose value was 1 is now a nonbasic variable and the original nonbasic variable x_2 is now a basic variable with value $\frac{1}{2}$. The variable x_4 is said to be the *leaving variable* or to *leave the basis* and the variable x_2 is said to be the *entering variable* or to *enter the basis.*

Before presenting Step 3 in general, we call attention to another possible situation. Suppose that we have an example such that instead of (15) we arrived at a system

$$\begin{aligned} x_1 &= 1 + \alpha x_2 + x_3, \\ x_4 &= 1 + \beta x_2 + x_3, \\ x_5 &= 3 + \gamma x_2 + x_3, \end{aligned}$$

with z as in (16) and where α, β, and γ are all nonnegative. If we set $x_3 = 0$ and let $x_2 = t$, then we obtain feasible solutions for all $t > 0$. If we let $t \to \infty$, we get that $z \to \infty$. Thus, the objective function is unbounded and the problem has no solution.

We now describe Step 3, which we break up into two substeps, for the general problem.

Step 3(i). Choose the nonbasic variable that will enter the new basis.

Let $\mathbf{d}_N$ be as in (7). Since the current basic feasible solution is not optimal, at least one of the components of $\mathbf{d}_N$ is positive. Choose the component of $\mathbf{d}_N$ that is the largest. If there is more than one such component, choose the one with the smallest index. Suppose that the component of $\mathbf{d}_N$ that is chosen is the qth component of $\mathbf{x} = (x_1, \ldots, x_n)$. The variable x_q will enter the basis. The column of A corresponding to x_q is $\mathbf{A}_q$.

Step 3(ii). Determine that the problem has no solution or choose the basic variable that will leave the basis.

Let $\mathbf{v}_q$ be the column of $B^{-1}N$ corresponding to x_q. Then $\mathbf{v}_q = B^{-1}\mathbf{A}_q$. Let

$$\mathbf{x}_B = (x_{i_1}, \ldots, x_{i_m}) \qquad \boldsymbol{\xi}_B = (\xi_{i_1}, \ldots, \xi_{i_m}) \tag{17}$$

and let $\mathbf{v}_q = (v_{1q}, \ldots, v_{mq})$. Set $x_q = t$, where $t \geqslant 0$ and let the other components of $\mathbf{x}_N$ remain at zero. For the resulting vector to be a feasible solution, $\mathbf{x}_B$ must satisfy the conditions (4) and $\mathbf{x}_B \geqslant \mathbf{0}$. Equation (4) becomes

$$\mathbf{x}_B = \boldsymbol{\xi}_B - t\mathbf{v}_q. \tag{18}$$

Thus, the condition $\mathbf{x}_B \geqslant \mathbf{0}$ imposes the restrictions

$$tv_{rq} \leqslant \xi_{i_r}, \qquad r = 1, \ldots, m. \tag{19}$$

From (8) we get that

$$z = \zeta + d_q t. \tag{20}$$

Since $\boldsymbol{\xi}$ is feasible, $\boldsymbol{\xi}_B \geqslant \mathbf{0}$, and so $\xi_{i_r} \geqslant 0$ for all $r = 1, \ldots, m$. Hence if $v_{rq} \leqslant 0$ for all $r = 1, \ldots, m$, then (18) and (19) hold for all $t \geqslant 0$. Since $d_q > 0$, it follows that z is unbounded above and the problem has no solution.

If not all of the v_{rq} are nonpositive, let

$$t^* = \min\left\{\frac{\xi_{i_r}}{v_{rq}} : r = 1, \ldots, m \text{ and } v_{rq} > 0\right\}. \tag{21}$$

Let ρ denote the smallest index r at which the minimum is achieved. Let $x_q = t^*$ and let

$$\begin{aligned} \xi'_{i_r} &= \xi_{i_r} - t^* v_{rq}, \qquad i_r \neq i_\rho, \\ \xi'_q &= t^* \\ \xi'_{i_\rho} &= 0, \\ \xi'_i &= 0 \quad \text{all other indices.} \end{aligned} \tag{22}$$

Let $\boldsymbol{\xi}'$ denote the vector whose coordinates are given by (22). Thus

$$\boldsymbol{\xi}' = (\xi'_{i_1}, \ldots, \xi'_{i_{\rho-1}}, 0, \xi'_{i_{\rho+1}}, \ldots, \xi'_{i_m}, 0, \ldots, 0, t^*, 0, \ldots, 0),$$

where t^* is the qth coordinate of $\boldsymbol{\xi}'$. It follows from the way $\boldsymbol{\xi}'$ was constructed that $\boldsymbol{\xi}'$ is a feasible solution. We shall show that $\boldsymbol{\xi}'$ is also basic.

We can write $\boldsymbol{\xi}'$ as $\boldsymbol{\xi}' = (\xi'_{B'}, \xi'_N)$ with

$$\boldsymbol{\xi}'_{B'} = (\xi'_{i_1}, \ldots, \xi'_{i_{\rho-1}}, t^*, \xi'_{i_{\rho+1}}, \ldots, \xi'_{i_m}), \qquad \boldsymbol{\xi}_{N'} = \mathbf{0}.$$

We shall show that the coordinate variables of

$$\mathbf{x}_{B'} = (x_{i_1}, \ldots, x_{i_{\rho-1}}, x_q, x_{i_{\rho+1}}, \ldots, x_{i_m}) \tag{23}$$

are a basis. Since $\mathbf{x}_{B'}$ is obtained from $\mathbf{x}_B$ by replacing x_{i_ρ} by x_q, this will also justify our saying that x_q enters the basis and x_{i_ρ} leaves the basis.

From (17) we get that the matrix I_m in (2) and (9) is

$$(\mathbf{e}_{i_1} \ldots \mathbf{e}_{i_{\rho-1}} \mathbf{e}_{i_\rho} \mathbf{e}_{i_{\rho+1}} \ldots \mathbf{e}_{i_m}),$$

where $\mathbf{e}_{i_j}$ is the vector whose i_jth component is 1 and whose other components are zero. Let

$$B' = (\mathbf{e}_{i_1} \ldots \mathbf{e}_{i_{\rho-1}} \mathbf{v}_q \mathbf{e}_{i_{\rho+1}} \ldots \mathbf{e}_{i_m}).$$

That is, B' is the matrix obtained from I_m by replacing the i_ρth column by $\mathbf{v}_q$. Let N' be the matrix obtained from $B^{-1}N$ by replacing column $\mathbf{v}_q$ by $\mathbf{e}_{i_p}$. Let $\mathbf{x}_{B'}$ be as in (23) and let $\mathbf{x}_{N'}$ be the vector obtained from $\mathbf{x}_N$ by replacing x_q by x_{i_p}. Let d_q denote the qth coordinate of the vector $\mathbf{d}_N$ defined in (7) and let $\mathbf{d}_{N'}$ denote the vector obtained from $\mathbf{d}_N$ by replacing the qth coordinate of $\mathbf{d}_N$ by 0. The system of equations

$$\begin{pmatrix} \mathbf{e}_{i_1} & \cdots & \mathbf{e}_{i_{p-1}} & \mathbf{v}_q & \mathbf{e}_{i_{p+1}} & \cdots & \mathbf{e}_{i_m} & N' & \mathbf{0}_m \\ 0 & \cdots & 0 & d_q & 0 & \cdots & 0 & \mathbf{d}_{N'} & -1 \end{pmatrix} \begin{pmatrix} \mathbf{x}_{B'} \\ \mathbf{x}_{N'} \\ z \end{pmatrix} = \begin{pmatrix} B^{-1}\mathbf{c} \\ -\zeta \end{pmatrix} \tag{24}$$

is equivalent to the system whose augmented matrix is (11), since (24) is obtained from (11) by a reordering of the variables.

We assert that the matrix B' is a basis and hence that the variables $\mathbf{x}_{B'}$ are basic. To prove this, we must show that the columns of B' are linearly independent. By (21) and the definition of the index ρ, the i_ρth component of $\mathbf{v}_q$ is positive. The i_ρth component of each of the other columns of B' is zero. Hence $\mathbf{v}_q$ cannot be written as a linear combination of the other columns of B'. Since the remaining columns of B' are linearly independent, it follows that all the columns of B' are linearly independent. Hence the coordinate variables of

$\mathbf{x}_{B'}$ are a basis, and the vector $\boldsymbol{\xi}'$ is a basic feasible solution. From (20) we get that

$$\zeta' = \zeta + d_q t^*.$$

Since $d_q \geqslant 0$ and $t \geqslant 0$, it follows that the value of the objective function is not decreased.

We say that in Step 3 we have *updated* the unprimed quantities to the primed quantities. The primed, or updated, quantities are the "new" or "updated current" quantities.

Remark 4.3. In Step 3(i) we choose the entering variable x_q to be the variable corresponding to the largest component of $\mathbf{d}_N$. Upon examining Step 3, it should be clear that the choice of entering variable could be any variable corresponding to a positive component of $\mathbf{d}_N$. The choice of largest component was motivated by the fact that at a given iteration this choice would appear to result in the largest increase in the value of z.

Step 4. Return to Step 1 using the updated current quantities and start the next iteration.

We now illustrate Steps 3 and 4 using Example 4.1.

Step 3(i). From (14) we see that $\mathbf{d}_N = (d_2, d_3)$ and that $d_2 > 0$ and $d_3 < 0$. Thus, we take x_2 as the entering variable. The entering column $\mathbf{v}_q$ is the second column of (14), and $d_q = d_2 = 3$.

Step 3(ii). By Remark 4.2, the entries in the last column of (14) give

$$(\xi_1, \xi_4, \xi_5) = (\xi_{i_1}, \xi_{i_2}, \xi_{i_3}) = (1, 1, 3), \qquad \zeta = -1.$$

From this and the fact that the first three entries in column 2 are positive, we get that

$$\frac{\xi_{i_1}}{v_1} = \frac{\xi_1}{v_1} = 1, \qquad \frac{\xi_{i_2}}{v_2} = \frac{\xi_4}{v_2} = \frac{1}{2}, \qquad \frac{\xi_{i_3}}{v_3} = \frac{\xi_5}{v_3} = 3.$$

Hence $t^* = \frac{1}{2}$ and x_4 is the leaving variable. The new, or updated, basic variables are (x_1, x_2, x_5) and the updated nonbasic variables are x_3 and x_4. From (22) we get the updated basic feasible solution:

$$\xi'_{i_1} = \xi'_1 = 1 - \tfrac{1}{2}(1) = \tfrac{1}{2}, \qquad \xi'_q = \xi'_2 = \tfrac{1}{2}$$

$$\xi'_3 = 0, \qquad \xi'_{i_2} = \xi'_4 = 0, \qquad \xi'_{i_3} = \xi'_5 = 3 - \tfrac{1}{2}(1) = \tfrac{5}{2}.$$

From (20) we get

$$\zeta' = (-1) + 3(\tfrac{1}{2}) = \tfrac{1}{2}.$$

These calculations, however, are *unnecessary* as we will get them automatically in Step 1 of the next iteration.

Step 4. We go to Step 1 and start the next iteration using the updated data. We do so dropping the primes that indicated updated data. The variables $x_1, \ldots, x_5, z$ satisfy the system of equations represented by (14). We rewrite (14) with the labeling of the new basic variables:

$$\begin{array}{c} \\ x_1 \\ x_2 \\ x_5 \\ \\ \end{array} \begin{pmatrix} x_1 & x_2 & x_3 & x_4 & x_5 & z & \\ 1 & 1 & -1 & 0 & 0 & 0 & 1 \\ 0 & 2 & -1 & 1 & 0 & 0 & 1 \\ 0 & 1 & 1 & 0 & 1 & 0 & 3 \\ 0 & 3 & -1 & 0 & 0 & -1 & 1 \end{pmatrix}. \tag{25}$$

Step 1 requires us to transform (25) into a form corresponding to (11). The matrix B is now

$$B = \begin{pmatrix} 1 & 1 & 0 \\ 0 & 2 & 0 \\ 0 & 1 & 1 \end{pmatrix}.$$

Thus, to achieve the desired result, we apply the following sequence of transformations to (25).

(i) Multiply row 2 by $\frac{1}{2}$.
(ii) Add the negative of the new row 2 to row 1.
(iii) Add the negative of the new row 2 to row 3.
(iv) Multiply the new row 2 by -3 and add the resulting row to row 4.

We get

$$\begin{array}{c} \\ x_1 \\ x_2 \\ x_5 \\ \\ \end{array} \begin{pmatrix} x_1 & x_2 & x_3 & x_4 & x_5 & z & \\ 1 & 0 & -\frac{1}{2} & -\frac{1}{2} & 0 & 0 & \frac{1}{2} \\ 0 & 1 & -\frac{1}{2} & \frac{1}{2} & 0 & 0 & \frac{1}{2} \\ 0 & 0 & \frac{3}{2} & -\frac{1}{2} & 1 & 0 & \frac{5}{2} \\ 0 & 0 & \frac{1}{2} & -\frac{3}{2} & 0 & -1 & -\frac{1}{2} \end{pmatrix}. \tag{26}$$

From (26) we can read off, if we desire, the basic variables $(x_1, x_2, x_5) = (x_{i_1}, x_{i_2}, x_{i_3})$ and z in terms of the nonbasic variables (x_3, x_4). By Remark 4.3 the last column of (26) gives

$$\xi_1 = \tfrac{1}{2}, \qquad \xi_2 = \tfrac{1}{2}, \qquad \xi_5 = \tfrac{5}{2}, \qquad \zeta = \tfrac{1}{2}$$

for $\boldsymbol{\xi}_B$, the current basic feasible solution, and ζ, the current value of the objective functions. Note that we have agreement with the previous calculation of these quantities. The current basic feasible solution corresponds to the point $(\frac{1}{2}, \frac{1}{2})$ in Figure 6.1. The current value $\zeta = \frac{1}{2}$ is greater than the previous current value of -1.

Before proceeding to Step 2, we introduce some terminology and comment on the carrying out of Step 1 in general. The entry a_{22} in row 2, column 2 in this example is called the *pivot* and the position in the matrix is called the *pivot position.* In performing the sequence of elementary operations used to obtain (26) from (25), we say that we have *pivoted on the element* a_{22}.

In our example the pivot was the entry in the row and column of the entering variable. We now show that in each iteration after the initial iteration we always pivot on the element in the row and column of the entering variable. The system (24) gives the relationship among the variables $\mathbf{x}_{B'}$, $\mathbf{x}_{N'}$, and z at the end of Step 4 of each iteration. It expresses z and each component of

$$\mathbf{x}_{B'} = (x_{i_1} \ldots x_{i_{p-1}}, x_q, x_{i_{p+1}} \ldots x_{i_m})$$

except x_q in terms of the components of $\mathbf{x}_{N'}$ and the variable x_q. Step 1 of the current iteration requires that $\mathbf{x}_{B'}$ and z be expressed in terms of the components of $\mathbf{x}_{N'}$. This will be achieved if we express x_q in terms of the components of $\mathbf{x}_{N'}$ and do so in a way that does not destroy the already established expressions of z and the x_{i_j}, $j = 1, \ldots, m$, $j \neq \rho$, in terms of x_q and $x_{N'}$. We assert that pivoting on row i_ρ, the entering row, and on column i_ρ (column $\mathbf{v}_q$), the entering column, achieves this.

By the definition of ρ and (21), we have that $v_{i_\rho} > 0$. Hence in (24) we may multiply row i_ρ by $1/v_{i_\rho}$ and obtain a row with entry 1 in the (i_ρ, i_ρ) position. If we now multiply row i_ρ by $-v_{1q}$ and add to row 1, the resulting first row will have a zero in the i_ρth position; that is, the column vector $\mathbf{v}_q$ now has a 1 in the i_ρ entry and a zero in the first position. Since each of the vectors $\mathbf{e}_{i_j}$, $j = 1, \ldots, m, j \neq \rho$, has a zero in the i_ρth entry, the first entries of these vectors are not changed. Similarly, if for each $i = 2, \ldots, m$, $i \neq \rho$, we multiply row i_ρ by $-v_{iq}$ and add the resulting row to row i, then the resulting ith row will have a zero in the i_ρth position. Each of the vectors $\mathbf{e}_{i_j}$, $j = 1, \ldots, m, j \neq \rho$, will have ith entries unchanged. Finally, if we multiply row i_ρ by $-d_q$ and add the resulting row to the last row of (24), then we shall get a system

$$\begin{pmatrix} I_m & N'' & \mathbf{0}_m \\ \mathbf{0}_{B'} & \mathbf{d}''_N & -1 \end{pmatrix} \begin{pmatrix} \mathbf{x}_{B'} \\ \mathbf{x}_{N'} \\ z \end{pmatrix} = \begin{pmatrix} \mathbf{c}'' \\ -\zeta' \end{pmatrix}.$$

This system expresses $\mathbf{x}_{B'}$ and z in terms of $\mathbf{x}_{N'}$.

We now return to Step 2 and (26) of the current iteration. Since $\mathbf{d}_N = (\frac{1}{2}, -\frac{3}{2})$, the current basic feasible solution is not optimal. The current value $\zeta = \frac{1}{2}$ is an improvement, however.

Step 3(i). Since $d_3 > 0$ and $d_4 < 0$, we take x_3 to be the entering variable.

Step 3(ii). The column corresponding to x_3 is the third column of (26). Since the only positive entry in this column is the entry in row 3 corresponding to x_5, the variable x_5 will be leaving variable. We rewrite the matrix (26) with the updated basic variables indicated on the left:

$$\begin{array}{c} \\ x_1 \\ x_2 \\ x_3 \\ \\ \end{array}\begin{pmatrix} x_1 & x_2 & x_3 & x_4 & x_5 & z & \\ 1 & 0 & -\frac{1}{2} & -\frac{1}{2} & 0 & 0 & \frac{1}{2} \\ 0 & 1 & -\frac{1}{2} & \frac{1}{2} & 0 & 0 & \frac{1}{2} \\ 0 & 0 & \frac{3}{2} & -\frac{1}{2} & 1 & 0 & \frac{5}{2} \\ 0 & 0 & \frac{1}{2} & -\frac{3}{2} & 0 & -1 & -\frac{1}{2} \end{pmatrix}.$$

We now follow Step 4 and begin another iteration with Step 1. We follow our pivot rule and pivot on $\frac{3}{2}$ in row 3, column 3. This will result in a matrix

$$\begin{array}{c} \\ x_1 \\ x_2 \\ x_3 \\ \\ \end{array}\begin{pmatrix} x_1 & x_2 & x_3 & x_4 & x_5 & z & \\ 1 & 0 & 0 & -\frac{2}{3} & \frac{1}{3} & 0 & \frac{4}{3} \\ 0 & 1 & 0 & \frac{1}{3} & \frac{1}{3} & 0 & \frac{4}{3} \\ 0 & 0 & 1 & -\frac{1}{3} & \frac{2}{3} & 0 & \frac{5}{3} \\ 0 & 0 & 0 & -\frac{4}{3} & -\frac{1}{3} & -1 & -\frac{4}{3} \end{pmatrix}.$$

Step 2. $\mathbf{d}_N = (-\frac{4}{3}, -\frac{1}{3})$. All components of $\mathbf{d}_N$ are negative, so the current basic feasible solution is optimal. From the last column we get that

$$\boldsymbol{\xi}_B = (\xi_1, \xi_2, \xi_3) = (\tfrac{4}{3}, \tfrac{4}{3}, \tfrac{5}{3})$$

and that the value of the objective function is $\zeta = \frac{4}{3}$. This solution corresponds to the point $(\frac{4}{3}, \frac{4}{3})$ in Figure 6.1.

5. TERMINATION AND CYCLING

In Section 4 we showed that each iteration of the simplex method results in a value of the objective function that is greater than or equal to the value of the objective function at the preceding iteration. If we could show that each iteration produces a value of the objective function that is strictly greater than the value at the preceding iteration, then the simplex method would terminate at an optimal solution in a finite number of steps, for then no basic feasible solution could appear in different iterations and there are a finite number of basic feasible solutions. Unfortunately, it is not true that each iteration produces a value of the objective function that is strictly greater than the preceding value, as the following simple example shows.

Let us suppose that at the end of Step 2 of an iteration we had arrived at a system given by

$$\begin{array}{c} \\ x_1 \\ x_4 \\ x_5 \\ \\ \end{array}\begin{pmatrix} x_1 & x_2 & x_3 & x_4 & x_5 & z & \\ 1 & 1 & 1 & 0 & 0 & 0 & 1 \\ 0 & 2 & -1 & 1 & 0 & 0 & 0 \\ 0 & 1 & 1 & 0 & 1 & 0 & 3 \\ 0 & 3 & -1 & 0 & 0 & -1 & 1 \end{pmatrix}.$$

Then the entering variable is x_2 and the entering column $\mathbf{v}_q$ is the second column. Recalling that the last column gives $(\boldsymbol{\xi}_B, -\zeta)$, we get that

$$\frac{\xi_{i_1}}{v_1} = \frac{\xi_1}{v_1} = \frac{1}{1}, \qquad \frac{\xi_{i_2}}{v_2} = \frac{\xi_4}{v_2} = \frac{0}{2} = 0, \qquad \frac{\xi_{i_3}}{v_3} = \frac{\xi_5}{v_3} = \frac{3}{1}. \tag{1}$$

Thus $t^* = 0$ and the leaving variable is x_4. It then follows from (4.20) that $\zeta' = \zeta$ and the updated value of z is not changed. From (4.22) we get that the updated basic feasible solution is

$$\boldsymbol{\xi}_{B'} = (\xi'_{i_1}, \xi'_{i_2}, \xi'_{i_3}) = (\xi'_1, \xi'_2, \xi'_5) = (1, 0, 3).$$

Looking at (1), we see that $t^* = 0$ because $\xi_4 = 0$. In other words, $t^* = 0$ is a consequence of the current basic feasible solution being degenerate. In this example, the updated basic solution is also degenerate. In the next example we show that a degenerate basic solution need not lead to another degenerate solution with no increase in the value of z.

Let us now suppose that at the end of Step 2 we arrived at the system described by

$$\begin{array}{c} \\ x_1 \\ x_4 \\ x_5 \\ \\ \end{array}\begin{pmatrix} x_1 & x_2 & x_3 & x_4 & x_5 & z & \\ 1 & 1 & 0 & 0 & 0 & 0 & 1 \\ 0 & -2 & -1 & 1 & 0 & 0 & 0 \\ 0 & 1 & 1 & 0 & 1 & 0 & 3 \\ 0 & 3 & -1 & 0 & 0 & -1 & 1 \end{pmatrix}.$$

The current basic feasible solution $(\xi_1, \xi_4, \xi_5) = (1, 0, 3)$ is again degenerate. The entering variable is again x_2. Now, however, the ratios used to determine t^* are

$$\frac{\xi_{i_1}}{v_1} = \frac{\xi_1}{v_1} = 1, \qquad \frac{\xi_{i_3}}{v_3} = \frac{\xi_5}{v_3} = 3.$$

Thus $t^* = 1$ and the leaving variable is x_1. Since $d_q = 3$, $\zeta' = -1$, and $t^* = 1$, from (4.20) we get

$$\zeta' = -1 + 3(1) = 2.$$

Thus the value of z is increased. From (4.22) we calculate the new basic feasible solution to be

$$\xi_{B'} = (\xi'_{i_1}, \xi'_{i_2}, \xi'_{i_3}) = (\xi'_2, \xi'_4, \xi'_5) = (1, 2, 2),$$

which is nondegenerate.

We now consider the general situation. Let ξ_B be a basic degenerate feasible solution with zero component ξ_{i_p}. If the corresponding component v_p of the leaving column $\mathbf{v}$ is positive, then $t^* = 0$, the updated basic feasible solution will be degenerate, and the updated value of the objective function will be unchanged. This follows from (4.21), (4.22), and (4.20).

The following question now arises: Can there exist a sequence of iterations $I_k, I_{k+1}, \ldots, I_{k+p}$ such that the basic feasible solutions $\xi_{B,k+j}, j = 0, 1, \ldots, p$, are all degenerate and such that $\xi_{B,k+p} = \xi_{B,k}$? If this occurs, then all current values $\zeta_k, \zeta_{k+1}, \ldots, \zeta_{k+p}$ of the objective will be equal and the sequence of degenerate basic solutions $\xi_{B,k}, \xi_{B,k+1}, \ldots, \xi_{B,k+p}$ will repeat infinitely often. Thus, the simplex method will not terminate. The phenomenon just described is called *cycling* and the simplex method is said to *cycle*.

The bad news is that cycling can occur. The good news is that cycling appears to be extremely rare in problems arising in applications, and there are modifications to the simplex method that prevent cycling.

The following example due Beale [1955] cycles in six iterations.

Example 5.1. Minimize $z = 22x_1 - 93x_2 - 21x_2 + 24x_5$ subject to

$$\begin{aligned}
-4x_1 + 8x_2 + 2x_3 \qquad + x_5 - 9x_6 \qquad &= 0,\\
\tfrac{1}{2}x_1 - \tfrac{3}{2}x_2 - \tfrac{1}{2}x_3 + x_4 \qquad + x_6 + x_7 &= 0,\\
x_1 \qquad\qquad\qquad\qquad + x_7 &= 1,
\end{aligned}$$

$$x_i \geqslant 0, \qquad i = 1, \ldots, 7.$$

From the preceding discussion and from Section 4 we have the following:

LEMMA 5.1. *Phase II of the simplex method does one of the following:*

(i) *determines an optimal solution in a finite number of steps,*
(ii) *determines that no solution exists in a finite number of steps, or*
(iii) *cycles.*

We shall present a very simple rule for the selection of entering and leaving variables that makes cycling impossible. This rule is the *smallest index rule* or *Bland's rule*, named after its originator R. G. Bland [1977]. Other methods for preventing cycling are the *perturbation method* and the related *lexicographic ordering rule*. For a discussion of these methods see Chvátal [1983], Dantzig [1963], and Gass [1985].

Before stating the smallest index rule, we recall that we pointed out in Remark 4.3 that the choice of entering variable could be any variable corresponding to a positive component of $\mathbf{d}_N$. Also, it is possible to choose any variable x_{i_r} at which t^* in (21) of Section 4 is attained.

Definition 5.1 (Smallest Index Rule). From among all possible entering variables, choose the one with smallest index. From among all possible leaving variables, choose the one with smallest index.

Remark 5.1. According to the smallest index rule, the choice of entering variable will, in general, be different from the choice of index such that d_i is a maximum. The choice of leaving variable is the same as that previously used.

THEOREM 5.1. *If at each iteration of the simplex method the smallest index rule is used, then cycling will not occur.*

Proof. We shall prove the theorem by showing that the use of the smallest index rule and the assumption of cycling lead to a contradiction. The argument that we shall give follows the one by Chvátal [1983].

Let us assume that cycling occurs and that $B_k, B_{k+1}, \ldots, B_{k+p}$ is a sequence of bases corresponding to iterations $I_k, I_{k+1}, \ldots, I_{k+p}$ and such that $B_{k+p} = B_k$. Since cycling occurs, all of these bases are degenerate.

Let J denote the set of variables that are basic in some iterations and nonbasic in others. Let x_t denote the variable in J with the greatest index. If x_t is basic in I_k, then there must exist an iteration I_{k+i} with $0 < i < p$ such that x_t leaves B_{k+i} and is nonbasic at the next iteration. If x_t is nonbasic in I_k, then there must exist a basis B_{k+i} with $0 < i < p$ such that x_t enters B_{k+i}. Since $B_{k+p} = B_k$, there exists a basis B_{k+j} with $i < j < p$ such that x_t leaves B_{k+j} and is nonbasic at the next iteration. In conclusion, we have shown that there is a basis B in the sequence $B_k, \ldots, B_{k+p}$ such that x_t leaves B at the iteration I corresponding to B and becomes nonbasic in the next basis. Let x_s denote the entering variable at the iteration I. Thus, $x_s \notin B$ and x_s is in the next basis. Since cycling occurs, there is basis B^* at an iteration I^* beyond I and in the sequence $B_k, B_{k+1}, \ldots, B_{k+p}, B_{k+p+1}, \ldots, B_{k+2p}$ such that x_t enters the basis at I^*. That is, x_t is not in B^* but is in the succeeding basis.

We shall write $i \in B$ to mean that x_i is a basic variable in the basis B and $i \notin B$ to mean that x_i is nonbasic. If we denote the elements of the matrix $B^{-1}N$

in relation (4) of Section 4 by v_{ij}, then we can write (4) and (8) of Section 4 as

$$\begin{aligned} x_i &= \xi_i - \sum_{j \notin B} v_{ij} x_j, \qquad i \in B, \\ z &= \zeta + \sum_{j \notin B} d_j x_j. \end{aligned} \tag{2}$$

Since x_s enters at B, $s \notin B$. A solution of (2) is then given by

$$\begin{aligned} x_s &= y, \qquad y \text{ arbitrary}, \\ x_i &= 0, \qquad i \notin B, \qquad i \neq s, \\ x_i &= \xi_i - v_{is} y, \qquad i \in B, \\ z &= \zeta + d_s y. \end{aligned} \tag{3}$$

For the basis B^* equations (4) and (8) in Section 4 become

$$\begin{aligned} \mathbf{x}_{B^*} &= \boldsymbol{\xi}_{B^*} - (B^*)^{-1} N^* x_N^*, \\ z &= \zeta^* + \sum_{j=1}^{n} d_j^* x_j, \end{aligned} \tag{4}$$

where $d_j^* = 0$ for $j \in B^*$.

The systems (2) and (4) are equivalent; that is, they have the same solution sets. In particular, (3) is a solution of (4). Since all of the bases from B to B^* are degenerate, the value of the objective function does not change from iteration to iteration. Hence, $\zeta^* = \zeta$. Equating the expressions for z in (2) and (4) and using (3) and $\zeta^* = \zeta$ give

$$\zeta + d_s y = \zeta + d_s^* y + \sum_{i \in B} d_i^* (\xi_i - v_{is} y).$$

Hence

$$\left(d_s - d_s^* + \sum_{i \in B} d_i^* v_{is} \right) y = \sum_{i \in B} d_i^* \xi_i \quad \text{for all } y.$$

Since the right-hand side is constant and the left-hand side holds for all y, we must have

$$d_s - d_s^* + \sum_{i \in B} d_i^* v_{is} = 0. \tag{5}$$

Since x_s enters B at iteration I, $d_s > 0$. Also, since x_t is leaving, $s \neq t$. Hence $s < t$. Since x_s is not entering at I^* and $s < t$, it follows that $d_s^* \leqslant 0$. For otherwise, by the smallest index rule, x_t would not be the entering variable at I^*. It then follows from (5) that there exists an index r in B such that

$$d_r^* v_{rs} < 0. \tag{6}$$

Since $d_r^* \neq 0$, the variable x_r is not in the basis B^*. But $r \in B$, so $x_r \in J$. Hence $r \leqslant t$.

It follows from (21) in Section 4 that since x_t is leaving in B and x_s is entering, $v_{ts} > 0$. Since x_t is entering at B^*, it follows that $d_t^* > 0$. Hence $d_t^* v_{ts} > 0$. Comparing this with (6) shows that $r \neq t$. Hence $r < t$. Since $r < t$, since the variable x_r is not in B^*, and since x_t enters B^*, we conclude from the smallest index rule that $d_r^* \leqslant 0$. From (6) we conclude that $d_r^* < 0$ and

$$v_{rs} > 0.$$

Since $x_r \notin B^*$, the basic feasible solution $\boldsymbol{\xi}^*$ at B^* has component $\xi_r^* = 0$. Let ξ_r denote the rth component of the basic feasible solution $\boldsymbol{\xi}$ corresponding to the basis B. We assert that $\xi_r = 0$. To prove this assertion, we note that because all the bases $B \ldots B^*$ are degenerate, it follows from (21) of Section 4 that $t^* = 0$ at all the iterations $I \ldots I^*$. Hence if $\xi_r > 0$, then x_r would not leave any of the bases $B \ldots B^*$. This contradicts the fact that $x_r \notin B^*$, and this proves the assertion.

Since $v_{rs} > 0$ and $\xi_r = 0$, it follows from (21) of Section 4 that x_r is a candidate for leaving the basis B. Yet, we chose x_t with $t > r$ to leave the basis, in contradiction to the smallest index rule.

Remark 5.2. Computational experience indicates that using the smallest subscript rule or the lexicographic rule instead of the $\max d_i$ rule considerably increases the time required for the simplex method or the revised simplex method to find an optimal solution. Since very few instances of cycling have been reported, some computer codes do not incorporate any anticycling routines. Other computer codes introduce an anticycling routine only after a sequence of specified length of degenerate bases occurs. The anticycling routine will then lead to a nondegenerate basis, at which point the use of the largest coefficient rule can be resumed.

Exercise 5.1. Carry out the iterations of the simplex method for Example 5.1 and show that cycling occurs at the sixth iteration. *Note:* $\boldsymbol{\xi}_B = (\xi_4, \xi_5, \xi_7) = (0, 0, 1)$ is a basic feasible solution.

Exercise 5.2. Solve the problem in Example 5.1 using the simplex method and Bland's rule.

6. PHASE I OF SIMPLEX METHOD

As noted in Section 3, phase I of the simplex method either determines a basic feasible solution or determines that none exists, in which case the problem has no optimal solution.

We begin with the linear programming problem in standard form (SLP):

$$\begin{aligned} &\text{Maximize } \langle \mathbf{b}, \mathbf{x} \rangle \\ &\text{subject to } A\mathbf{x} \leqslant \mathbf{c}, \qquad \mathbf{x} \geqslant \mathbf{0}. \end{aligned} \tag{1}$$

Let us first suppose that the vector $\mathbf{c}$ is nonnegative. We then put the problem in canonical form (CLP) by adjoining a vector of slack variables,

$$\mathbf{x}_S = (x_{n+1}, \ldots, x_{n+m}), \qquad x_{n+i} \geqslant 0, \qquad i = 1, \ldots, m,$$

and write the problem as

$$\begin{aligned} &\text{Maximize } \langle \mathbf{b}, \mathbf{x} \rangle \\ &\text{subject to } (A, I)\begin{pmatrix} \mathbf{x} \\ \mathbf{x}_S \end{pmatrix} = \mathbf{c}, \qquad \mathbf{x} \geqslant \mathbf{0}, \qquad \mathbf{x}_S \geqslant \mathbf{0}. \end{aligned} \tag{2}$$

Since $\mathbf{c} \geqslant 0$, the vector $(\mathbf{x}, \mathbf{x}_s) = (\mathbf{0}_n, \mathbf{c})$, where $\mathbf{0}_n$ is the zero vector in $\mathbb{R}^n$, is a basic feasible solution for the problem in (2). We then immediately pass to phase II with initial basic feasible solution $(\xi, \xi_S) = (\mathbf{0}_n, \mathbf{c})$.

If some of the components of $\mathbf{c}$ are negative, the vector $(\mathbf{0}_n, \mathbf{c})$ is still a basic solution of the system of equations in (2), but it is not feasible. We therefore resort to another device and introduce an auxiliary problem.

Auxiliary Problem

Minimize x_0 subject to

$$\begin{gathered} (A, I, -\mathbf{e})\begin{pmatrix} \mathbf{x} \\ \mathbf{x}_S \\ x_0 \end{pmatrix} = \mathbf{c} \\ \mathbf{x} \geqslant \mathbf{0}, \qquad \mathbf{x}_S \geqslant \mathbf{0}, \qquad x_0 \geqslant 0, \end{gathered} \tag{3}$$

where $\mathbf{e} = (1, \ldots, 1)$ and I is the $m \times m$ identity matrix.

In component form the system in (3) is

$$\begin{gathered} \sum_{j=1}^{n} a_{ij} x_j + x_{n+i} - x_0 = c_i, \qquad i = 1, \ldots, m, \\ x_j \geqslant 0, \qquad j = 0, 1, \ldots, n+m. \end{gathered}$$

A vector $(\boldsymbol{\xi}, \boldsymbol{\xi}_S)$ is feasible for relation (2) if and only if $(\boldsymbol{\xi}, \boldsymbol{\xi}_S, 0)$ is feasible for the auxiliary problem. Therefore, $(\boldsymbol{\xi}, \boldsymbol{\xi}_S)$ *is feasible for relation* (2) *if and only if* $(\boldsymbol{\xi}, \boldsymbol{\xi}_S, 0)$ *is optimal for the auxiliary problem.*

If we write the objective function of the auxiliary problem as "Maximize $w = -x_0$," then the auxiliary problem becomes a linear programming problem in canonical form to which we can apply phase II of the simplex method, provided we can get an initial basic feasible solution. A basic solution to the system (3) is $(\boldsymbol{\xi}, \boldsymbol{\xi}_S, \xi_0) = (\mathbf{0}, \mathbf{c}, 0)$. This solution, however, is not feasible since some components of $\mathbf{c}$ are negative. We can obtain a basic feasible solution from this solution by replacing an appropriate basic variable by x_0. Before we do this in general, we shall illustrate using Example 4.1.

In solving Example 4.1, we obtained an initial feasible solution from Figure 6.1. We now ignore the figure and write the auxiliary problem in canonical form: Maximize $w = -x_0$ subject to

$$\begin{aligned} -x_1 - x_2 + x_3 \qquad\qquad - x_0 &= -1, \\ -x_1 + x_2 \qquad + x_4 \qquad - x_0 &= 0, \\ x_1 + 2x_2 \qquad\qquad + x_5 - x_0 &= 4, \\ x_i \geqslant 0, \qquad i = 0, 1, \ldots, 5. & \end{aligned} \tag{4}$$

The vector $(\xi_0, \xi_1, \xi_2, \xi_3, \xi_4, \xi_5) = (0, 0, 0, -1, 0, 4)$ is a basic solution but is not feasible. The only negative entry on the right-hand side is -1 and occurs in the first equation. Therefore, if in the augmented coefficient matrix of (4) we pivot on x_0 in row 1, then the right-hand side of the resulting system should be positive. Performing the indicated pivot gives

$$\begin{pmatrix} x_1 & x_2 & x_3 & x_4 & x_5 & x_0 & \mathbf{c} \\ 1 & 1 & -1 & 0 & 0 & 1 & 1 \\ 0 & 2 & -1 & 1 & 0 & 0 & 1 \\ 2 & 3 & -1 & 0 & 1 & 0 & 5 \end{pmatrix} \tag{5}$$

Thus, $(\xi_0, \xi_1, \xi_2, \xi_3, \xi_4, \xi_5) = (1, 0, 0, 0, 1, 5)$ is a basic feasible solution. From (5) we also get that

$$w = -x_0 = -1 + x_1 + x_2 - x_3. \tag{6}$$

We can now start phase II for the problem of minimizing (6) subject to the constraining equations given by (5). We leave this to the reader and return to the general situation.

We now obtain a system equivalent to (3) and having a basic feasible solution. Let c_p denote the most negative component of $\mathbf{c}$. Thus,

$$c_p = \min\{c_i : i = 1, \ldots, m\}.$$

If there are more than one such components, choose the one with smallest index. Next, pivot on x_0 in the pth row of the system (3). The result will be an equivalent system

$$(A', \mathbf{e}_1, \ldots, \mathbf{e}_{p-1}, -\mathbf{e}, \mathbf{e}_{p+1}, \ldots, \mathbf{e}_m, \mathbf{e}_p)\begin{pmatrix} \mathbf{x} \\ \mathbf{x}_S \\ x_0 \end{pmatrix} = \mathbf{c}', \tag{7}$$

where $\mathbf{e} = (1, \ldots, 1)$ and

$$c'_p = -c_p \geqslant 0, \qquad c'_i = c_i - c_p \geqslant 0, \qquad i = 1, \ldots, p-1, p+1, \ldots, m.$$

The system (7) in component form is

$$\sum_{j=1}^{n} a'_{ij}x_j - x_{n+p} + x_{n+i} = c'_i, \qquad i \neq p,$$

$$\sum_{j=1}^{n} a'_{pj}x_j - x_{n+p} + x_0 = c'_p.$$

Thus a basic feasible solution to (7) is

$$\begin{aligned} \xi_0 = c'_p, \qquad \xi_{n+i} = c'_i, \qquad & i = 1, \ldots, p-1, p+1, \ldots, m, \\ \xi_i = 0, \qquad & i = 1, \ldots, n, n+p, \end{aligned} \tag{8}$$

and

$$w = -x_0 = -c'_p - x_{n+p} + \sum_{j=1}^{n} a'_{pj}x_j. \tag{9}$$

The basic variables are

$$x_{n+1}, \ldots, x_{n+p-1}, x_0, x_{n+p+1}, \ldots, x_{n+m}. \tag{10}$$

We have transformed the auxiliary problem (3) to the following problem. Maximize w as given in (9) subject to (8) and $x_i \geqslant 0$, $i = 0, 1, \ldots, n+m$. For this problem we have determined a basic feasible solution given by (8) with basic variables given by (10). We can now apply phase II to the problem in this form. If we incorporate an anticycling subroutine in our implementation of Phase II, then according to Lemma 5.1, we shall, in a finite number of steps, determine that the auxiliary problem either has no solution or determine an optimal solution to the auxiliary problem. *If the auxiliary problem has no solution, then the original linear programming problem* (1) *has no feasible solution.* For, as we saw, a vector $(\boldsymbol{\xi}, \boldsymbol{\xi}_S)$ is feasible for the original problem if and only if $(\boldsymbol{\xi}, \boldsymbol{\xi}_S, 0)$ is optimal for the auxiliary problem.

The remaining alternative is that the auxiliary problem has a solution. We now explore the possibilities for this alternative. The variable x_0 is basic initially. In applying phase II of the simplex method to the auxiliary problem, we note that our rule for choosing the leaving variable is such that if at some iteration x_0 is a candidate for leaving the basis, then x_0 will be chosen to be the leaving variable. If at some iteration I_k the variable x_0 leaves the basis, then at the iteration I_{k+1}, the variable x_0 will have the value zero. The solution (ξ, ξ_S, ξ_0) of the constraining equations at I_{k+1} will also be a solution of (3). Hence if $\xi_0 = 0$, then $w = 0$ and (ξ, ξ_S) is optimal for relation (1). Thus, if x_0 ever leaves the basis at an iteration I_k, at the next iteration I_{k+1}, we will have a solution to the auxiliary problem and hence a basic feasible solution for relation (2). We can then begin phase II with this basic feasible solution.

An optimal solution could also conceivably occur under the following circumstances:

(i) x_0 basic, $w = 0$, and
(ii) x_0 basic, $w \neq 0$.

If (ii) occurs, then the original problem has no feasible solution. We now show that (i) is impossible. If (i) did occur, then $x_0 = -w = 0$. Since the previous iteration did not end in an optimal solution, $x_0 \neq 0$ at the start of the previous iteration and became zero in the previous iteration. If this happened, then x_0 was a candidate for leaving the basis but did not do so. This contradicts the rule for determining whether x_0 leaves the basis.

The next lemma follows from the discussion in this section.

LEMMA 6.1. *If an anticycling routine is included, then phase I of the simplex method either determines that no feasible solution exists or determines a basic feasible solution.*

Lemmas 5.1 and 6.1 imply the following.

THEOREM 6.1. *The simplex method incorporating an anticycling routine applied to a linear programming problem determines one of the following in a finite number of iterations:*

(i) No feasible solutions exist.
(ii) The problem does not have an optimal solution.
(iii) An optimal solution.

Exercise 6.1. Complete the solution of the illustrative example.

Exercise 6.2. Let the linear programming problem be given originally in canonical form:

$$\text{Maximize } \langle \mathbf{b}, \mathbf{x} \rangle$$
$$\text{subject to } A\mathbf{x} = \mathbf{c}, \qquad \mathbf{x} \geqslant \mathbf{0}.$$

Show that an appropriate auxiliary problem for phase I is

$$\text{Maximize } w = -\sum_{i=1}^{m} y_i$$

$$\text{subject to } (A, I)\begin{pmatrix}\mathbf{x}\\ \mathbf{y}\end{pmatrix} = \mathbf{c}, \qquad \mathbf{x} \geqslant \mathbf{0}, \qquad \mathbf{y} \geqslant \mathbf{0}.$$

Show how one obtains an initial basic feasible solution for the auxiliary problem.

Exercise 6.3. Solve the following linear programming problems using the simplex method:

(a) Maximize $x_1 + 3x_3$ subject to

$$\begin{aligned} x_1 + 2x_2 + 7x_3 &\leqslant 4, \\ x_1 + 3x_2 + x_3 &\leqslant 5, \\ x_i \geqslant 0, \qquad i &= 1, 2, 3. \end{aligned}$$

(b) Maximize $3x_2 - x_1$ subject to

$$\begin{aligned} x_1 + 2x_2 &\geqslant 2, \\ 3x_1 + x_2 &\leqslant 3, \\ x_i \leqslant 4, \qquad x_i \geqslant 0. \quad i &= 1, 2, 3, 4. \end{aligned}$$

(c) Maximize $-2x_1 - x_2 + x_3 + x_4$ subject to

$$\begin{aligned} x_1 - x_2 + 2x_3 - x_4 &= 2, \\ 2x_1 + x_2 - 3x_3 + x_4 &= 6, \\ x_1 + x_2 + x_3 + x_4 &= 7. \end{aligned}$$

7. REVISED SIMPLEX METHOD

The revised simplex method does not differ from the simplex method in principle, that is, in going from one basic solution to another in such a way that the objective function is not decreased but differs in the manner in which some of the computations are carried out. In describing the revised simplex method, we assume, as we did in our description of the simplex method in Section 4, that the problem has been formulated as in (4) of Section 3. The reader will recall from Section 4 that at each iteration a basis B is determined and the calculation of $B^{-1}N$ is used to determine the entering and leaving variable. The revised simplex method differs from the version of the simplex method given in Section 4 in that a system (1) in Section 4 with current B is not reduced to (11) in Section 4 and the current basis not given by (22) in Section 4. In the revised simplex method the entering variable x_q is chosen as before. If $\mathbf{A}_q$ is the column of A corresponding to x_q, then the updated basis B' is obtained from B by replacing the column A_{i_ρ} of the leaving variable x_{i_ρ} by $\mathbf{A}_q$. This procedure enables us to calculate $(B')^{-1}$ from B^{-1} very easily and expedites other calculations, as we shall show.

Before we present the steps in an iteration of the revised simplex method, we take up the calculation of $(B')^{-1}$, the updated inverse. At the current iteration I let the basis B be given by

$$B = (\mathbf{A}_{i_1} \ldots \mathbf{A}_{i_m}), \tag{1}$$

where the $\mathbf{A}_{i_i}$ are columns of B. We then write

$$N = (\mathbf{A}_{i_{m+1}} \ldots \mathbf{A}_{i_n}).$$

Let x_q be the entering variable at I, determined as in Section 4, and let $\mathbf{A}_q$ denote the column of N corresponding to x_q. Let the leaving variable x_{i_ρ} be determined as in Section 4. Let B' denote the matrix obtained from B by replacing the column $\mathbf{A}_{i_\rho}$ by $\mathbf{A}_q$ and let N' denote the matrix obtained from N by replacing column $\mathbf{A}_q$ by $\mathbf{A}_{i_\rho}$. Thus

$$\begin{aligned} B' &= (\mathbf{A}_{i_1} \ldots \mathbf{A}_{i_{\rho-1}} \mathbf{A}_q \mathbf{A}_{i_{\rho+1}} \ldots \mathbf{A}_{i_m}), \\ N' &= (\mathbf{A}_{i_{m+1}} \ldots \mathbf{A}_{i_{q-1}} A_{i_\rho} A_{i_{q+1}} \ldots A_{i_n}). \end{aligned} \tag{2}$$

Hence

$$\begin{aligned} B^{-1}B' &= B^{-1}(\mathbf{A}_{i_1} \ldots \mathbf{A}_{i_{\rho-1}} \mathbf{A}_q A_{i_{\rho+1}} \ldots A_{i_m}) \\ &= \begin{pmatrix} 1 & 0 & \cdots & 0 & v_{1q} & 0 & \cdots & 0 \\ 0 & 1 & \cdots & 0 & v_{2q} & 0 & \cdots & 0 \\ \vdots & \vdots & \ddots & \vdots & \vdots & \vdots & \ddots & \vdots \\ 0 & 0 & \cdots & 0 & v_{mq} & 0 & \cdots & 1 \end{pmatrix} \\ &= (\mathbf{e}_1 \ldots \mathbf{e}_{\rho-1} \mathbf{v}_q \mathbf{e}_{\rho+1} \ldots \mathbf{e}_m), \end{aligned} \tag{3}$$

where $\mathbf{v}_q = B^{-1}\mathbf{A}_q$. Note that $\mathbf{v}_q$ is the qth column of $B^{-1}N$ associated with x_q and thus is as in Step 3(ii) of an iteration in the simplex method. From the way ρ is determined we have that $v_{\rho q} > 0$, and hence the matrix on the right in (3) is nonsingular. If we multiply both sides of (3) by B on the left, we get that

$$B' = B(\mathbf{e}_1 \ldots \mathbf{e}_{\rho-1} \mathbf{v}_q \mathbf{e}_{\rho+1} \ldots \mathbf{e}_m).$$

Since B' is the product of nonsingular matrices, it is nonsingular and so is a basis. We also get

$$(B')^{-1} = (\mathbf{e}_1 \dots \mathbf{e}_{\rho-1}\mathbf{v}_q\mathbf{e}_{\rho+1} \dots \mathbf{e}_m)^{-1}B^{-1}. \tag{4}$$

It is readily verified that multiplying the matrix on the right-hand side of the following by the matrix on the right-hand side of (3) gives the $m \times m$ identity matrix. Hence

$$\begin{pmatrix} 1 & 0 & \cdots & 0 & v_{1q} & 0 & \cdots & 0 \\ 0 & 1 & \cdots & 0 & v_{2q} & 0 & \cdots & 0 \\ \vdots & \vdots & \ddots & \vdots & \vdots & \ddots & & \\ 0 & 0 & \cdots & 0 & v_{mq} & & & \\ & & & & \uparrow & & & \\ & & & & \text{col } \rho & & & \end{pmatrix}^{-1} = \begin{pmatrix} 1 & 0 & \cdots & 0 & \eta_{1q} & 0 & \cdots & 0 \\ 0 & 1 & \cdots & 0 & \eta_{2q} & 0 & \cdots & 0 \\ \vdots & \vdots & \ddots & \vdots & \vdots & \vdots & \ddots & \\ 0 & 0 & \cdots & 0 & \eta_{mq} & 0 & \cdots & 1 \\ & & & & \uparrow & & & \\ & & & & \text{col } \rho & & & \end{pmatrix}, \tag{5}$$

where

$$\eta_{iq} = -\frac{v_{iq}}{v_{\rho q}}, \qquad i = 1, \dots, m, \qquad i \neq \rho, \qquad \eta_{\rho q} = \frac{1}{v_{\rho q}}.$$

The matrix on the right in (5) is called an *eta matrix* and will be denoted by E_1. Thus we may write (4) as

$$(B')^{-1} = E_1 B^{-1}, \tag{6}$$

which updates B^{-1} for the next iteration.

We summarize this discussion in the following lemma

LEMMA 7.1. *Let B be a basic matrix as in* (1) *and let B' be a matrix as in* (2). *Then B' is basic and* (6) *holds, where E_1 is the matrix on the right in* (5).

Remark 7.1. In the various calculations in the simplex method the essential matrix is B^{-1}, not B. The matrix B is needed, in principle, only to calculate B^{-1}. Equation (6) enables us to calculate $(B')^{-1}$ from B^{-1} *without knowing B* or B'. It is this fact that underlies the revised simplex method. Also note that an eta matrix differs from an identity matrix in one column.

We now present a generic revised simplex method. Most computer codes for solving linear programming problems use the revised simplex method. Codes used in practice incorporate various modifications to improve accuracy, reduce the number of iterations needed, decrease the running time, and so on. For details the reader is referred to Chvátal [1983], Dantzig [1963], Gass [1985], and Murtagh [1981].

As in our presentation of the simplex method we assume that the problem is in canonical form: Maximize z subject to

$$A\mathbf{x} = \mathbf{c}, \qquad \langle \mathbf{b}, \mathbf{x} \rangle - z = 0, \qquad \mathbf{x} \geqslant 0. \tag{7}$$

Let the initial stage or iteration of phase II be called the zeroth iteration. At the kth iteration, $k = 0, 1, 2, \ldots$, let $B(k)$ denote the basis, let $N(k)$ denote the submatrix of A complementary to $B(k)$, let $\mathbf{x}_{B(k)}$ denote the vector of basic variables, let $\mathbf{x}_{N(k)}$ denote the vector of nonbasic variables, let $\xi(k) = (\xi_{B(k)}, \mathbf{0}_{N(k)})$ be the corresponding basic feasible solution, and let $\zeta(k)$ be the current value of the objective function. Equations (4)–(6) from Section 4 yield, for $k = 0, 1, 2, \ldots$,

$$\begin{aligned} \mathbf{x}_{B(k)} &= \xi_{B(k)} - [B(k)]^{-1} N(k) \mathbf{x}_{N(k)}, \\ \zeta(k) &= \langle \mathbf{b}_{B(k)}, \xi_{B(k)} \rangle, \\ z &= \zeta(k) - \langle \mathbf{b}_{N(k)} - \mathbf{b}^t_{B(k)} [B(k)]^{-1} N(k),\ \mathbf{x}_{N(k)} \rangle. \end{aligned} \tag{8}$$

From Lemma 7.1 we get that

$$[B(k+1)]^{-1} = E_{k+1} [B(k)]^{-1}, \qquad k = 0, 1, 2, \ldots,$$

where E_{k+1} is an approximate eta matrix. Proceeding inductively, we get

$$[B(k+1)]^{-1} = E_{k+1}(E_k(\ldots E_2(E_1[B(0)]^{-1}) \ldots)). \tag{9}$$

A sequential file in the computer storing the matrices $E_{k+1}, E_k, \ldots, E_1$ is called an *eta file*. The calculation as indicated in (9) (from right to left) is called a *forward transformation*. The matrix $[B(0)]^{-1}$ is calculated using elementary matrices as in (11) from Section 3.

We now present the steps in the kth iteration, $k \geqslant 1$, of our generic revised simplex algorithm. In essence, the steps mirror the steps of the simplex method as presented in Section 4. Some of the calculations are different and Step 1 of the simplex method is deleted. For typographical convenience we will replace the designations $B(k)$, $N(k)$, $\mathbf{x}_{B(k)}$, $\mathbf{x}_{N(k)}, \ldots$ of the current quantities by B, N, $\mathbf{x}_B$, $\mathbf{x}_N, \ldots$. The updated quantities $B(k+1)$, $N(k+1)$, $\mathbf{x}_{B(k+1)}$, $\mathbf{x}_{N(k+1)}, \ldots$ will be designated as B', N', $\mathbf{x}_{B'}$, $\mathbf{x}_{N'}, \ldots$. As in (7) from Section 4 let

$$\mathbf{d}_N = \mathbf{b}_N - \mathbf{b}^t_B B^{-1} N = (d_{i_{m+1}} \ldots d_{i_n}), \tag{10}$$

where $i_{m+1}, \ldots, i_n$ are the components of $\mathbf{x}$ that constitute $\mathbf{x}_N$. We may then write the last equation in (8) as

$$\mathbf{z} = \zeta + \langle \mathbf{d}_N, \mathbf{x}_N \rangle = \zeta + \sum_{p=m+1}^{n} d_{i_p} x_{i_p}.$$

Step 1. Calculate $\mathbf{d}_N$. This is done in two steps.

(i) Calculate

$$\boldsymbol{\pi} = \mathbf{b}_B^t E_k E_{k-1} \dots E_1 [B(0)]^{-1}$$

from left to right, that is, as

$$(\dots((\mathbf{b}_B^t E_k) E_{k-1}) \dots) E_1)[B(0)]^{-1}.$$

This *backward transformation* involves repeated multiplications of a vector by a matrix, whereas the forward calculation (from right to left) involves repeated multiplications of matrices. Clearly, the backward transformation involves fewer operations.

(ii) Calculate

$$d_{i_j} = b_{i_j} - \langle \boldsymbol{\pi}, \mathbf{A}_{i_j} \rangle, \qquad j = m + 1, \dots, n,$$

where the i_j are the indices of the nonbasic variables at the kth stage and the $\mathbf{A}_{i_j}$ are the corresponding columns of N.

Step 2. If all the d_{i_j} are less than or equal to zero, the current basic feasible solution is optimal and the algorithm terminates. If the algorithm terminates, then we use (8) to calculate ζ, the optimal value.

If the algorithm does not terminate at Step 2, we go to Step 3.

Step 3. Choose an entering variable.

Choose an index i_j such that $d_{i_j} > 0$. This may be done according to the smallest index rule, the max d_{i_j} rule, or any other rule, as long as $d_{i_j} > 0$. We denote the index i_j by q and take x_q to be the entering variable. The column $\mathbf{A}_q$ will be the entering column for the basis at the $(k + 1)$st iteration.

Step 4. Either determine that the problem is unbounded or determine the leaving variable.

(i) Calculate

$$\mathbf{v}_q = E_k E_{k-1} \dots E_1 [B(0)]^{-1} \mathbf{A}_q$$

from right to left. Note that $\mathbf{v}_q$ is as in Section 4.

(ii) If all components of $\mathbf{v}_q$ are nonpositive, then the problem is unbounded and no solution exists. The justification for this conclusion is the same as the one given in Step 3(ii) of Section 4.

(iii) If $\mathbf{v} = (v_{1q} \ldots v_{mq})$ has at least one positive component, then as in (21) of Section 4 let ρ denote the smallest index r at which

$$t^* = \min\left\{\frac{\xi_{i_r}}{v_{r_q}} : v_{r_q} > 0\right\}$$

is attained. Choose x_{i_ρ} to be the leaving variable.

Step 5. Calculate E_{k+1} as given by the matrix on the right in (5).

Step 6. Update the data for the $(k+1)$st iteration.

(i) Update $\mathbf{x}_B$ to $\mathbf{x}_{B'}$ by replacing x_{i_p} by x_q.
(ii) Update $\mathbf{x}_N$ to $\mathbf{x}_{r'}$ by replcing x_q by x_{i_p}.
(iii) Update B to B' by replacing $\mathbf{A}_{i_\rho}$ by $\mathbf{A}_q$.
(iv) Update N to N' by replacing $\mathbf{A}_q$ by $\mathbf{A}_{i_\rho}$.
(v) Update (ξ_B, ξ_N) to $(\xi_{B'}, \xi_{N'})$ using (22) from Section 4.
(vi) Update B^{-1} to $(B')^{-1}$ using (9).
(vii) Update $\mathbf{b}_N$ and $\mathbf{b}_B$ to $\mathbf{b}_{N'}$ and $\mathbf{b}_{B'}$.

Step 7. Begin iteration $k+1$ using the updated quantities.

Remark 7.2. If the initial basis consists of the slack variables, then $B(0) = I$ and (9) becomes

$$[B(k+1)]^{-1} = E_{k+1}E_k \ldots E_1 I.$$

Remark 7.3. The eta file can become very long. In some computer codes, once the eta file reaches a predetermined length, say $k+1$, then $B(k+1)$ is set as the initial basis $B(0)$, the matrix $[B(k+1)]^{-1}$ is set as $[B(0)]^{-1}$, and a new eta file is constructed.

Exercise 7.1. Solve the linear programming problem in Example 1.1 using the revised simplex method. Start with phase I.

Exercise 7.2. Solve the linear programming problem in Exercises 6.3(a), (b), and (c) using the revised simplex method.

BIBLIOGRAPHY

Bartle, R. G. and D. R. Sherbert, *The Elements of Real Analysis*, 3rd ed., John Wiley & Sons, New York, 1999.

Bazaraa, M. S., H. D. Sherali, and C. M. Shetty, *Nonlinear Programming, Theory and Algorithms*, 2nd ed., John Wiley & Sons, New York, 1993.

Beale, E. M. L., "Cycling in the Dual Simplex Algorithm," *Naval Research Logistics Quarterly* **2** (1955), 269–275.

Bland, R. G., "New Finite Pivoting Rules for the Simplex Method," *Mathematics of Operations Research* **2** (1977), 103–107.

Chvátal, V., *Linear Programming*, W. H. Freeman, New York, 1983.

Dantzig, G. B., *Linear Programming and Extensions*, Princeton University Press, Princeton, N.J., 1963.

Dresher, M., *Games of Strategy: Theory and Applications*, Prentice-Hall, Englewood Cliffs, N.J., 1961.

Farkas, J., "Über die Theorie der einfachen Ungleichungen," *Journal für die reine und angewandte Mathematik* **124** (1901), 1–27.

Fiacco, A. V. and G. P. McCormick, *Nonlinear Programming: Sequential Unconstrained Minimization Techniques*, John Wiley & Sons, New York, 1968.

Fleming, W., *Functions of Several Variables*, 2nd ed., Springer, New York, 1977.

Gale, D., *The Theory of Linear Economic Models*, McGraw-Hill, New York, 1960.

Gass, S. I., *Linear Programming, Methods and Applications*, 5th ed., McGraw-Hill, New York, 1985.

Golub, G. H. and C. F. Van Loan, *Matrix Computations*, Johns Hopkins University Press, Baltimore, 1996.

Gordan, P., "Über die Anflösungen linearer Gleichungen mit reelen coefficienten," *Mathematische Annalen* **6** (1873), 23–28.

Hahn, S-P. and O. L. Mangasarian, "Exact Penalty Functions in Nonlinear Programming," *Mathematical Programming* **17** (1979), 251–269.

John, F., *Extremum Problems with Inequalities as Subsidiary Conditions*, Studies and Essays: Courant Anniversary Volume (K. O. Friedricks, O. E. Neugebauer, and J. J. Stoker, Ed.), Interscience, New York, 1948, pp. 187–204.

Karush, W., "Minima of Functions of Several Variables with Inequalities as Side Conditions," Masters' Dissertation, University of Chicago, Chicago, IL, Dec. 1939.

Kuhn, H. W. and A. W. Tucker, *Nonlinear Programming*, Proceedings of the Second Berkeley Symposium on Mathematical Statistics and Probability (J. Neyman, Ed.), University of California Press, Berkeley, CA, 1951, pp. 481–492.

Mangasarian, O. L. *Nonlinear Programming*, McGraw-Hill, New York, 1969.

McShane, E. J., "The Lagrange Multiplier Rule," *American Mathematics Monthly* **80** (1973), 922–924.

Motzkin, T. S., "Béitrage zur Theorie Linearen Ungleichungen," Inaugural Dissertation, Basel, Jerusalem, 1936.

Murtagh, B. A., *Advanced Linear Programming: Computation and Practice*, McGraw-Hill, New York, 1981.

Pennisi, L. L., "An Indirect Sufficiency Proof for the Problem of Lagrange with Differential Inequalities as Side Conditions," *Transactions of the American Mathematical Society* **74** (1953), 177–198.

Rockafellar, R. T., *Convex Analysis*, Princeton University Press, Princeton, N.J., 1970.

Rudin, W., *Principles of Mathematical Analysis*, 3rd ed., McGraw-Hill, New York, 1976.

INDEX

PURE AND APPLIED MATHEMATICS

A Wiley-Interscience Series of Texts, Monographs, and Tracts

Founded by RICHARD COURANT
Editors: MYRON B. ALLEN III, DAVID A. COX, PETER LAX
Editors Emeriti: PETER HILTON, HARRY HOCHSTADT, JOHN TOLAND

ADÁMEK, HERRLICH, and STRECKER—Abstract and Concrete Catetories
ADAMOWICZ and ZBIERSKI—Logic of Mathematics
AINSWORTH and ODEN—A Posteriori Error Estimation in Finite Element Analysis
AKIVIS and GOLDBERG—Conformal Differential Geometry and Its Generalizations
ALLEN and ISAACSON—Numerical Analysis for Applied Science
*ARTIN—Geometric Algebra
AUBIN—Applied Functional Analysis, Second Edition
AZIZOV and IOKHVIDOV—Linear Operators in Spaces with an Indefinite Metric
BERG—The Fourier-Analytic Proof of Quadratic Reciprocity
BERMAN, NEUMANN, and STERN—Nonnegative Matrices in Dynamic Systems
BERKOVITZ—Convexity and Optimization in $\mathbb{R}^n$
BOYARINTSEV—Methods of Solving Singular Systems of Ordinary Differential Equations
BURK—Lebesgue Measure and Integration: An Introduction
*CARTER—Finite Groups of Lie Type
CASTILLO, COBO, JUBETE, and PRUNEDA—Orthogonal Sets and Polar Methods in Linear Algebra: Applications to Matrix Calculations, Systems of Equations, Inequalities, and Linear Programming
CASTILLO, CONEJO, PEDREGAL, GARCIÁ, and ALGUACIL—Building and Solving Mathematical Programming Models in Engineering and Science
CHATELIN—Eigenvalues of Matrices
CLARK—Mathematical Bioeconomics: The Optimal Management of Renewable Resources, Second Edition
COX—Primes of the Form $x^2 + ny^2$: Fermat, Class Field Theory, and Complex Multiplication
*CURTIS and REINER—Representation Theory of Finite Groups and Associative Algebras
*CURTIS and REINER—Methods of Representation Theory: With Applications to Finite Groups and Orders, Volume I
CURTIS and REINER—Methods of Representation Theory: With Applications to Finite Groups and Orders, Volume II
DINCULEANU—Vector Integration and Stochastic Integration in Banach Spaces
*DUNFORD and SCHWARTZ—Linear Operators
Part 1—General Theory
Part 2—Spectral Theory, Self Adjoint Operators in Hilbert Space
Part 3—Spectral Operators
FARINA and RINALDI—Positive Linear Systems: Theory and Applications
FOLLAND—Real Analysis: Modern Techniques and Their Applications
FRÖLICHER and KRIEGL—Linear Spaces and Differentiation Theory
GARDINER—Teichmüller Theory and Quadratic Differentials
GREENE and KRANTZ—Function Theory of One Complex Variable

*Now available in a lower priced paperback edition in the Wiley Classics Library.
†Now available in paperback.

*GRIFFITHS and HARRIS—Principles of Algebraic Geometry
GRILLET—Algebra
GROVE—Groups and Characters
GUSTAFSSON, KREISS and OLIGER—Time Dependent Problems and Difference Methods
HANNA and ROWLAND—Fourier Series, Transforms, and Boundary Value Problems, Second Edition
*HENRICI—Applied and Computational Complex Analysis
Volume 1, Power Series—Integration—Conformal Mapping—Location of Zeros
Volume 2, Special Functions—Integral Transforms—Asymptotics—Continued Fractions
Volume 3, Discrete Fourier Analysis, Cauchy Integrals, Construction of Conformal Maps, Univalent Functions
*HILTON and WU—A Course in Modern Algebra
*HOCHSTADT—Integral Equations
JOST—Two-Dimensional Geometric Variational Procedures
KHAMSI and KIRK—An Introduction to Metric Spaces and Fixed Point Theory
*KOBAYASHI and NOMIZU—Foundations of Differential Geometry, Volume I
*KOBAYASHI and NOMIZU—Foundations of Differential Geometry, Volume II
KOSHY—Fibonacci and Lucas Numbers with Applications
LAX—Linear Algebra
LOGAN—An Introduction to Nonlinear Partial Differential Equations
McCONNELL and ROBSON—Noncommutative Noetherian Rings
MORRISON—Functional Analysis: An Introduction to Banach Space Theory
NAYFEH—Perturbation Methods
NAYFEH and MOOK—Nonlinear Oscillations
PANDEY—The Hilbert Transform of Schwartz Distributions and Applications
PETKOV—Geometry of Reflecting Rays and Inverse Spectral Problems
*PRENTER—Splines and Variational Methods
RAO—Measure Theory and Integration
RASSIAS and SIMSA—Finite Sums Decompositions in Mathematical Analysis
RENELT—Elliptic Systems and Quasiconformal Mappings
RIVLIN—Chebyshev Polynomials: From Approximation Theory to Algebra and Number Theory, Second Edition
ROCKAFELLAR—Network Flows and Monotropic Optimization
ROITMAN—Introduction to Modern Set Theory
*RUDIN—Fourier Analysis on Groups
SENDOV—The Averaged Moduli of Smoothness: Applications in Numerical Methods and Approximations
SENDOV and POPOV—The Averaged Moduli of Smoothness
*SIEGEL—Topics in Complex Function Theory
Volume l—Elliptic Functions and Uniformization Theory
Volume 2—Automorphic Functions and Abelian Integrals
Volume 3—Abelian Functions and Modular Functions of Several Variables
SMITH and ROMANOWSKA—Post-Modern Algebra
STAKGOLD—Green's Functions and Boundary Value Problems, Second Editon
*STOKER—Differential Geometry
*STOKER—Nonlinear Vibrations in Mechanical and Electrical Systems
*STOKER—Water Waves: The Mathematical Theory with Applications
WESSELING—An Introduction to Multigrid Methods
†WHITHAM—Linear and Nonlinear Waves
†ZAUDERER—Partial Differential Equations of Applied Mathematics, Second Edition

*Now available in a lower priced paperback edition in the Wiley Classics Library.
†Now available in paperback.